Industrial Coating and Lining Application Specialist

Level One

Trainee Guide
First Edition

PEARSON

NCCER

New York, New York
Columbus, Ohio

NCCER

President: Don Whyte
Director of Curriculum Revision and Development: Daniele Dixon
Industrial Coating and Lining Application Specialist Project Manager: Daniele Dixon/Angela Jonas
Production Manager: Tim Davis
Quality Assurance Coordinator: Debie Ness
Editors: Rob Richardson and Matt Tischler
Desktop Publishing Coordinator: James McKay
Production Assistant: Laura Wright

NACE International

Executive Director: Tony Keane, CAE
Senior Director: Helena Seelinger, CAE
Course Development Staff: Daniel Kurth, Karen North-Hurst, Paula McCordic, and Trey Moore

This information is general in nature and intended for training purposes only. Actual performance of activities described in this manual requires compliance with all applicable operating, service, maintenance, and safety procedures under the direction of qualified personnel. References in this manual to patented or proprietary devices do not constitute a recommendation of their use.

PEARSON

10 9 8 7 6 5 4 3 2 1
ISBN 0-13-604508-1
ISBN 13 978-0-13-604508-3

PREFACE

TO THE TRAINEE

By deciding to begin training as an industrial coatings application specialist, you are selecting a career that offers variety and unlimited opportunity. Industrial coatings applicators specialize in painting industrial structures to prevent deterioration. They may apply protective coatings to steel bridges to fight corrosion, or they may coat interior and exterior facilities and equipment such as storage tanks, plant buildings, lockers, piping, structural steel, and ships.

Industrial coatings specialists play a very important role. Each year, the total cost of corrosion to the U.S. economy is $276 billion dollars, or approximately 3 percent of the nation's GDP (gross domestic product). That's a lot of deterioration and much of it can be prevented by industrial coatings specialists. By applying industrial coatings, specialists are adding years to the life expectancy of industrial structures.

Opportunities for industrial painters should be excellent as the positions available are greater than the pool of qualified individuals to fill them (source: BLS). Industrial structures that require painting are everywhere, too. Having a job in this field could mean working on maintenance painting of bridges in the U.S or Sweden, a new skyscraper in Shanghai or Toronto, a ship in port in Japan or Korea, or a new pipeline installation in Argentina or Kazikstan. There is a high international demand for qualified coatings specialists as well. How you become qualified is the next step, and by opening this book, you've just taken it.

Welcome to the NCCER-NACE Industrial Coatings Applicator Training Program. The primary goal of this program is to set a globally-recognized standard for training of paint applicator personnel so that both employers and employees will benefit. A trainee who successfully progresses through this NCCER-accredited NCCER-NACE Industrial Coatings Applicator Program will become part of a national registry for craft workers and will receive a certificate for each level of training completed. If the trainee applies for a job with a participating sponsor, a transcript of training will be available to the sponsor. The continuity of this program is portable—a trainee who begins employment with another participating sponsor may continue his or her NCCER-NACE training with an uninterrupted transcript in the registry.

We wish you success as you embark on your first year of training and hope that you'll continue through the two next skill levels of the program. There are close to a half a million painters working in the industry and, as most can tell you, there are many opportunities awaiting those with the skills and the desire to move forward.

We also invite you to visit the NCCER website at **www.nccer.org** for the latest releases, training information, newsletter, and much more. You can also reference the Pearson product catalog online at **www.crafttraining.com**. Your feedback is welcome. You may email your comments to **curriculum@nccer.org** or send general comments and inquiries to **info@nccer.org**.

NCCER STANDARDIZED CURRICULA

NCCER is a not-for-profit 501(c)(3) education foundation established in 1996 by the world's largest and most progressive construction companies and national construction associations. It was founded to address the severe workforce shortage facing the industry and to develop a standardized training process and curricula. Today, NCCER is supported by hundreds of leading construction and maintenance companies, manufacturers, and national associations. The NCCER Standardized Curricula was developed by NCCER in partnership with Pearson Education, Inc., the world's largest educational publisher.

Some features of NCCER's Standardized Curricula are as follows:

- An industry-proven record of success
- Curricula developed by the industry for the industry
- National standardization providing portability of learned job skills and educational credits
- Compliance with the Office of Apprenticeship requirements for related classroom training (CFR 29:29)
- Well-illustrated, up-to-date, and practical information

NCCER also maintains a Registry that provides transcripts, certificates, and wallet cards to individuals who have successfully completed modules of NCCER's Standardized Curricula. *Training programs must be delivered by an NCCER Accredited Training Sponsor in order to receive these credentials.*

NACE INTERNATIONAL

NACE International has more than 19,000 members in 100 countries networking on many corrosion control technologies. NACE's largest program is the NACE Coating Inspector Program, providing comprehensive training and certification for coating inspectors who often work alongside applicators to ensure lasting success of protective coating systems. Paint applicators considering a move into the coating inspection field are encouraged to consider coating inspection training and certification, described at **www.nace.org**. NACE International also is a leader in the development of written standards for the coatings industry, which can be accessed at no charge, by NACE members, at **www.nace.org**.

Contents

NCCER Standardized Curricula

NCCER's training programs comprise over 80 construction, maintenance, and pipeline areas and include skills assessments, safety training, and management education.

Boilermaking
Cabinetmaking
Carpentry
Concrete Finishing
Construction Craft Laborer
Construction Technology
Core Curriculum: Introductory Craft Skills
Drywall
Electrical
Electronic Systems Technician
Heating, Ventilating, and Air Conditioning
Heavy Equipment Operations
Highway/Heavy Construction
Hydroblasting
Industrial Coating and Lining Application Specialist
Industrial Maintenance Electrical and Instrumentation Technician
Industrial Maintenance Mechanic
Instrumentation
Insulating
Ironworking
Masonry
Millwright
Mobile Crane Operations
Painting
Painting, Industrial
Pipefitting
Pipelayer
Plumbing
Reinforcing Ironwork
Rigging
Scaffolding
Sheet Metal
Signal Person
Site Layout
Sprinkler Fitting
Tower Crane Operator
Welding

MARITIME

Maritime Industry Fundamentals
Maritime Pipefitting
Maritime Structural Fitter

GREEN/SUSTAINABLE CONSTRUCTION

Building Auditor
Fundamentals of Weatherization
Introduction to Weatherization
Sustainable Construction Supervisor
Weatherization Crew Chief
Weatherization Technician
Your Role in the Green Environment

ENERGY

Alternative Energy
Introduction to the Power Industry
Introduction to Solar Photovoltaics
Introduction to Wind Energy
Power Industry Fundamentals
Power Generation Maintenance Electrician
Power Generation I&C Maintenance Technician
Power Generation Maintenance Mechanic
Power Line Worker
Power Line Worker: Distribution
Power Line Worker: Substation
Power Line Worker: Transmission
Solar Photovoltaic Systems Installer
Wind Turbine Maintenance Technician

PIPELINE

Control Center Operations, Liquid
Corrosion Control
Electrical and Instrumentation
Field Operations, Liquid
Field Operations, Gas
Maintenance
Mechanical

SAFETY

Field Safety
Safety Orientation
Safety Technology

SUPPLEMENTAL TITLES

Applied Construction Math
Tools for Success

MANAGEMENT

Fundamentals of Crew Leadership
Project Management
Project Supervision

SPANISH TITLES

Acabado de concreto: nivel uno
Aislamiento: nivel uno
Albañilería: nivel uno
Andamios
Carpintería: Formas para carpintería, nivel tres
Currículo básico: habilidades introductorias del oficio
Electricidad: nivel uno
Herrería: nivel uno
Herrería de refuerzo: nivel uno
Instalación de rociadores: nivel uno
Instalación de tuberías: nivel uno
Instrumentación: nivel uno, nivel dos, nivel tres, nivel cuatro
Orientación de seguridad
Paneles de yeso: nivel uno
Seguridad de campo

Acknowledgments

This curriculum was revised as a result of the farsightedness and leadership of the following sponsors:

Basic Industries
Protherm
MARK 10 Resource Group, Inc
CSI Coating Systems
Poncio Coating Consultants

This curriculum would not exist were it not for the dedication and unselfish energy of those volunteers who served on the Authoring Team. A sincere thanks is extended to the following:

David Hurtado
Richard McGinty
Michael O'Brien
Steve Poncio
Gil Rogers
Tom Swan
John Troeger

NCCER PARTNERS

American Fire Sprinkler Association
Associated Builders and Contractors, Inc.
Associated General Contractors of America
Association for Career and Technical Education
Association for Skilled and Technical Sciences
Construction Industry Institute
Construction Users Roundtable
Construction Workforce Development Center
Design Build Institute of America
GSSC – Gulf States Shipbuilders Consortium
ISN
Manufacturing Institute
Mason Contractors Association of America
Merit Contractors Association of Canada
NACE International
National Association of Minority Contractors
National Association of Women in Construction
National Insulation Association
National Technical Honor Society
National Utility Contractors Association
NAWIC Education Foundation
North American Crane Bureau
North American Technician Excellence
Pearson
Pearson Qualifications International
Prov
SkillsUSA®
Steel Erectors Association of America
U.S. Army Corps of Engineers
University of Florida, M. E. Rinker School of Building Construction
Women Construction Owners & Executives, USA

Industrial Coating and Lining Application Specialist

Modules from the

Core Curriculum: Introductory Craft Skills

00101-04
Basic Safety

00106-04
Basic Rigging

A Message About

Core Curriculum: Introductory Craft Skills

The task training modules in this section are from NCCER's *Core Curriculum: Introductory Craft Skills*. In the NCCER-NACE Industrial Coating and Lining Application Specialist training program, two modules (35 instructional hours) from the *Core* — 00101-04: *Basic Safety*; and 00106-04: *Basic Rigging* — are considered prerequisite to the trade-specific training you will receive in Level One.

For most of NCCER's craft training programs, a full 72.5 instructional hours from the *Core Curriculum* are prerequisite to any Level One completion or credential. This is because the majority of NCCER's material is oriented to the building trades. The NCCER-NACE Industrial Coating and Lining Application Specialist training program, however, applies primarily to the industrial workplace.

The authoring team who oversaw the development of the NCCER-NACE Industrial Coating and Lining Application Specialist training program determined that the following two *Core* modules were integral to your training for the following reasons:

- In *Basic Safety*, the safety obligations of workers, supervisors, and managers and the role of OSHA regulations in maintaining a safe workplace are defined. Your recognition of job-site hazards and your awareness of safety procedures are vital to your training as an industrial coating and lining application specialist.
- *Basic Rigging* explains how ropes, chains, hoists, loaders and cranes are used to move material and equipment from one location to another on a job site. Depending upon your job location, a knowledge of rigging principles and rigging inspection criteria could be essential to getting the job done.

After successful completion of these two important building blocks, you will be ready to begin your trade-specific training in industrial coating and lining application.

Basic Safety
00101-04

**Build America Merit Winner—
Building New**

The McCormick Tribune Campus Center and the renovation of the historic Illinois Institute of Technology Commons Building in Chicago, Illinois, was completed by the Gilbane Building Company. This bold and innovative project included the replacement of the Chicago Transit Authority subway track supports, the construction of an elliptical acoustic tube around the elevated subway tracks, and the construction and renovation of the campus center and commons building.

00101-04
Basic Safety

Topics to be presented in this module include:

Overview

Construction safety has advanced a great deal over the years. There was a time in the industry when very few safety precautions were taken. Workers often worked without hard hats, gloves, eye protection, fall protection, or safety shoes. This resulted in illnesses, injuries, and deaths. Today, with the help of the Occupational Safety and Health Administration (OSHA) and state and local rules and regulations, the construction industry is becoming safer.

Rules and regulations are only a part of what contributes to a safe work environment. Construction workers have the ultimate responsibility for their safety and the safety of others. One way to remain safe is always to use appropriate personal protective equipment (PPE). The type of personal protective equipment will vary depending on your job. For example, a heavy machine operator is not required to wear the same face shield a welder is required to wear.

Proper training is also an essential part of creating a safe work environment. Training must be provided to all workers to ensure that they perform work and use and maintain all tools and equipment safely. When a worker does not know how to use a tool or equipment or has not been properly trained to use personal protective equipment, he or she should immediately tell a supervisor and ask for the necessary training. Safe behavior and safe equipment help to prevent accidents and injury.

Objectives

When you have completed this module, you will be able to do the following:

1. Explain the role that safety plays in the construction crafts.
2. Describe the meaning of job-site safety.
3. Describe the characteristics of a competent person and a qualified person.
4. Explain the appropriate safety precautions to take around common job-site hazards.
5. Demonstrate the use and care of appropriate personal protective equipment (PPE).
6. Properly don and remove personal protective equipment (safety goggles, hard hat, and personal fall protection).
7. Follow the safety procedures required for lifting heavy objects.
8. Describe safe behavior on and around ladders and scaffolds.
9. Explain the importance of hazard communications (HazCom) and material safety data sheets (MSDSs).
10. Describe fire prevention and firefighting techniques.
11. Define safe work procedures to use around electrical hazards.

Key Trade Terms

Apparatus
Arc
Arc welding
Combustible
Competent person
Concealed receptacle
Confined space
Cross-bracing
Dross
Electrical distribution panel
Excavation
Extension ladder
Flammable
Flash
Flashback
Flash burn
Flash goggles
Flash point
Ground
Guarded
Hand line
Hazard Communication Standard (HazCom)
Lanyard
Lockout/tagout
Management system
Material safety data sheet (MSDS)
Maximum intended load
Mid-rail
Occupational Safety and Health Administration (OSHA)
Permit-required confined space
Personal protective equipment (PPE)
Planked
Proximity work
Qualified person
Respirator
Scaffold
Shoring
Signaler
Slag
Stepladder
Straight ladder
Switch enclosure
Toeboard
Top rail
Trench
Trencher
Welding shield
Wind sock

Required Trainee Materials

1. Sharpened pencils and paper
2. Appropriate personal protective equipment

Prerequisites

There are no prerequisites for this module.

This course map shows all of the modules in *Core Curriculum: Introductory Craft Skills.* The suggested training order begins at the bottom and proceeds up. Skill levels increase as you advance on the course map. The local Training Program Sponsor may adjust the training order.

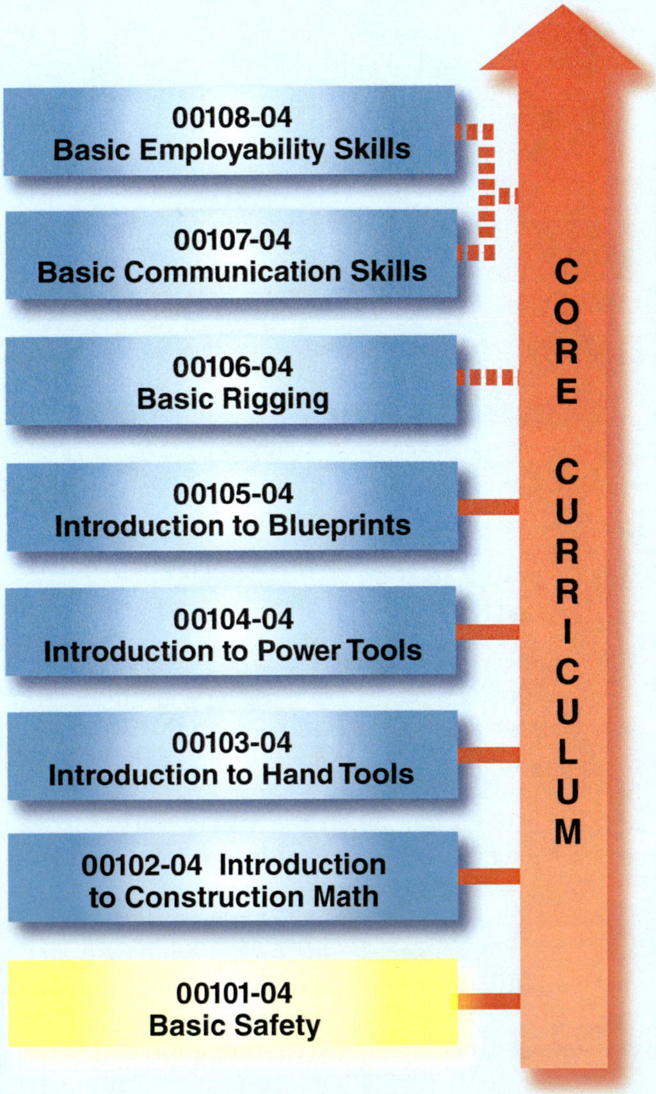

101CMAP.EPS

1.0.0 ◆ INTRODUCTION

When you take a job, you have a safety obligation to your employer, co-workers, family, and yourself. In exchange for your wages and benefits, you agree to work safely. You are also obligated to make sure anyone you work with is working safely. Your employer is likewise obligated to maintain a safe workplace for all employees. The ultimate responsibility for on-the-job safety, however, rests with you. Safety is part of your job. In this module, you will learn to ensure your safety and that of the people you work with by adhering to the following rules:

- Follow safe work practices and procedures.
- Inspect safety equipment before use.
- Use safety equipment properly.

To take full advantage of the wide variety of training, job, and career opportunities the construction industry offers, you must first understand the importance of safety. Successful completion of this module will be your first step toward achieving this goal. Other modules offer more detailed explanations of safety procedures and opportunities to practice them.

DID YOU KNOW?

Safety training is required for all activities. Never operate tools, machinery, or equipment without prior training. Always refer to the manufacturer's instructions.

2.0.0 ◆ ACCIDENTS: CAUSES AND RESULTS

Your boss might say, "I want my company to have a perfect safety record." What does that mean? A safety record is more than the number of days a company has worked without an accident. Safety is a learned behavior and attitude. Safety is a way of working. The time you spend learning and practicing safety procedures can save your life and the lives of others.

What causes accidents? Both poor behavior and poor working conditions can cause accidents. You can help prevent accidents by learning safe work habits and understanding what causes accidents. Accidents cost billions of dollars each year and cause much needless suffering. The National Safety Council estimates that the organized safety movement has saved more than 4.2 million lives since it began in 1913. This section examines why accidents happen and how you can help prevent them.

The lessons you will learn in this module will help you work safely. You will be able to spot and avoid hazardous conditions on the job site. By following safety procedures and being aware of the need for safety, you will help keep your workplace free from accidents and protect yourself and your co-workers from injury or even death.

2.1.0 What Causes Accidents?

You may already know some of the main causes of accidents. They include the following:

- Failure to communicate
- Poor work habits
- Alcohol or drug abuse
- Lack of skill
- Intentional acts
- Unsafe acts
- Rationalizing risks
- Unsafe conditions
- **Management system** failure

We will discuss each of these causes in the following sections.

2.1.1 Failure to Communicate

Many accidents happen because of a lack of communication. For example, you may learn how to do things one way on one job, but what happens when you go to a new job site? You need to communicate with the people at the new job site to find out whether they do things the way you have learned to do them. If you do not communicate clearly, accidents can happen. Remember that different people, companies, and job sites do things in different ways.

If you think that people know something without talking with them about it, then you are assuming that they know. Assuming that other people know and will do what you think they will do can cause accidents.

All work sites have specific markings and signs to identify hazards and provide emergency information (see *Figure 1*). Learn to recognize these types of signs:

- Informational
- Safety
- Caution
- Danger
- Temporary warnings

Informational markings or signs provide general information. These signs are blue. The following are considered informational signs:

- No Admittance
- No Trespassing
- For Employees Only

Safety signs give general instructions and suggestions about safety measures. The background on these signs is white; most have a green panel with white letters. These signs tell you where to find such important areas as the following:

- First-aid stations
- Emergency eye-wash stations
- Evacuation routes
- **Material safety data sheet (MSDS)** stations
- Exits (usually have white letters on a red field)

Caution markings or signs tell you about potential hazards or warn against unsafe acts. When you see a caution sign, protect yourself against a possible hazard. Caution signs are yellow and have a black panel with yellow letters. They may give you the following information:

- Hearing and eye protection are required.
- **Respirators** are required.
- Smoking is not allowed.

Danger markings or signs tell you that an immediate hazard exists and that you must take certain precautions to avoid an accident. Danger signs are red, black, and white. They may indicate the presence of the following:

- Defective equipment
- **Flammable** liquids and compressed gases
- Safety barriers and barricades
- Emergency stop button
- High voltage

Safety tags are temporary warnings of immediate and potential hazards. They are not designed to replace signs or to serve as permanent means of protection. Learn to recognize the standard accident prevention signs and tags (see *Table 1*). Other important guidelines regarding communication are covered in the module, *Basic Communication Skills*.

101F01A.EPS

INFORMATION SIGN

101F01B.EPS

SAFETY SIGN

101F01C.EPS

CAUTION SIGN

101F01D.EPS

DANGER SIGN

Figure 1 ◆ Communication tags/signs.

Table 1 Tags and Signs

Basic Stock (background)	Safety Colors (ink)	Message(s)
White	Red panel with white or gray letters	Do Not Operate Do Not Start
White	Black square with a red oval and white letters	Danger Unsafe Do Not Use
Yellow	Black square with yellow letters	Caution
White	Black square with white letters	Out of Order Do Not Use
Yellow	Red/magenta (purple) panel with black letters and a radiation symbol	Radiation Hazard
White	Fluorescent orange square with black letters and a biohazard symbol	Biological Hazard

WARNING!

Never assume anything! It never hurts to ask questions, but disaster can result if you don't ask. For example, do not assume that an electrical current is turned off. First ask whether the current is turned off, then check it yourself to be completely safe.

2.1.2 Poor Work Habits

Poor work habits can cause serious accidents. Examples of poor work habits are procrastination, carelessness, and horseplay. Procrastination, or putting things off, is a common cause of accidents. For example, delaying the repair, inspection, or cleaning of equipment and tools can cause accidents. If you try to push machines and equipment beyond their operating capacities, you risk injuring yourself and your co-workers.

Machines, power tools, and even a pair of pliers can hurt you if you don't use them safely. It is your responsibility to be careful. Tools and machines don't know the difference between wood or steel and flesh and bone.

Work habits and work attitudes are closely related. If you resist taking orders, you may also resist listening to warnings. If you let yourself be easily distracted, you won't be able to concentrate. If you aren't concentrating, you could cause an accident.

Your safety is affected not only by how you do your work, but also by how you act on the job site. This is why most companies have strict policies for employee behavior. Horseplay and other inappropriate behavior are forbidden. Workers who engage in horseplay and other inappropriate behavior on the job site may be fired.

These strict policies are for your protection. There are many hazards on construction sites. Each person's behavior—at work, on a break, or at lunch—must follow the principles of safety.

DID YOU KNOW?

Dull blades cause more accidents than sharp ones. If you do not keep your cutting tools sharpened, they won't cut very easily. When you have a hard time cutting, you exert more force on the tool. When that happens, something is bound to slip. And when something slips, you can get cut.

The man pouring the water on his co-worker in *Figure 2* may look like he's just having fun by playing a prank on his co-worker. In fact, what he's doing could cause his co-worker serious, even fatal, injury. If you horse around on the job, play pranks, or don't concentrate on what you are doing, you are showing a poor work attitude. That can lead to a serious accident.

DID YOU KNOW?

Quick Quiz

Look at the man who's trying to play a prank on his co-worker in *Figure 2*. How many things are wrong in this picture? What other safety rules is he breaking?

- No hard hat
- No safety goggles
- No gloves
- Breaking shop rules

DID YOU KNOW?

Stress creates a chemical change in your body. Although stress may heighten your hearing, vision, energy, and strength, long-term stress can harm your health.

Not all stress is job-related; some stress develops from the pressures of dealing with family and friends and daily living. In the end, your ability to handle and manage your stress determines whether stress hurts or helps you. Use common sense when you are dealing with stressful situations. For example, consider the following:

- Keep daily occurrences in perspective. Not everything is worth getting upset, angry, or anxious about.
- When you have a particularly difficult workday scheduled, get plenty of rest the night before.
- Manage your time. The feeling of always being behind creates a lot of stress. Waiting until the last minute to finish an important task adds unnecessary stress.
- Talk to your supervisor. Your supervisor may understand what is causing your stress and may be able to suggest ways to manage it better.

101F02.EPS

Figure 2 ◆ Horseplay can be dangerous.

2.1.3 Alcohol and Drug Abuse

Alcohol and drug abuse costs the construction industry millions of dollars a year in accidents, lost time, and lost productivity. The true cost of alcohol and drug abuse is much more than just money, of course. Abuse can cost lives. Just as drunk driving kills thousands of people on our highways every year, alcohol and drug abuse kills on the construction site. Examine the person in *Figure 3*. Would you want to be like him or be working near him?

Using alcohol or drugs creates a risk of injury for everyone on a job site. Many states have laws that prevent workers from collecting insurance benefits if they are injured while under the influence of alcohol or illegal drugs.

Would you trust your life to a crane operator who was high on drugs? Would you bet your life on the responses of a co-worker using alcohol or drugs? Alcohol and drug abuse have no place in the construction industry. A person on a construction site who is under the influence of alcohol or drugs is an accident waiting to happen—possibly a fatal accident.

People who work while using alcohol or drugs are at risk of accident or injury, and their co-workers are at risk as well. That's why your employer probably has a formal substance abuse policy. You should know that policy and follow it for your own safety.

You don't have to be abusing illegal drugs such as marijuana, cocaine, or heroin to create a job hazard. Many prescription and over-the-counter drugs, taken for legitimate reasons, can affect your ability to work safely. Amphetamines, barbiturates, and antihistamines are only a few of the legal drugs that can affect your ability to work safely or to operate machinery.

CAUTION

If your doctor prescribes any medication that you think might affect your job performance, ask about its effects. Your safety and the safety of your co-workers depend on everyone being alert on the job.

Figure 3 ◆ How many violations can you identify?

Do yourself and the people you work with a big favor. Be aware of and follow your employer's substance abuse policy. Avoid any substances that can affect your job performance. The life you save could be your own.

DID YOU KNOW?

Quick Quiz

If the man in *Figure 3* doesn't kill himself first, he will almost certainly kill someone else eventually. How many violations can you identify? In this scene alone, the man is

- Consuming alcohol on the job site
- Not following proper safety procedures for operating a motorized vehicle near hazardous materials
- Not using both hands to drive the vehicle
- Not wearing a seat/safety belt
- Not wearing a hard hat
- Not wearing safety glasses or goggles

2.1.4 Lack of Skill

You should learn and practice new skills under careful supervision. Never perform new tasks alone until you've been checked out by a supervisor.

Lack of skill can cause accidents quickly. For example, suppose you are told to cut some 2 × 8s with a circular saw, but you aren't skilled with that tool. A basic rule of circular saw operation is never to cut without a properly functioning guard. Because you haven't been trained, you don't know this. You find that the guard on the saw is slowing you down. So you jam the guard open with a small block of wood. The result could be a serious accident. Proper training can prevent this type of accident.

WARNING!

Never operate a power tool until you have been trained to use it. You can greatly reduce the chances of accidents by learning safety rules for each task you perform.

2.1.5 Intentional Acts

When someone purposely causes an accident, it is called an intentional act. Sometimes an angry or dissatisfied employee may purposely create a situation that leads to property damage or personal injury. If someone you are working with threatens to get even or pay back someone, let your supervisor know at once.

2.1.6 Unsafe Acts

An unsafe act is a change from an accepted, normal, or correct procedure that usually causes an accident. It can be any conduct that causes unnecessary exposure to a job-site hazard or that makes an activity less safe than usual. Here are examples of unsafe acts:

- Failing to use **personal protective equipment (PPE)**
- Failing to warn co-workers of hazards
- Lifting improperly
- Loading or placing equipment or supplies improperly
- Making safety devices (such as saw guards) inoperable
- Operating equipment at improper speeds
- Operating equipment without authority
- Servicing equipment in motion
- Taking an improper working position
- Using defective equipment
- Using equipment improperly

2.1.7 Rationalizing Risk

Everybody takes risks every day. When you get in your car to drive to work, you know there is a risk of being involved in an accident. Yet when you drive using all the safety practices you have learned, you know that there is a good chance that you will arrive at your destination safely. Driving is an appropriate risk because you have some control over your own safety and that of others.

Some risks are not appropriate. On the job, you must never take risks that endanger yourself or others just because you can make an excuse for doing so. This is called rationalizing risk. Rationalizing risk means ignoring safety warnings and practices. For example, because you are late for work, you might decide to run a red light. By trying to save time, you could cause a serious accident.

The following are common examples of rationalized risks on the job:

- Crossing boundaries because no activity is in sight
- Not wearing gloves because it will take only a minute to make a cut
- Removing your hard hat because you are hot and you cannot see anyone working overhead
- Not tying off your fall protection because you only have to lean over by about a foot

Think about the job before you do it. If you think that it is unsafe, then it is unsafe. Stop working until the job can be done safely. Bring your concerns to the attention of your supervisor. Your health and safety, and that of your co-workers, make it worth taking extra care.

DID YOU KNOW?

Most workers who die from falls were wearing harnesses but had failed to tie off properly. Always follow the manufacturer's instructions when wearing a harness. Know and follow your company's safety procedures when working on roofs, ladders, and other elevated locations.

2.1.8 Unsafe Conditions

An unsafe condition is a physical state that is different from the acceptable, normal, or correct condition found on the job site. It usually causes an accident. It can be anything that reduces the degree of safety normally present. The following are some examples of unsafe conditions:

- Congested workplace
- Defective tools, equipment, or supplies
- Excessive noise
- Fire and explosive hazards
- Hazardous atmospheric conditions (such as gases, dusts, fumes, and vapors)
- Inadequate supports or guards
- Inadequate warning systems
- Poor housekeeping
- Poor lighting
- Poor ventilation
- Radiation exposure
- Unguarded moving parts such as pulleys, drive chains, and belts

2.1.9 Management System Failure

Sometimes the cause of an accident is failure of the management system. The management system should be designed to prevent or correct the acts and conditions that can cause accidents. If the management system did not do these things, that system failure may have caused the accident.

What traits could mean the difference between a management system that fails and one that succeeds? A few important traits of a good management system follow:

- The company puts safety policies and procedures in writing.
- The company distributes written safety policies and procedures to each employee.
- The company reviews safety policies and procedures periodically.
- The company enforces all safety policies and procedures fairly and consistently.
- The company evaluates supplies, equipment, and services to see whether they are safe.
- The company provides regular, periodic safety training for employees.

2.2.0 Housekeeping

In construction, housekeeping means keeping your work area clean and free of scraps or spills. It also means being orderly and organized. You must store your materials and supplies safely and label them properly. Arranging your tools and equipment to permit safe, efficient work practices and easy cleaning is also important.

If the work site is indoors, make sure it is well lighted and ventilated. Don't allow aisles and exits to be blocked by materials and equipment. Make sure that flammable liquids are stored in safety cans. Oily rags must be placed only in approved, self-closing metal containers. Remember that the major goal of housekeeping is to prevent accidents. Good housekeeping reduces the chances for slips, fires, explosions, and falling objects.

Here are some good housekeeping rules:

- Remove all scrap material and lumber with nails sticking out from work areas.
- Clean up spills to prevent falls.
- Remove all **combustible** scrap materials regularly.
- Make sure you have containers for the collection and separation of refuse. Containers for flammable or harmful refuse must be covered.
- Dispose of wastes often.
- Store all tools and equipment when you're finished using them.

Another term for good housekeeping is pride of workmanship. If you take pride in what you are doing, you won't let trash build up around you. The old saying "A place for everything and everything in its place" is the right idea on the job site.

2.3.0 Company Safety Policies and OSHA Regulations

The mission of the **Occupational Safety and Health Administration (OSHA)** is to save lives, prevent injuries, and protect the health of America's workers. To accomplish this, federal and state governments work in partnership with the 111 million working men and women and their 7 million employers who are covered by the Occupational Safety and Health Act (OSH Act) of 1970.

Nearly every worker in the nation comes under OSHA's jurisdiction. There are some exceptions, such as miners, transportation workers, many public employees, and the self-employed.

2.3.1 The Code of Federal Regulations

The *Code of Federal Regulations* (CFR) *Part 1910* covers the OSHA standards for general industry. *CFR Part 1926* covers the OSHA standards for the construction industry. Either or both may apply to you, depending on where you are working and what you are doing. If a job-site condition is covered in the CFR book, then that standard must be used. However, if a more stringent requirement is listed in *CFR 1910*, it should also be met. Check with your supervisor to find out which standards apply to your job.

If a standard does not specifically address a hazard, the general duty clause must be invoked. This clause, found in *Section 5(a)1* of the OSH Act, summarizes the intent of the law:

> *Each Employer—shall furnish to each of his employees employment and a place of employment which are free from recognized hazards that are causing or are likely to cause death or serious physical harm to his employees.*

DID YOU KNOW?

All of OSHA's safety requirements in the *Code of Federal Regulations* apply to residential as well as commercial construction. In the past, OSHA enforced safety only at commercial sites. The increasing rate of accidents at residential sites led OSHA to enforce safety guidelines for the building of houses and townhouses. Today, however, OSHA still focuses its enforcement efforts on commercial construction.

2.3.2 Violations

Employers who violate OSHA regulations can be fined. The fines are not always high, but they can harm a company's reputation for safety. Fines for serious safety violations can be as high as $70,000 for each violation that was done willfully. In 2002, more than 78,000 fines were levied at a cost of $70,000 per violation.

2.3.3 Compliance

Just as employers are responsible to OSHA for compliance, employees must comply with their company's safety policies and rules. Employers are required to identify hazards and potential hazards within the workplace and eliminate them, control them, or provide protection from them. This can only be done through the combined efforts of the employer and employees. Employers must provide written programs and training on hazards, and employees must follow the procedures. You, as the employee, must read and understand the OSHA poster at your job site explaining your rights and responsibilities. If you are unsure where the OSHA poster is, ask your supervisor.

To help employers provide a safe workplace, OSHA requires companies to provide a **competent person** to ensure the safety of the employees. In *OSHA 29 CFR 1926*, OSHA defines a competent person as follows:

> *A person who can identify working conditions or surroundings that are unsanitary, hazardous, or dangerous to employees and who has authorization to correct or eliminate these conditions promptly.*

In comparison, *OSHA 29 CFR 1926* defines a **qualified person** as follows:

> *Someone who, by possession of a recognized degree, certificate, or professional standing, or who by extensive knowledge, training, and experience, has successfully demonstrated his ability to solve or resolve problems relating to the subject matter, work, or the project.*

In other words, a competent person is experienced and knowledgeable about the specific operation and has the authority from the employer to correct the problem or shut down the operation until it is safe. A qualified person has the knowledge and experience to handle problems. A competent person is not necessarily a qualified person.

These terms will be an important part of your career. It is important for you to know who the competent person is on your job site. OSHA requires a competent person for many of the tasks you may be assigned to perform, such as **confined space** entry, ladder use, and trenching. Different individuals may be assigned as a competent person for different tasks, according to their expertise. To ensure safety for you and your co-workers, work closely with your competent person and supervisor.

2.4.0 Reporting Injuries, Accidents, and Incidents

There are three categories of on-the-job events: injuries, accidents, and incidents. An injury is anything that requires treatment, even minor first aid. An accident is anything that causes an injury or property damage. An incident is anything that could have caused an injury or damage but, because it was caught in time, did not.

You must report all on-the-job injuries, accidents, or incidents, no matter how minor, to your supervisor (see *Figure 4*). Some workers think they will get in trouble if they report minor injuries. That's not true. Small injuries, like cuts and scrapes, can later become big problems because of infection and other complications.

101F04.EPS

Figure 4 ◆ All accidents, injuries, or incidents must be reported to your supervisor.

Many employers are required to maintain a log of significant work-related injuries and illnesses using OSHA Form 300. Employee names can be kept confidential in certain circumstances. A summary of these injuries must be posted at certain intervals, although employers do not need to submit it to OSHA unless requested. Employers can calculate the total number of injuries and illnesses and compare it with the average national rates for similar companies.

By analyzing accidents, companies and OSHA can improve safety policies and procedures. By reporting an accident, you can help keep similar accidents from happening in the future.

Table 2 shows an analysis of the causes of fatal accidents in the construction industry for 1997 through 1998.

Table 2 Causes of Fatal Accidents

Cause	Percentage of Fatalities
Falls from elevation	33%
Struck by . . .	27%
Caught in or caught between . . .	16%
Electrical shock	14%

All other accidents combined accounted for only 10 percent of the total. Because of these findings, OSHA developed its Focused Inspection Program to target these four high-hazard areas. OSHA hopes to reduce accidents, injuries, and fatalities in the construction industry.

Here are explanations of the four leading hazard groups:

- Falls from elevation are accidents involving failure of, failure to provide, or failure to use appropriate fall protection.
- Struck-by accidents involve unsafe operation of equipment, machinery, and vehicles, as well as improper handling of materials, such as through unsafe rigging operations.
- Caught-in or caught-between accidents involve unsafe operation of equipment, machinery, and vehicles, as well as improper safety procedures at **trench** sites and in other confined spaces.
- Electrical shock accidents involve contact with overhead wires, use of defective tools, failure to disconnect power source before repairs, or improper **ground** fault protection.

2.5.0 Evacuation Procedures

In many work environments, specific evacuation procedures are needed. These procedures go into effect when dangerous situations arise, such as fires, chemical spills, and gas leaks. In an emergency, you must know the evacuation procedures. You must also know the signal (usually a horn or siren) that tells workers to evacuate.

When you hear the evacuation signal, follow the evacuation procedures exactly. That usually means taking a certain route to a designated assembly area and telling the person in charge that you are there. If hazardous materials are released into the air, you may have to look at the **wind sock** to see which way the wind is blowing. Different evacuation routes are planned for different wind directions. Taking the right route will keep you from being exposed to the hazardous material.

Review Questions

Section 2.0.0

1. __________ is (are) *not* a main cause of accidents.
   a. Unsafe acts
   b. Alcohol or drug abuse
   c. Weather
   d. Poor work habits

2. Blue signs or markings that provide general information such as No Trespassing are __________ signs.
   a. caution
   b. informational
   c. warning
   d. safety

3. White and green signs or markings that give general instructions or suggestions about first-aid stations, exits, and evacuation routes are __________ signs.
   a. safety
   b. danger
   c. information
   d. MSDS

4. Many states have laws that prevent workers from collecting insurance benefits if they are injured while under the influence of alcohol or illegal drugs.
   a. True
   b. False

5. __________ is an example of an unsafe act.
   a. Taking an improper working position
   b. Radiation exposure
   c. Excessive noise
   d. Using a respirator

6. All of the following are good housekeeping guidelines *except* __________.
   a. keep aisles and exits clear
   b. clean up spills
   c. place oily rags in an uncovered container
   d. dispose of wastes often

7. The __________ summarizes the intent of the *OSHA Act of 1970.*
   a. competent person clause
   b. general duty clause
   c. hazardous duty clause
   d. qualified person clause

8. If a sign has a white background and a red panel with white or gray letters, you might see __________ on it.
   a. Out of Order
   b. Danger
   c. Biological Hazard
   d. Do Not Start

9. __________ must be reported to the employer.
   a. Only major injuries
   b. Only incidents and major injuries
   c. All injuries and incidents
   d. Only incidents in which a death occurred

10. An __________ is anything that could have caused an injury or damage if it hadn't been caught in time.
    a. incident
    b. accident
    c. injury
    d. intentional act

3.0.0 ◆ HAZARDS ON THE CONSTRUCTION JOB SITE

It's impossible to list all the hazards that can exist on a construction job site. This section describes some of the more common hazards and explains how to deal with them. You may want to make a list of other hazards you think could be present on your job site and discuss them with your instructor or supervisor.

For your safety, you must know the specific hazards where you are working and how to prevent accidents and injuries. If you have questions specific to your job site, consult your supervisor.

3.1.0 Welding

Even if you're not welding, you can be injured when you are around a welding operation. The oxygen and acetylene used in gas welding are very dangerous. The cylinders containing oxygen and acetylene must be transported, stored, and handled very carefully. Always follow these safety guidelines:

- Keep the work area clean and free from potentially hazardous items such as combustible materials and grease or petroleum products.

> **WARNING!**
> Keep oxygen away from sources of flame and combustible materials, especially substances containing oil, grease, or other petroleum products. Compressed oxygen mixed with oil or grease will explode. Never use petroleum-based products around fittings that serve compressed oxygen lines.

- Use great caution when you handle compressed gas cylinders.
- Store cylinders in an upright position and separate them by metal.

> **WARNING!**
> Do not remove the protective cap unless a cylinder is secured. If the cylinder falls over and the nozzle breaks off, the cylinder will shoot off like a rocket, injuring or killing anyone in its path.

- Never look at an **arc welding** operation without wearing the proper eye protection. The **arc** will burn your eyes.

> **WARNING!**
> In an arc welding operation, even a reflected arc can harm your eyes. It is extremely important to follow proper safety procedures at and around all welding operations. Serious eye injury or even blindness can result from unsafe conditions.

- If you are welding, use the proper PPE (see *Figure 5*), including the following:
  - Full face shield with proper lens
  - Earplugs to prevent flying sparks from entering your ears
  - All-leather, gauntlet-type welder's gloves
  - High-top leather boots to prevent **slag** from dropping inside your boots
  - Cuffless trousers that cover your ankles and boot tops
  - A respirator, if necessary

> **CAUTION**
> When welding on construction job sites, a hard hat with a full face shield may be required.

- If you are welding and other workers are in the area around your work, set up **welding shields.** Make sure everyone wears **flash goggles.** These goggles protect the eyes from the **flash,** which is the sudden bright light associated with starting a welding operation.
- A welder must be protected when the welding shield on the welder's headgear is down, because the shield restricts the welder's field of vision. A helper or monitor must watch the welder and the surrounding area in case of a fire or similar emergency, or rope off the area to prevent collisions and keep other workers away from the area.
- Welded material is hot! Mark it with a sign and stay clear for a while after the welding has been completed.

> **WARNING!**
> Post a fire watch when you are welding or cutting. One person other than the welding or cutting operator must constantly scan the work area for fires. Fire watch personnel should have ready access to fire extinguishers and alarms and know how to use them. Welding and cutting operations should never be performed without a fire watch. The area where welding is done must be monitored afterwards until there is no longer a risk of fire.

Figure 5 ◆ Personal protective equipment for welding.

Pay special attention to the safety guidelines about never looking at the arc without proper eye protection. Even a brief exposure to the ultraviolet light from arc welding can cause a **flash burn** and damage your eyes badly. You may not notice the symptoms until some time after the exposure. Here are some symptoms of flash burns to the eye:

- Headache
- Feeling of sand in your eyes
- Red or weeping eyes
- Trouble opening your eyes
- Impaired vision
- Swollen eyes

If you think you may have a flash burn to your eyes, seek medical help at once.

WARNING!

Never wear contact lenses while you are welding. The ultraviolet rays may dry out the moisture beneath the contact lens, causing it to stick to your eye.

3.2.0 Flame Cutting

Many of the safety guidelines for welding apply to flame cutting as well. Cutting is not dangerous as long as you follow safety precautions. Here are some of the precautions:

- Wear appropriate PPE, including a welding hood with a filter.

- Never open the valve of an acetylene cylinder near an open flame.
- Store oxygen cylinders separately from fuel gas cylinders.
- Store acetylene cylinders in an upright position.
- Always use a friction striker to light a cutting torch.

WARNING!

Never cut galvanized metal without proper ventilation. The zinc oxide fumes given off as the galvanized material is cut are hazardous. Also, use a respirator when you are cutting galvanized material.

The cutting process results in oxides that mix with molten iron and produce **dross.** The dross is blown from the cut by the jet of cutting oxygen. Hot dross can cause severe injury or can start fires on contact with flammable materials.

Before and during welding and cutting operations, you must follow certain safety procedures. *Figure 6* shows a typical oxyacetylene welding/cutting outfit. As the operator, you must check three things:

- Hoses
- Regulators
- Work area

3.2.1 Hoses and Regulators

Use the proper hose. The fuel gas hose is usually red (sometimes black) and has a left-hand threaded nut for connecting to the torch. The oxygen hose is green and has a right-hand threaded nut for connecting to the torch.

Hoses with leaks, burns, worn places, or other defects that make them unfit for service must be repaired or replaced. When inspecting hoses, look for charred sections close to the torch. These may have been caused by **flashback,** which is the result of a welding flame flaring up and charring the hose near the torch connection. Flashback is caused by improperly mixed fuel. Also check that hoses are not taped up to cover leaks (see *Figure 7*).

WARNING!

If the torch goes out and begins to hiss, shut off the gas supply to the torch immediately. Otherwise, a flashback could occur. Never relight a torch from hot metal. Doing so could cause an explosion.

New hoses contain talc and loose bits of rubber. These materials must be removed from the hoses before the torch is connected. If they are not removed, they will clog the torch needle valves. Common industry practice is to use compressed air to blow these materials out of the hose. Always

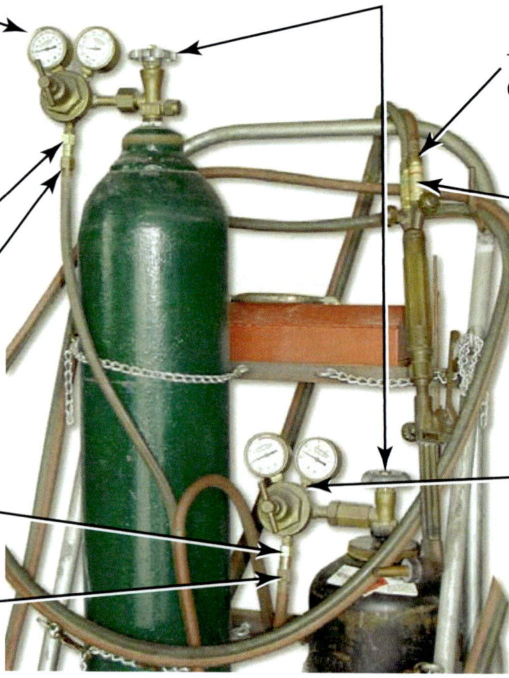

Figure 6 ◆ Typical oxyacetylene welding/cutting outfit.

make sure that the regulator valve is turned down to minimal pressures before using compressed air to clean a hose.

WARNING!
Never point a compressed air hose toward anyone. Flying debris and particles of dirt may cause serious injury.

Regulators are attached to the cylinder valve. They lower the high cylinder pressures to the required working pressures and maintain a steady flow of gas from the cylinder.

To prevent damage to regulators, always follow these guidelines:

- Never jar or shake regulators, because that can damage the equipment beyond repair.
- Always check that the adjusting screw is released before the cylinder valve is turned on.
- Always open cylinder valves slowly.

WARNING!
When opening valves, always stand to the side. Dirt that is stuck in the valve may fly out and cause serious injury.

- Once cutting or welding has been completed, fully release the adjusting screw to relieve line pressure.
- Never use oil to lubricate a regulator, because that can cause an explosion.
- Never operate fuel regulators on oxygen cylinders or oxygen regulators on fuel gas cylinders.
- Never use a defective regulator. If a regulator is not working properly, shut off the gas supply and have a qualified person repair the regulator.
- Never operate the fuel regulator above the recommended safe operating pressure.
- Never use pliers or channel locks to install or remove regulators.

3.2.2 Work Area

Before beginning a cutting or welding operation, check the area for fire hazards. Cutting sparks can fly 30 feet or more and can fall several floors. Remove any flammable material in the area or cover it with an approved fire blanket. Have an approved fire extinguisher available before starting your work.

WARNING!
The slag and products that result from cutting and welding operations can start fires and cause severe injuries. Always wear appropriate personal protective equipment, including gloves and eye protection, when cutting or welding. Do not wear clothes made of polyester when welding or cutting. Observe the safety instructions of both the manufacturer and your shop.

Always perform cutting operations in a well-ventilated area. Heating and cutting metals with an oxyfuel torch can create toxic fumes.

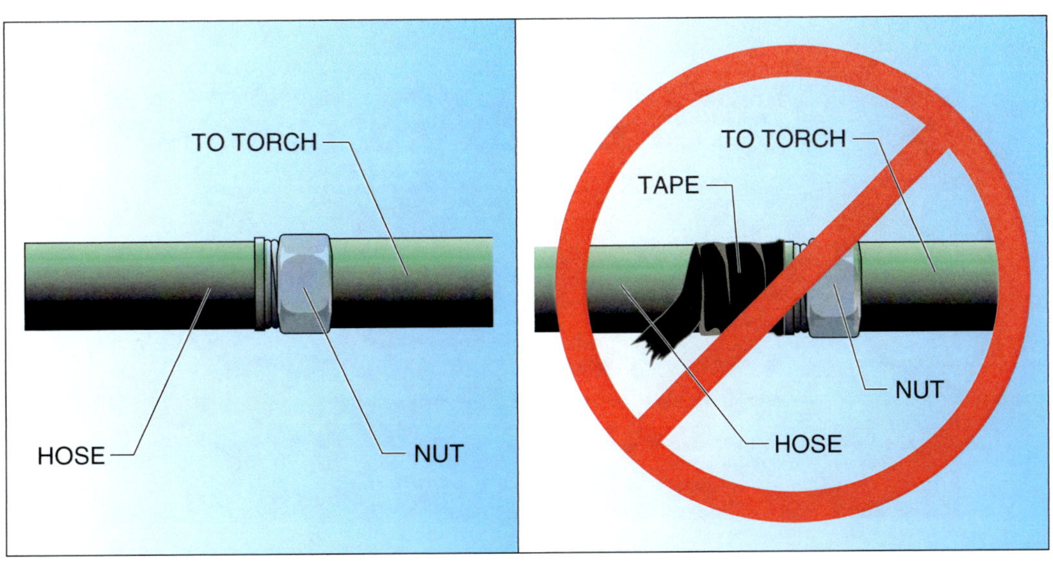

Figure 7 ◆ Proper hose connection.

Maintaining a clean and neat work area promotes safety and efficiency. When you are finished welding, be sure to do the following:

- Pick up cutting scraps.
- Sweep up any scraps or debris around the work area.
- Return cylinders and equipment to the proper places.
- Prevent fires by making sure that cut metals and dross are cooled before disposing of them.

3.3.0 Trenches and Excavations

In many construction jobs you will need to work in trenches or **excavations.** Cave-ins and falling objects are hazards in these areas. Obey the following safety rules when working around trenches and excavations:

- Never put tools, materials, or loose dirt or rocks within 2 feet of the edge of a trench. They can easily fall in and injure the people in the trench. Also, too much weight near the edge of a trench can cause a cave-in.
- Always walk around a trench; never jump over it or straddle it. You could lose your footing and fall in, or your weight could cause a cave-in.
- Never jump into a trench. Always use a ladder to get in and out.
- A stairway, ladder, or ramp must be provided for exit from any trench that is 4 or more feet deep.
- Put barricades around all trenches, as shown in *Figure 8*.
- Always follow OSHA regulations and your employer's procedures for **shoring** up a trench to prevent a cave-in. Never work beyond the shoring.
- Always follow OSHA regulations for determining the maximum allowable slope.

WARNING!

Never work on the face of either sloped excavations or excavations in which concrete slabs were used to flatten the surrounding area at levels above other workers. The workers below you may not be adequately protected from the hazard of falling, rolling, or sliding materials or equipment.

A competent person will inspect excavations daily and decide whether cave-ins or failures of protective systems could occur, and whether there are any other hazardous conditions. The competent person will conduct the inspection before any work begins and as needed throughout the shift. He or she will also inspect the excavations after every rainstorm or other hazard-increasing incident.

You cannot work in excavations that have standing water or in excavations where water is coming in unless you take precautions to protect yourself. A competent person will know what these precautions are. Always ask the competent person on site or your immediate supervisor if you have any questions about proper safety practices.

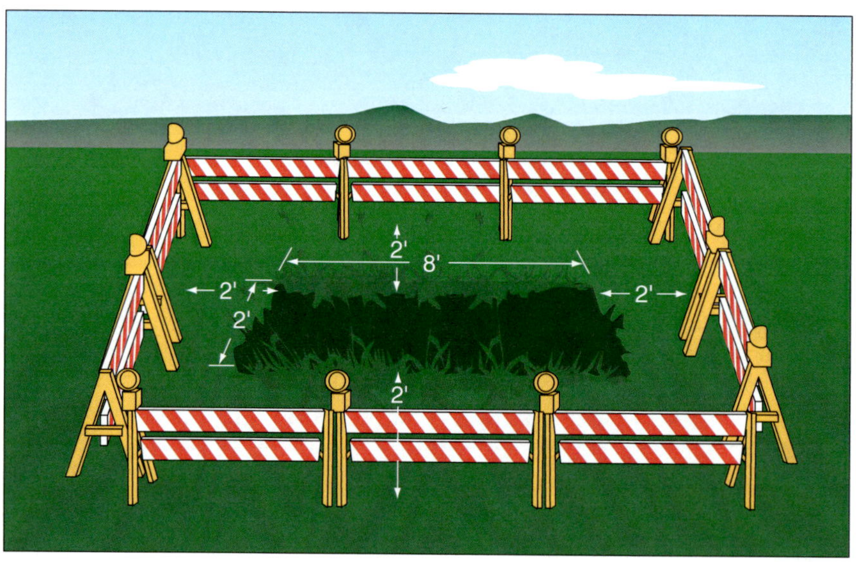

101F08.EPS

Figure 8 ◆ Barricade around a trench.

3.4.0 Proximity Work

Work that is done near a hazard but not in direct contact with it is called **proximity work.** Proximity work requires extra caution and awareness. The hazard may be hot piping, energized electrical equipment, or running motors or machinery (see *Figure 9*). You must do your work so that you do not come into contact with the nearby hazard.

You may need to put up barricades to prevent accidental contact. Lifting and rigging operations may have to be done in a way that minimizes the risk of dropping things on the hazard. A monitor may watch you and alert you if you are in danger of touching the hazard while you work.

Energized electrical equipment is very hazardous. Regulations and policies will tell you the minimum safe working distance from energized electrical conductors. You'll learn more about working with and around energized electrical equipment later in this module.

3.4.1 Pressurized or High-Temperature Systems

In many construction jobs, you must work close to tanks, piping systems, and pumps that contain pressurized or high-temperature fluids. Be aware of these two possible hazards:

- Touching a container of high-temperature fluid can cause burns (see *Figure 10*). Many industrial processes involve fluids that are as hot as several thousand degrees.
- If a container holding pressurized fluids is damaged, it may leak and spray dangerous fluids.

Any work around pressurized or high-temperature systems is proximity work. Barricades, a monitor, or both may be needed for safety (see *Figure 11*).

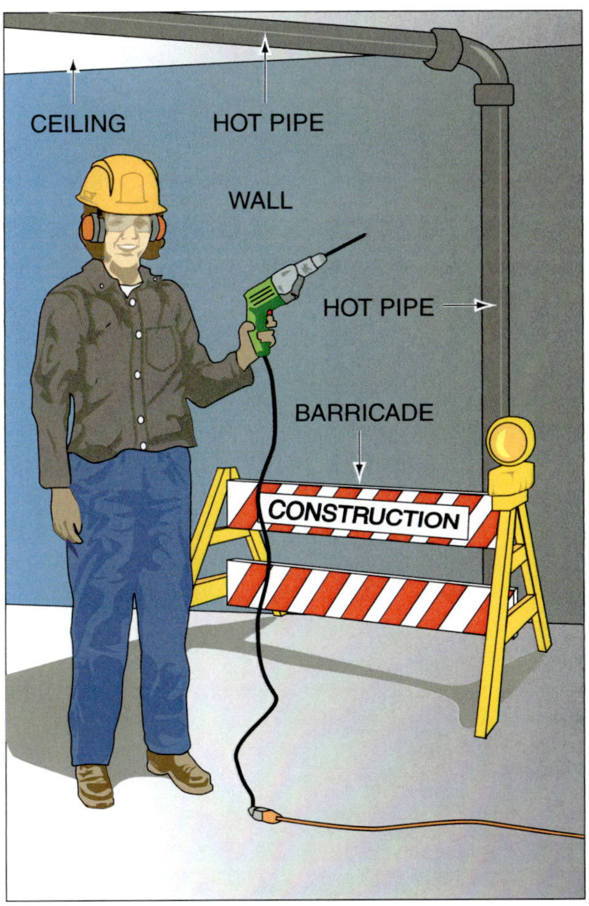

Figure 9 ◆ Proximity work.

Figure 10 ◆ Avoid touching high-temperature components.

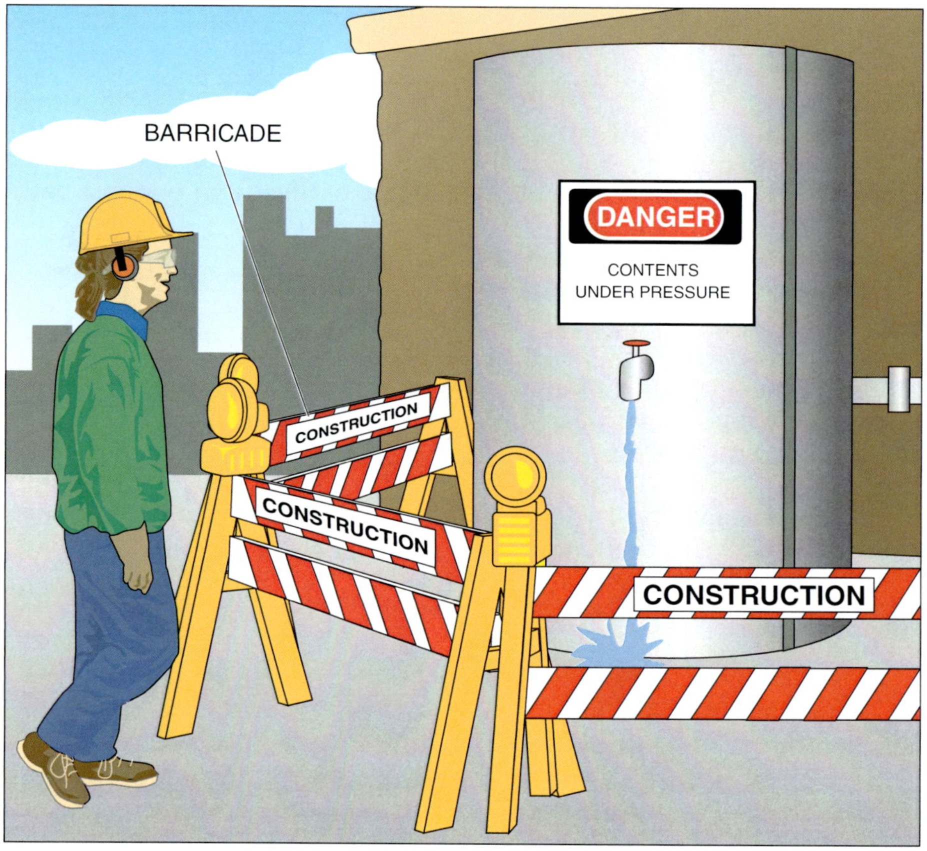

101F11.EPS

Figure 11 ◆ Work safely near pressurized or high-temperature systems.

3.5.0 Confined Spaces

Construction and maintenance work isn't always done outdoors. A lot of it is done in confined spaces. A confined space is a space that is large enough to work in but that has limited means of entry or exit. A confined space is not designed for human occupancy, and it has limited ventilation. Examples of confined spaces are tanks, vessels, silos, storage bins, hoppers, vaults, and pits (see *Figure 12*).

A **permit-required confined space** is a type of confined space that has been evaluated by a qualified person and found to have actual or potential hazards (see *Figure 13*). You must have written authorization to enter a permit-required confined space.

When equipment is in operation, many confined spaces contain hazardous gases or fluids. In addition, the work you are doing may introduce hazardous fumes into the space. Welding is an example of such work. For safety, you must take special precautions both before you enter and leave a confined space and while you work there.

Until you have been trained to work in permit-required confined spaces and have taken the needed precautions, you must stay out of them. If you aren't sure whether a confined space requires a permit, ask your supervisor. You must always follow your employer's procedures and your supervisor's instructions. Confined space procedures may include getting clearance from a safety representative before starting the work. You will be told what kinds of hazards are involved and what precautions you need to take. You will also be shown how to use the required PPE. Remember, it is better to be safe than sorry, so ask!

WARNING!

Without proper training, no employee is allowed to enter a permit-required or non-permit-required confined space. Employers are required to have programs to control entry to and hazards in both types of confined spaces.

Never work alone in a confined space. OSHA requires an attendant to remain outside a permit-required confined space. The attendant monitors entry, work, and exit.

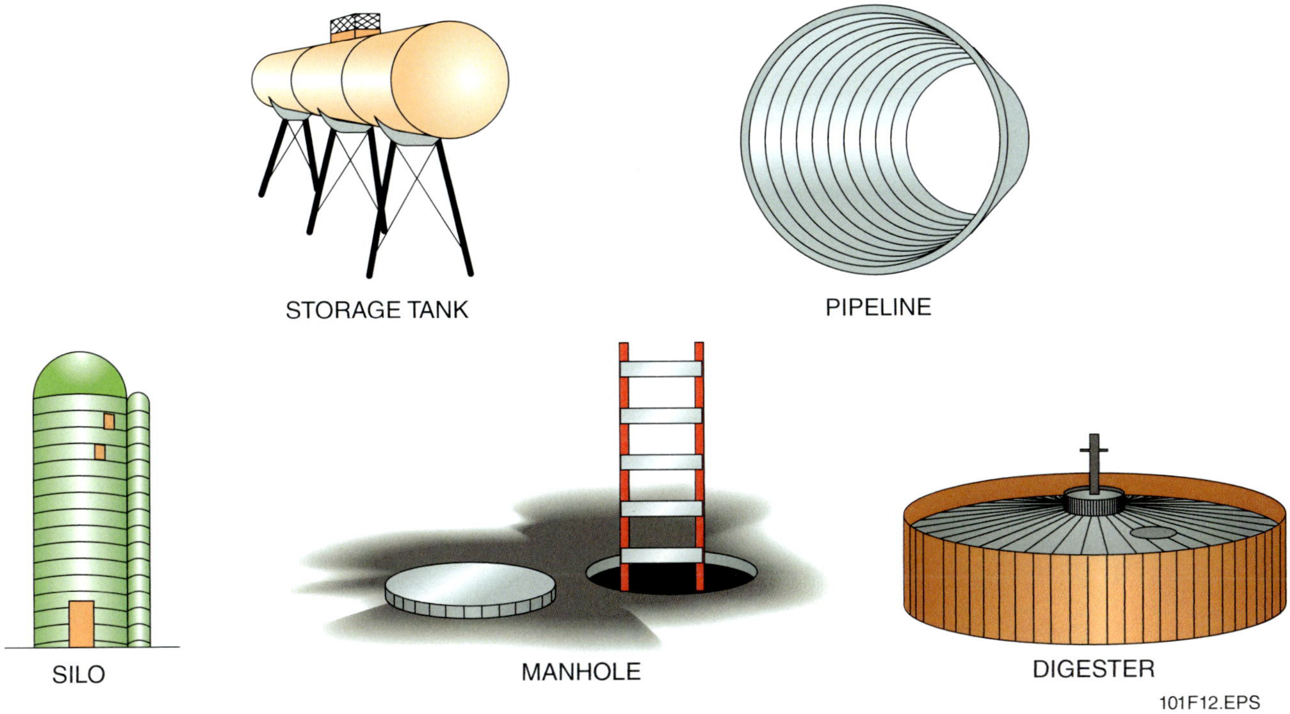

Figure 12 ◆ Examples of confined spaces.

DID YOU KNOW?

About 100 people die in confined spaces each year. Most of these deaths are caused by inhaling poisonous substances or being engulfed by a substance such as grain. The remainder of deaths are caused by lack of oxygen.

3.6.0 Motorized Vehicles

Motorized vehicles used on job sites include trucks, forklifts, backhoes, cranes, and **trenchers.** Operators must take care when driving vehicles. Helpers, riggers, and anyone else working nearby must also be careful.

If a vehicle is used indoors, ventilation of the work area is especially important. All internal combustion engines give off carbon monoxide as part of their exhaust. You cannot see, smell, or taste carbon monoxide, but it can kill you. Make sure there is good ventilation before you operate any motorized vehicles indoors.

The operator of any vehicle is responsible for the safety of passengers and the protection of the load. Follow these safety guidelines when you operate vehicles on a job site:

- Always wear a seat belt.
- Be sure that each person in the vehicle has a firmly secured seat and seat belt.
- Obey all speed limits. Reduce speed in crowded areas.
- Look to the rear and sound the horn before backing up. If your rear vision is blocked, get a **signaler** to direct you. (Hand signals are covered in the *Rigging* module.)
- Every vehicle must have a backup alarm. Make sure the backup alarm works.
- Always turn off the engine when you are fueling.
- Turn off the engine and set the brakes before you leave the vehicle.
- Never stay on or in a truck that is being loaded by excavating equipment.
- Keep windshields, rearview mirrors, and lights clean and functional.
- Carry road flares, fire extinguishers, and other standard safety equipment at all times.

WARNING!

Driving a vehicle indoors without good ventilation can make you sick or kill you because of the carbon monoxide given off by the exhaust. Carbon monoxide is especially dangerous because you cannot see, smell, or taste it.

101F13.EPS

Figure 13 ◆ Permit-required confined space.

Review Questions

Section 3.0.0

1. Oil or grease in contact with _________ will cause an explosion.
   a. a gas cylinder
   b. acetylene
   c. compressed oxygen
   d. combustible materials

2. Flash burns are caused by exposing your eyes to _________.
   a. ultraviolet light
   b. oxygen and acetylene
   c. petroleum products
   d. zinc oxide fumes

3. A _________ is attached to a cylinder valve to reduce the high cylinder pressure to the required lower working pressure.
   a. safety valve
   b. regulator
   c. torch assembly
   d. compression hose

4. A confined space _________.
   a. has a limited amount of ventilation
   b. has no means of entry
   c. is too small to work in
   d. may be entered by untrained employees

5. All internal combustion engines give off a deadly odorless, tasteless, invisible gas called _________ as part of their exhaust.
   a. carbon trioxide
   b. carbon monoxide
   c. carbon dioxide
   d. carbon cyanide

4.0.0 ◆ WORKING SAFELY WITH JOB HAZARDS

You can safely handle all the job hazards that you have learned about if you follow the rules. As long as everyone follows safety procedures, there is little risk of being hurt on the job site. In this section, you will learn about procedures and equipment used on construction sites to ensure worker safety.

4.1.0 Lockout/Tagout

A **lockout/tagout** system safeguards workers from hazardous energy while they work with machines and equipment. A lockout/tagout system protects workers from hazards such as the following:

- Acids
- Air pressure
- Chemicals
- Electricity
- Flammable liquids
- High temperatures
- Hydraulics
- Machinery
- Steam
- Other forms of energy

When people are working on or around any of these hazards, mechanical and other systems are shut down, drained, or de-energized. Tags and locks are placed on each switch, circuit breaker, valve, or other component to make sure that motors aren't started, valves aren't opened or closed, and no other changes are made that would endanger workers. Lockouts and tagouts protect workers from all possible sources of energy, including electrical, mechanical, hydraulic, thermal, pneumatic (air), and high temperature.

Generally, each lock has its own key, and the person who puts the lock on keeps the key. That person is the only one who can remove the lock. Tags have the words DANGER or CLEARANCE (see *Figure 14*).

Follow these rules for a safe lockout/tagout system:

- Never operate any device, valve, switch, or piece of equipment that has a lock or a tag attached to it.
- Use only tags that have been approved for your job site.
- If a device, valve, switch, or piece of equipment is locked out, make sure the proper tag is attached.

101F14.EPS

Figure 14 ◆ Typical safety tags.

- Lock out and tag all electrical systems when they are not in use.
- Lock out and tag pipelines containing acids, explosive fluids, or high-pressure steam during maintenance or repair.
- Tag motorized vehicles and equipment when they are being repaired and before anyone starts work. Also, disconnect or disable the starting devices.

The exact procedures for lockout/tagout may vary at different companies and job sites. Ask your supervisor to explain the lockout/tagout procedure on your job site. You must know and follow this procedure. This is for your safety and the safety of your co-workers. If you have any questions about lockout/tagout procedures, ask your supervisor.

4.2.0 Barriers and Barricades

Any opening in a wall or floor is a safety hazard. There are two types of protection for these openings: (1) they can be guarded or (2) they can be covered. Cover any hole in the floor when possible. When it is not practical to cover the hole, use barricades. If the bottom edge of a wall opening is fewer than 3 feet above the floor and would allow someone to fall 4 feet or more, then place guards around the opening.

The types of barriers and barricades used vary from job site to job site (see *Figure 15*). There may also be different procedures for when and how barricades are put up. Learn and follow the policies at your job site.

Several different types of guards are commonly used:

- Railings are used across wall openings or as a barrier around floor openings to prevent falls (see *Figure 15a*).
- Warning barricades alert workers to hazards but provide no real protection (see *Figure 15b*). Typical warning barricades are made of plastic tape or rope strung from wire or between posts. The tape or rope is color-coded:
  - Red means danger. No one may enter an area with a red warning barricade. A red barricade is used when there is danger from falling objects or when a load is suspended over an area.
  - Yellow means caution. You may enter an area with a yellow barricade, but be sure you know what the hazard is, and be careful. Yellow barricades are used around wet areas or areas containing loose dust. Yellow with black lettering warns of physical hazards such as bumping into something, stumbling, or falling.
  - Yellow and purple means radiation warning. No one may pass a yellow and purple barricade without authorization, training, and the appropriate personal protective equipment. These barricades are often used where piping welds are being X-rayed.
- Protective barricades give both a visual warning and protection from injury (see *Figure 15c*). They can be wooden posts and rails, posts and chain, or steel cable. People cannot get past protective barricades.
- Blinking lights are placed on barricades so they can be seen at night (see *Figure 15d*).
- Hole covers are used to cover open holes in a floor or in the ground (see *Figure 15e*). They must be secured and labeled. They must be strong enough to support twice the weight of anything that may be placed on top of them.

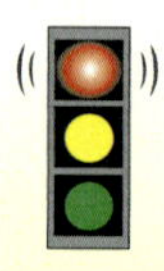

WARNING!

Never remove a barricade unless you have been authorized to do so. Follow your employer's procedures for putting up and removing barricades.

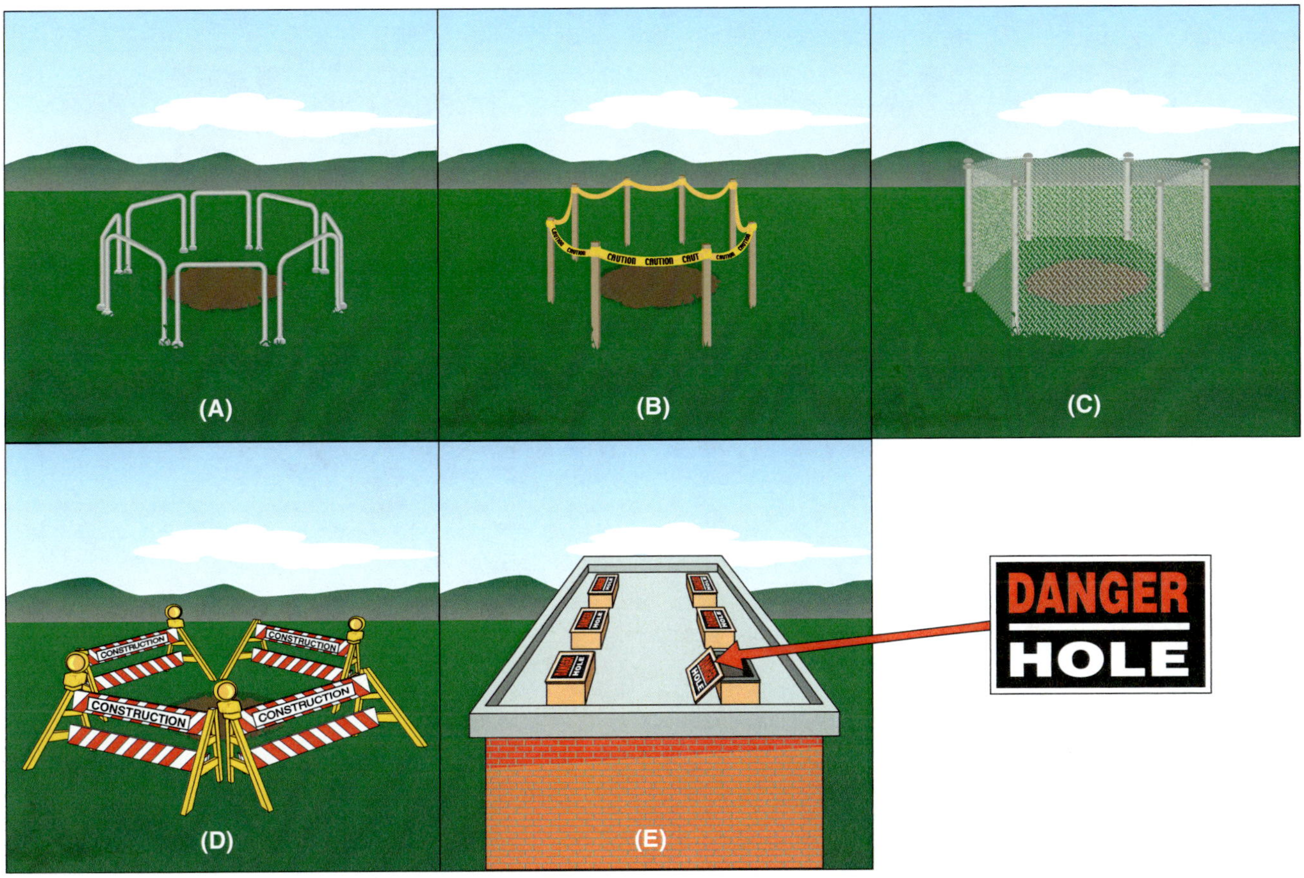

101F15.EPS

Figure 15 ◆ Common types of barriers and barricades.

5.0.0 ◆ PERSONAL PROTECTIVE EQUIPMENT

PPE is designed to protect you from injury. You must keep it in good condition and use it when you need to. Many workers are injured on the job because they are not using PPE.

5.1.0 Personal Protective Equipment Needs

You will not see all the potentially dangerous conditions just by looking around a job site. It's important to stop and consider what type of accidents could happen on any job that you are about to do. Using common sense and knowing how to use PPE will greatly reduce your chance of getting hurt.

5.2.0 Personal Protective Equipment Use and Care

The best protective equipment is of no use to you unless you do the following four things:

- Regularly inspect it.
- Properly care for it.
- Use it properly when it is needed.
- Never alter or modify it in any way.

The sections that follow describe protective equipment commonly used on construction sites and tell how to use and care for each piece of equipment. Be sure to wear the equipment according to the manufacturer's specifications.

5.2.1 Hard Hat

Figure 16 shows a typical hard hat. The outer shell of the hat can protect your head from a hard blow. The webbing inside the hat keeps space between the shell and your head. Adjust the headband so that the webbing fits your head and there is at least 1 inch of space between your head and the shell. Do not alter your hard hat in any way.

DID YOU KNOW?

Hard hats used to be made of metal. However, metal conducts electricity, so most hard hats are now made of reinforced plastic or fiberglass.

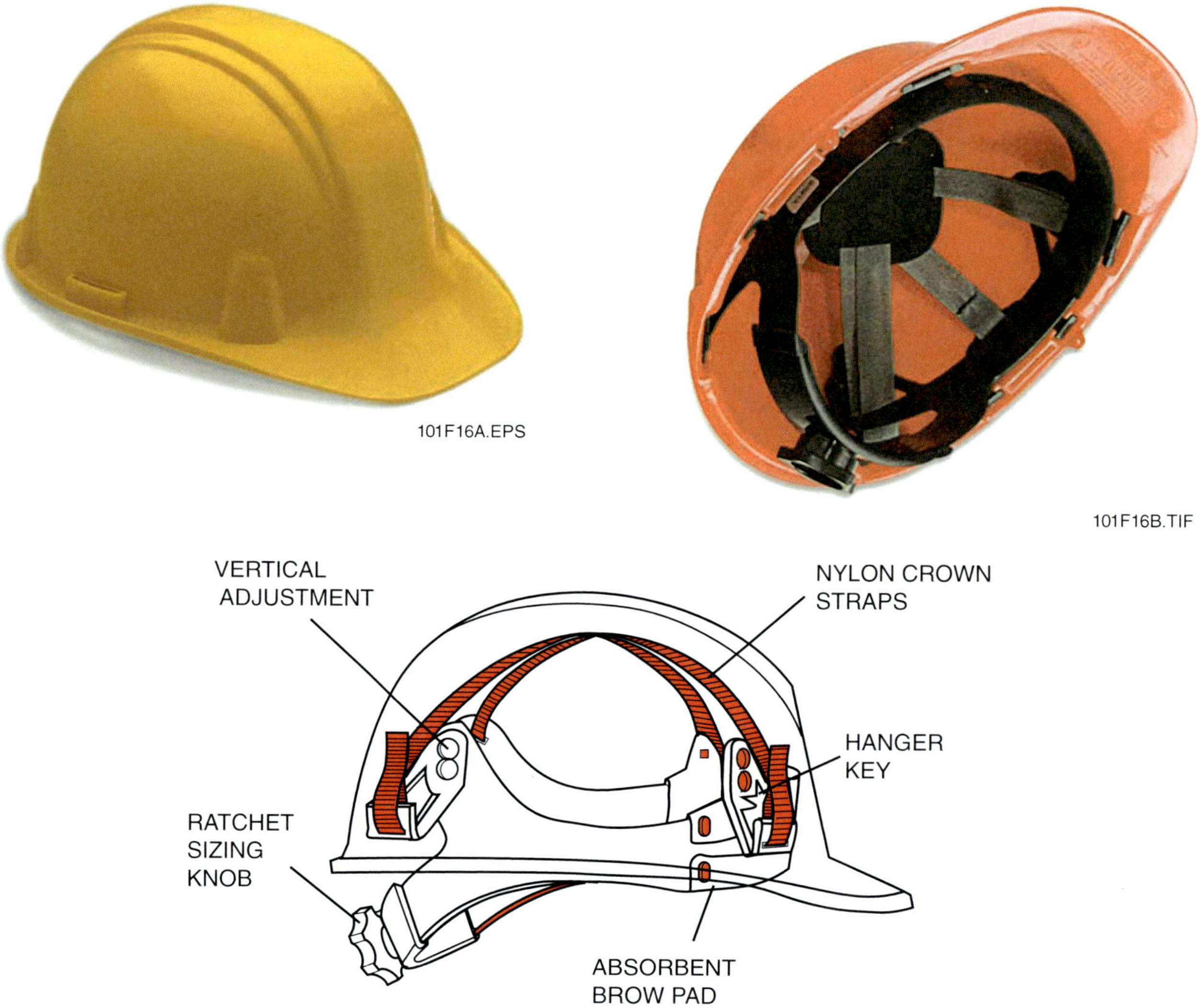

Figure 16 ◆ Typical hard hat.

ON-SITE

Hard Hat Care

Inspect your hard hat every time you use it. If there are any cracks or dents in the shell, or if the webbing straps are worn or torn, get a new hard hat. Wash the webbing and headband with soapy water as often as needed to keep them clean. Wear the hard hat only as the manufacturer recommends. Never wear anything under the hard hat.

5.2.2 Safety Glasses, Goggles, and Face Shields

Wear eye protection (see *Figure 17*) wherever there is even the slightest chance of an eye injury. Eye and face protection must meet the requirements specified in American National Standards Institute (ANSI) *Standard Z87.1-1968.* Areas where there are potential eye hazards from falling or flying objects are usually identified, but you should always be on the lookout for possible hazards.

Regular safety glasses will protect you from falling objects or from objects flying toward your face. You can add side shields for protection from the sides. In some cases, you may need a face shield. Safety goggles give your eyes the best protection from all directions.

Welders must use tinted goggles or welding hoods. The tinted lenses protect the eyes from the bright welding arc or flame.

> **WARNING!**
> Handle safety glasses and goggles with care. If they get scratched, replace them. The scratches will interfere with your vision. Clean the lenses regularly with lens tissues or a soft cloth.

Figure 17 ◆ Typical safety glasses, goggles, and full face shields.

5.2.3 Safety Harness

Safety harnesses, like the one in *Figure 18*, are extra-heavy-duty harnesses that buckle around your body. They have leg, shoulder, chest, and pelvic straps.

Safety harnesses have a D-ring attached to one end of a short section of rope called a **lanyard** (see *Figure 19*). The other end of the lanyard should be attached to a strong anchor point located above the work area. (A qualified person will tell you what a strong anchor point is.) The lanyard should be long enough to let you work but short enough to keep you from falling more than 6 feet.

Use a safety harness and lanyard when you are working in the following situations:

- More than 6 feet above ground or according to company policy
- Near a large opening in a floor
- Near a deep hole
- Near protruding rebar

101F18.EPS

Figure 18 ◆ Typical full-body safety harness.

WARNING!

Never use a safety harness and lanyard for anything except their intended purpose. Always follow the manufacturer's instructions for hooking up a safety harness or lanyard.

NOTE

The safety harness and lanyard are parts of a system that is known as the personal fall protection system. Workers must know how to properly inspect, don, and maintain their system.

Treat a safety harness as if your life depends on it, because it does! Carefully inspect the harness each time you use it. Check that the buckles and D-ring are not bent or deeply scratched. Check the harness for any cuts or rough spots. If you find any damage, turn in the harness for testing or replacement.

WARNING!

Always use a safety harness and lanyard properly. More than 70 percent of reported job-site accidents are caused by improper use of the lanyard and harness.

101F19A.TIF

101F19B.TIF

Figure 19 ◆ Lanyards.

5.2.4 Gloves

On many construction jobs, you must wear heavy-duty gloves to protect your hands (see *Figure 20*). Construction work gloves are usually made of cloth, canvas, or leather. Never wear cloth gloves around rotating or moving equipment. They can easily get caught in the equipment.

Gloves help prevent cuts and scrapes when you handle sharp or rough materials. Heat-resistant gloves are sometimes used for hot materials. Electricians use special rubber-insulated gloves when they work on or around live circuits.

Replace gloves when they become worn, torn, or soaked with oil or chemicals. Electrician's rubber-insulated gloves should be tested regularly to make sure they will protect the wearer.

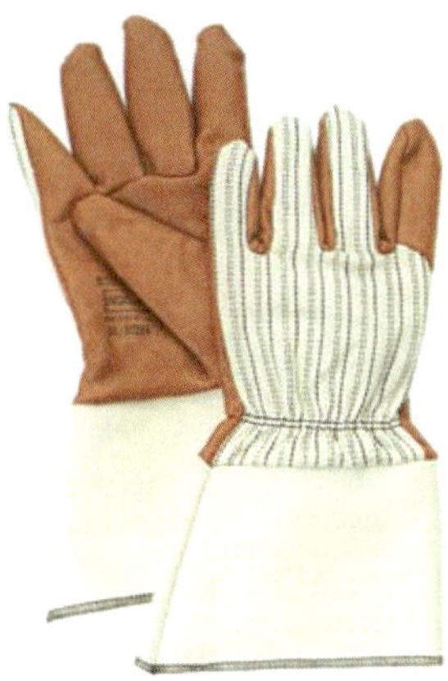

101F20.EPS

Figure 20 ◆ Work gloves.

5.2.5 Safety Shoes

The best shoes to wear on a construction site are ANSI-approved shoes (see *Figure 21*). The steel toe protects your toes from falling objects. The steel sole keeps nails and other sharp objects from puncturing your feet. The next best footwear material is heavy leather. Never wear canvas shoes or sandals on a construction site. They do not provide adequate protection.

Always replace boots or shoes when the sole tread becomes worn or the shoes have holes, even if the holes are on top. Don't wear oil-soaked shoes when you are welding, because of the risk of fire.

101F21.EPS

Figure 21 ◆ Safety shoe.

5.2.6 Hearing Protection

Damage to most parts of the body causes pain. But ear damage does not always cause pain. Exposure to loud noise over a long time can cause hearing loss, even if the noise is not loud enough to cause pain.

Most construction companies follow OSHA rules in deciding when hearing protection must be used. One type of hearing protection is specially designed earplugs that fit into your ears and filter out noise (see *Figure 22*). Clean earplugs regularly with soap and water to prevent ear infection.

Another type of hearing protection is earmuffs, which are large padded covers for the entire ear (see *Figure 23*). You must adjust the headband on earmuffs for a snug fit. If the noise level is very high, you may need to wear both earplugs and earmuffs.

Noise-induced hearing loss can be prevented by using noise control measures and personal protective devices. *Table 3* shows the recommended maximum length of exposure to sound levels rated 90 decibels and higher.

Table 3 Maximum Noise Levels

Sound Level (decibels)	Maximum Hours of Continuous Exposure per Day	Examples
90	8	Power lawn mower
92	6	Belt sander
95	4	Tractor
97	3	Hand drill
100	2	Chain saw
102	1.5	Impact wrench
105	1	Spray painter
110	0.5	Power shovel
115	0.25 or less	Hammer drill

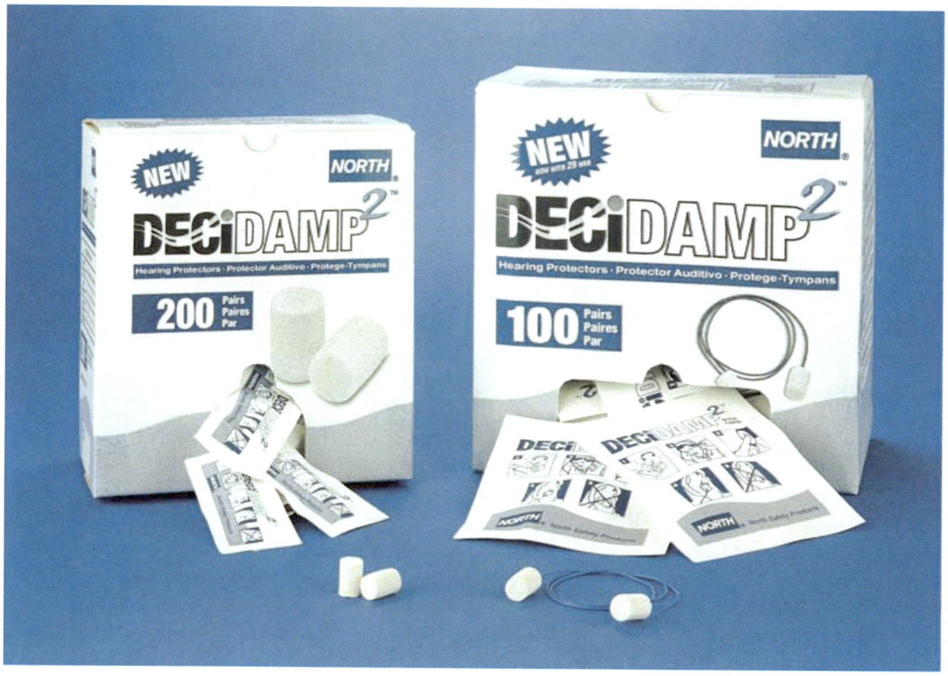

101F22.EPS

Figure 22 ◆ Earplugs for hearing protection.

101F23.EPS

Figure 23 ◆ Earmuffs for hearing protection.

5.2.7 Respiratory Protection

Wherever there is danger of an inhalation hazard, you must use a respirator.

Follow company and OSHA procedures when choosing the type of respirator for a particular job. Also be sure that it is safe for you to wear a respirator. Under OSHA's Respiratory Protection Standard, workers must fill out a questionnaire to determine any potential problems in wearing a respirator. Depending on the answers, a medical exam may be required.

Federal law specifies which type of respirator to use for different types of hazards. There are four general types of respirators (see *Figure 24*):

- Self-contained breathing **apparatus** (SCBA)
- Supplied air mask
- Full facepiece mask with chemical canister (gas mask)
- Half mask or mouthpiece with mechanical filter

An SCBA carries its own air supply in a compressed air tank. It is used where there is not enough oxygen or where there are dangerous gases or fumes in the air.

A supplied air mask uses a remote compressor or air tank. A hose supplies air to the mask. Supplied air masks can be used under the same conditions as SCBAs.

A full facepiece mask with chemical canisters is used to protect against brief exposure to dangerous gases or fumes.

A half mask or mouthpiece with a mechanical filter is used in areas where you might inhale dust or other solid particles. No medical exam is necessary to use this type of general respiratory protection.

101F24A.EPS

SELF-CONTAINED BREATHING APPARATUS

101F24B.EPS

SUPPLIED AIR MASK

101F24C.EPS

FULL FACEPIECE MASK

101F24D.EPS

HALF MASK

Figure 24 ◆ Examples of respirators.

It is very important to check a respirator carefully for damage and for proper fit. A leaking facepiece can be as dangerous as not wearing a respirator at all. All respirators must be fitted properly, and their facepiece-to-face seal must be checked each time the respirator is used. When conditions prevent a good seal, the respirator cannot be worn. The following conditions will interfere with the respirator's seal:

- Facial hair (such as sideburns or beards)
- Skullcaps that project under the facepiece
- Temple bars on glasses (especially when wearing full-face respirators)
- Absence of upper, lower, or all teeth
- Absence of dentures
- Gum and tobacco chewing

CAUTION

OSHA no longer bans the wearing of contact lenses with respirators (*29 CFR 1910.134 [g][1][ii]*). After sponsoring research and studies on the issue, OSHA recently concluded that wearing contact lenses with respirators does not pose an increased risk to the wearer's safety.

If you wear contact lenses, practice wearing a respirator with your contact lenses to see whether you have any problems. That way, you will identify any problems before you use the respirator under hazardous conditions.

Respirators used by only one person should be cleaned after each day of use, or more often if necessary. Those used by more than one person should be cleaned and disinfected (made germ-free) after each use.

When a respirator is not required, workers may voluntarily use a dust or particle mask for general protection. These masks do not require fit testing or a medical examination.

WARNING!

When a respirator is required, a personal monitoring device is usually also required. This device samples the air to measure the concentration of hazardous chemicals.

6.0.0 ◆ LIFTING

You may be surprised to learn that one-fourth of all occupational injuries happen when workers are handling or moving construction materials, especially lifting heavy objects. There is a right way and a wrong way to lift a heavy object. Lifting the wrong way can land you in the hospital. *Figure 25* shows the right way.

Step 1 Move close to the object you are going to lift. Position your feet in a forward/backward stride, with one foot at the side of the object.

Step 2 Bend your knees and lower your body, keeping your back straight and as upright as possible.

Step 3 Place your hands under the object, wrap your arms around it, or grasp the handles. To get your hands under an object that is flat on the floor, use both hands to lift one corner.
Slip one hand under that corner. With one hand under, tilt the object to get the other hand under the opposite side.

Step 4 Draw the object close to your body.

Step 5 Lift by slowly straightening your legs and keeping the object's weight over your legs as much as possible.

Step 6 Pick the object up facing the direction you are going to go, to avoid twisting your knees or back.

These steps let you use your strongest muscles (those in your legs) instead of your weakest ones (those in your back) to lift. Practice with light objects. Then, when you've got it down, move on to heavier ones.

Many employers supply back belts to help reduce back injuries. You should be trained in the right way to use a back belt. Remember, a back belt is no substitute for using proper lifting techniques.

WARNING!

Always check your company's rules and regulations regarding the amount of weight one person is allowed to carry at a time.

Figure 25 ◆ How to lift safely.

Review Questions

Sections 4.0.0–6.0.0

1. The ________ keeps the key to a lock used for lockout/tagout.
   a. site supervisor
   b. person who puts on the lock
   c. site safety manager
   d. OSHA inspector

2. You may operate a device that is tagged out if there is no imminent danger.
   a. True
   b. False

3. A yellow and purple warning barricade means ________.
   a. caution
   b. danger
   c. physical danger
   d. radiation hazard

4. Hole covers must be strong enough to support ________ the weight of anything that may be placed on top of them.
   a. exactly
   b. twice
   c. four times
   d. ten times

5. Adjust the webbing of a hard hat so that there is ________ between your head and the shell.
   a. no space
   b. as much space as possible
   c. at least 1 inch of space
   d. less than 1 inch of space

6. ________ provide the best all-around protection for your eyes.
   a. Welding hoods
   b. Face shields
   c. Safety goggles
   d. Strap-on glasses

7. A ________ has its own clean air supply.
   a. half mask
   b. mouthpiece with mechanical filter
   c. self-contained breathing apparatus
   d. full facepiece mask

8. Whenever there is danger of an inhalation hazard, you must use a respirator.
   a. True
   b. False

9. ________ do(es) *not* interfere with the respirator's seal.
   a. Facial hair
   b. Gum and tobacco chewing
   c. Dentures
   d. Temple bars on glasses

10. When lifting heavy objects, keep as much weight as possible over your ________.
   a. hips
   b. shoulders
   c. legs
   d. back

7.0.0 ◆ AERIAL WORK

Working in elevated locations is common in the construction industry. If it is done properly and the proper equipment is used, it is safe. But falls from heights can cause serious injuries or even death. You must always have your supervisor's permission before working in an elevated location. In this section, you will learn about the equipment used for aerial work. You'll learn how to use it, inspect it, and maintain it.

7.1.0 Ladders and Scaffolds

Ladders and **scaffolds** are used to perform work in elevated locations. Any time work is performed above ground level, there is a risk of accidents. You can reduce this risk by carefully inspecting ladders and scaffolds before you use them and by using them properly.

Overloading means exceeding the **maximum intended load** of a ladder. Overloading can cause ladder failure, which means that the ladder could buckle, break, or topple, among other possibilities. The maximum intended load is the total weight of all people, equipment, tools, materials, loads that are being carried, and other loads that the ladder can hold at any one time. Check the manufacturer's specifications to determine the maximum intended load. Ladders are usually given a duty rating that indicates their load capacity, as shown in *Table 4*.

> **WARNING!**
> When you use a ladder, be sure to maintain three-point contact with the ladder at all times. Three-point contact means that either two feet and one hand or one foot and two hands are always touching the ladder.

> **WARNING!**
> Use ladders and scaffolds for their intended uses only. Ladders are not interchangeable, and incorrect use can result in injury or damage.

Table 4 Ladder Duty Ratings and Load Capacities

Duty Ratings	Load Capacities
Type IA	300 lbs., extra-heavy duty/professional use
Type I	250 lbs., heavy duty/industrial use
Type II	225 lbs., medium duty/commercial use
Type III	200 lbs., light duty/household use

7.1.1 Portable Straight Ladders

Straight ladders consist of two rails, rungs between the rails, and safety feet on the bottom of the rails (see *Figure 26*). The straight ladders used in construction are made of wood or fiberglass.

ON-SITE

Fall Protection

Effective February 1995, fall protection must be provided for those working at an elevation above 6 feet or where a drop into a hole or trench is greater than 6 feet. Supervisors must ensure that all walking and working surfaces have the strength and structural integrity to support the workers. Work conducted on an otherwise unprotected side or edge that is 6 feet or more above a lower level must be protected by the use of guardrail systems, safety-net systems, or personal fall-arrest systems.

Effective January 1998, an acceptable personal fall-arrest system is a body harness with a lanyard attached to a D-ring in the center of the back. Body belts are not acceptable as part of a personal fall-arrest system because they unevenly distribute pressure on the wearer's midsection.

All companies are required by law to have a written Fall Protection Plan that addresses site-specific fall hazards and the steps taken to prevent each hazard. This information will normally be included in your training or regularly scheduled safety meetings. If you are unsure of what your company's plan is, or have questions about it, ask your supervisor before performing any aerial work.

ON-SITE

Three-Point Contact

When climbing or working from a ladder, you run the risk of falling. An important measure in safeguarding yourself against a fall is to maintain three-point contact with the ladder at all times. This means that you either have two hands and one foot or two feet and one hand touching the ladder constantly. Maintaining three-point contact with the ladder will help prevent you from falling and injuring yourself or a co-worker.

(A) 101SA01A.EPS

(B) 101SA01B.EPS

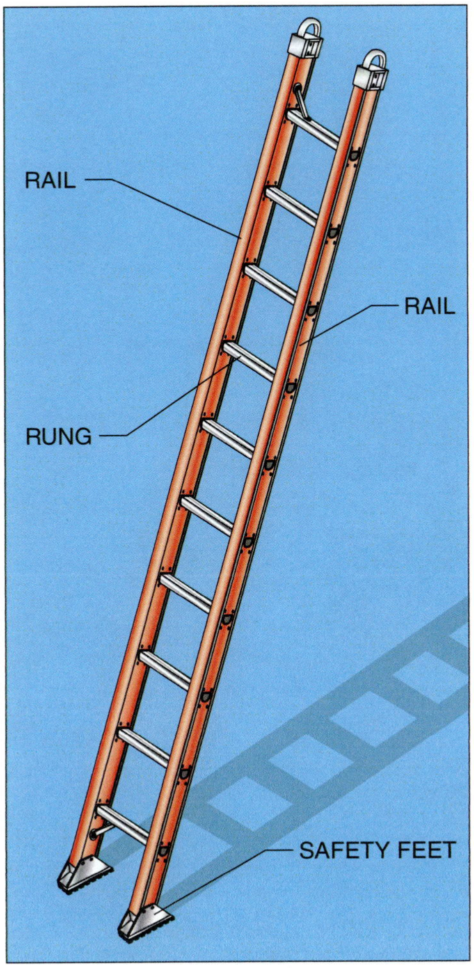

Figure 26 ◆ Portable straight ladder.

Metal ladders conduct electricity and should never be used around electrical equipment. Any portable metal ladder must have "Danger. Do Not Use Around Electrical Installations" stenciled on the rails in 2-inch, red letters. Ladders made of dry wood or fiberglass, neither of which conducts electricity, should be used around electrical equipment. Check that any wooden ladder is, in fact, completely dry before using it around electricity. Even a small amount of water will conduct electricity.

Different types of ladders should be used in different situations (see *Figure 27*). Aluminum ladders are corrosion-resistant and can be used where they might be exposed to the elements. They are also lightweight and can be used where they must frequently be lifted and moved. Wooden ladders, which are heavier and sturdier than fiberglass or aluminum ladders, can be used where heavy loads must be moved up and down. Fiberglass ladders are very durable, so they are useful where some amount of rough treatment is unavoidable. Both fiberglass and aluminum are easier to clean than wood.

7.1.2 Inspecting Straight Ladders

Always inspect a ladder before you use it. Check the rails and rungs for cracks or other damage. Also, check for loose rungs. If you find any damage, do not use the ladder. Check the entire ladder for loose nails, screws, brackets, or other hardware. If you find any hardware problems, tighten the loose parts or have the ladder repaired before you use it. OSHA requires regular inspections of all ladders and an inspection just before each use.

WARNING!

Wooden ladders should never be painted. The paint could hide cracks in the rungs or rails. Clear varnish, shellac, or a preservative oil finish will protect the wood without hiding defects.

Figure 28 shows the safety feet attached to a straight ladder. Make sure the feet are securely attached and that they are not damaged or worn down. Do not use a ladder if its safety feet are not in good working order.

7.1.3 Using Straight Ladders

It is very important to place a straight ladder at the proper angle before using it. A ladder placed at an improper angle will be unstable and could cause you to fall. *Figure 29* shows a properly positioned straight ladder.

The distance between the foot of a ladder and the base of the structure it is leaning against must be one-fourth of the distance between the ground and the point where the ladder touches the structure. For example, if the height of the wall shown in *Figure 29* is 16 feet, the base of the ladder should be 4 feet from the base of the wall. If you are going to step off a ladder onto a platform or roof, the top of the ladder should extend at least 3 feet above the point where the ladder touches the platform or roof.

Ladders should be used only on stable and level surfaces unless they are secured at both the bottom and the top to prevent any accidental movement (see *Figure 30*). Never try to move a ladder while you are on it. If a ladder must be placed in front of a door that opens toward the ladder, the door should be locked or blocked open. Otherwise, the door could be opened into the ladder.

Ladders are made for vertical use only. Never use a ladder as a work platform by placing it horizontally. Make sure the ladder you are about to climb or descend is properly secure before you do so. Check to make sure the ladder's feet are sol-

Figure 27 ◆ Different types of ladders and applications.

Figure 28 ◆ Ladder safety feet.

idly positioned on firm, level ground. Also check to make sure the top of the ladder is firmly positioned and in no danger of shifting once you begin your climb. Remember that your own weight will affect the ladder's steadiness once you mount it. So it is important to test the ladder first by putting some of your weight on it without actually beginning to climb. This way, you can be sure that the ladder will remain steady as you climb.

When climbing a straight ladder, keep both hands on the rails. Always keep your body's weight in the center of the ladder between the rails. Face the ladder at all times. Never go up or down a ladder while facing away from it (see *Figure 31*).

To carry a tool while you are on the ladder, use a **hand line** or tagline attached to the tool. Climb the ladder and then pull up the tool. Don't carry tools in your hands while you are climbing a ladder.

Figure 29 ◆ Proper positioning of a straight ladder.

Figure 30 ◆ Securing a ladder.

Figure 31 ◆ Moving up or down a ladder.

7.1.4 Extension Ladders

An **extension ladder** is actually two straight ladders. They are connected so you can adjust the overlap between them and change the length of the ladder as needed (see *Figure 32*).

7.1.5 Inspecting Extension Ladders

The same rules for inspecting straight ladders apply to extension ladders. In addition, you should inspect the rope that is used to raise and lower the movable section of the ladder. If the rope is frayed or has worn spots, it should be replaced before the ladder is used.

The rung locks (see *Figure 33*) support the entire weight of the movable section and the person climbing the ladder. Inspect them for damage before each use. If they are damaged, they should be repaired or replaced before the ladder is used.

7.1.6 Using Extension Ladders

Extension ladders are positioned and secured following the same rules as for straight ladders. When you adjust the length of an extension ladder, always reposition the movable section from the bottom, not the top, so you can make sure the rung locks are properly engaged after you make the adjustment. Check to make sure the section locking mechanism is fully hooked over the desired rung. Also check to make sure that all ropes used for raising and lowering the extension are clear and untangled.

Figure 33 ◆ Rung locks.

WARNING!
Extension ladders have a built-in extension stop mechanism. Do not remove it. This could cause the ladder to collapse under a load.

WARNING!
Haul materials up on a line rather than hand-carrying them up an extension ladder. Use extra caution when carrying anything on a ladder, because it affects your balance.

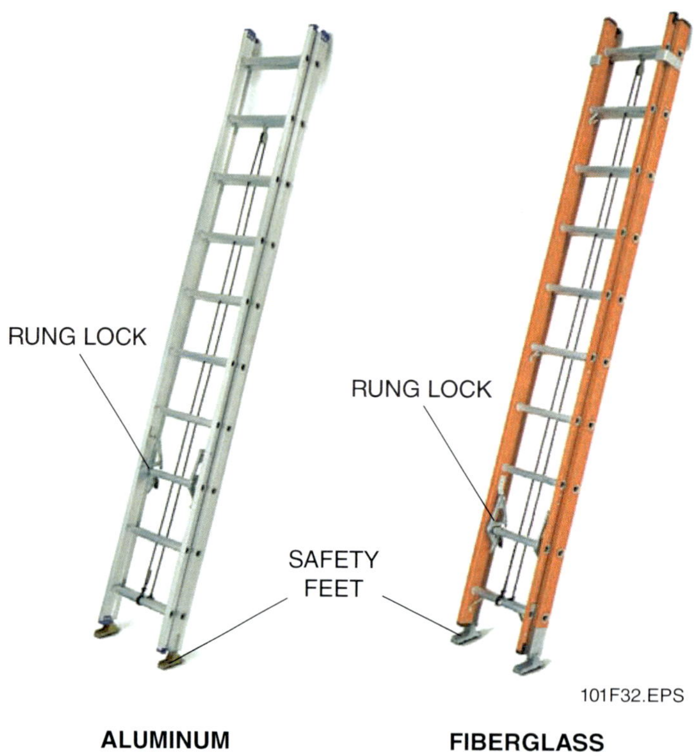

Figure 32 ◆ Typical extension ladders.

Never stand above the highest safe standing level on a ladder. The highest safe standing level on an extension ladder is the fourth rung from the top. If you stand higher, you may lose your balance and fall. Some ladders have colored rungs to show where you should not stand.

7.1.7 Stepladders

Stepladders are self-supporting ladders made of two sections hinged at the top (see *Figure 34*).

The section of a stepladder used for climbing consists of rails and rungs like those on straight ladders. The other section consists of rails and braces. Spreaders are hinged arms between the sections that keep the ladder stable and keep it from folding while in use.

7.1.8 Inspecting Stepladders

Inspect stepladders the way you inspect straight and extension ladders. Pay special attention to the hinges and spreaders to be sure they are in good repair. Also, be sure the rungs are clean. A stepladder's rungs are usually flat, so oil, grease, or dirt can build up on them and make them slippery.

7.1.9 Using Stepladders

When you position a stepladder, be sure that all four feet are on a hard, even surface. Otherwise, the ladder can rock from side to side or corner to corner when you climb it. With the ladder in position, be sure the spreaders are locked in the fully open position.

Figure 34 ◆ Typical fiberglass stepladder.

Never stand on the top step or the top of a stepladder. Putting your weight this high will make the ladder unstable. The top of the ladder is made to support the hinges, not to be used as a step. And, although the rear braces may look like rungs, they are not designed to support your weight. Never use the braces for climbing. And never climb the back of a stepladder. (For certain jobs, however, there are specially designed two-person ladders with steps on both sides.) *Figure 35* shows common dos and don'ts for using ladders.

7.2.0 Scaffolds

Scaffolds provide safe elevated work platforms for people and materials. They are designed and built to comply with high safety standards, but normal wear and tear or accidentally putting too much weight on them can weaken them and make them unsafe. That's why it is important to inspect every part of a scaffold before each use.

> **WARNING!**
> Never stand on a step with your knees higher than the top of a stepladder. You need to be able to hold on to the ladder with your hand. Also, keep your body centered between the side rails.

Two basic types of scaffolds—manufactured scaffolds and rolling scaffolds—are used in the construction industry. The rules for safe use apply to both of them.

7.2.1 Manufactured Scaffolds

Manufactured scaffolds are made of painted steel, stainless steel, or aluminum (see *Figure 36*). They are stronger and more fire-resistant than wooden scaffolds. They are supplied in ready-made, individual units, which are assembled on site.

7.2.2 Rolling Scaffolds

A rolling scaffold has wheels on its legs so that it can be easily moved (see *Figure 37*). The scaffold wheels have brakes so the scaffold will not move while workers are standing on it.

> **CAUTION**
> Only a competent person has the authority to supervise setting up, moving, and taking down scaffolding. Only a competent person can approve the use of scaffolding on the job site after inspecting the scaffolding.

DOs

- Be sure your ladder has been properly set up and is used in accordance with safety instructions and warnings.
- Wear shoes with non-slip soles.

- Keep your body centered on the ladder. Hold the ladder with one hand while working with the other. Never let your belt buckle pass beyond either ladder rail.

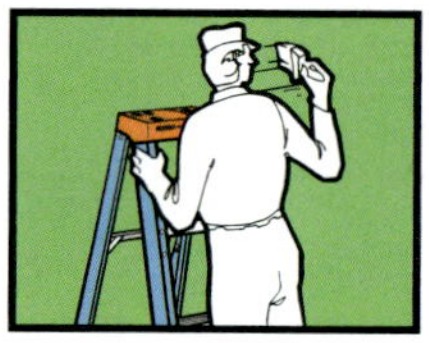

- Move materials with extreme caution. Be careful pushing or pulling anything while on a ladder. You may lose your balance or tip the ladder.

- Get help with a ladder that is too heavy to handle alone. If possible, have another person hold the ladder when you are working on it.

- Climb facing the ladder. Center your body between the rails. Maintain a firm grip.
- Always move one step at a time, firmly setting one foot before moving the other.

- Haul materials up on a line rather than carry them up an extension ladder.
- Use extra caution when carrying anything on a ladder.

Read ladder labels for additional information.

DON'Ts

- DON'T stand above the highest **safe standing level**.
- DON'T stand above the second step from the top of a stepladder and the 4th rung from the top of an extension ladder. A person standing higher may lose their balance and fall.

- DON'T climb a closed stepladder. It may slip out from under you.
- DON'T climb on the back of a stepladder. It is not designed to hold a person.

- DON'T stand or sit on a step-ladder top or pail shelf. They are not designed to carry your weight.
- DON'T climb a ladder if you are not physically and mentally up to the task.

- DON'T exceed the Duty Rating, which is the maximum load capacity of the ladder. Do not permit more than one person on a single-sided stepladder or on any extension ladder.

- DON'T place the base of an extension ladder too close to the building as it may tip over backward.
- DON'T place the base of an extension ladder too far away from the building, as it may slip out at the bottom. **Please refer to the 4 to 1 Ratio Box.**

- DON'T over-reach, lean to one side, or try to move a ladder while on it. You could lose your balance or tip the ladder. **Climb down and then reposition the ladder closer to your work!**

4 TO 1 Ratio

Place an extension ladder at a 75-1/2° angle. The set-back ("S") needs to be 1 ft. for each 4 ft. of length ("L") to the upper support point.

101F35.EPS

Figure 35 ◆ Ladder safety dos and don'ts.

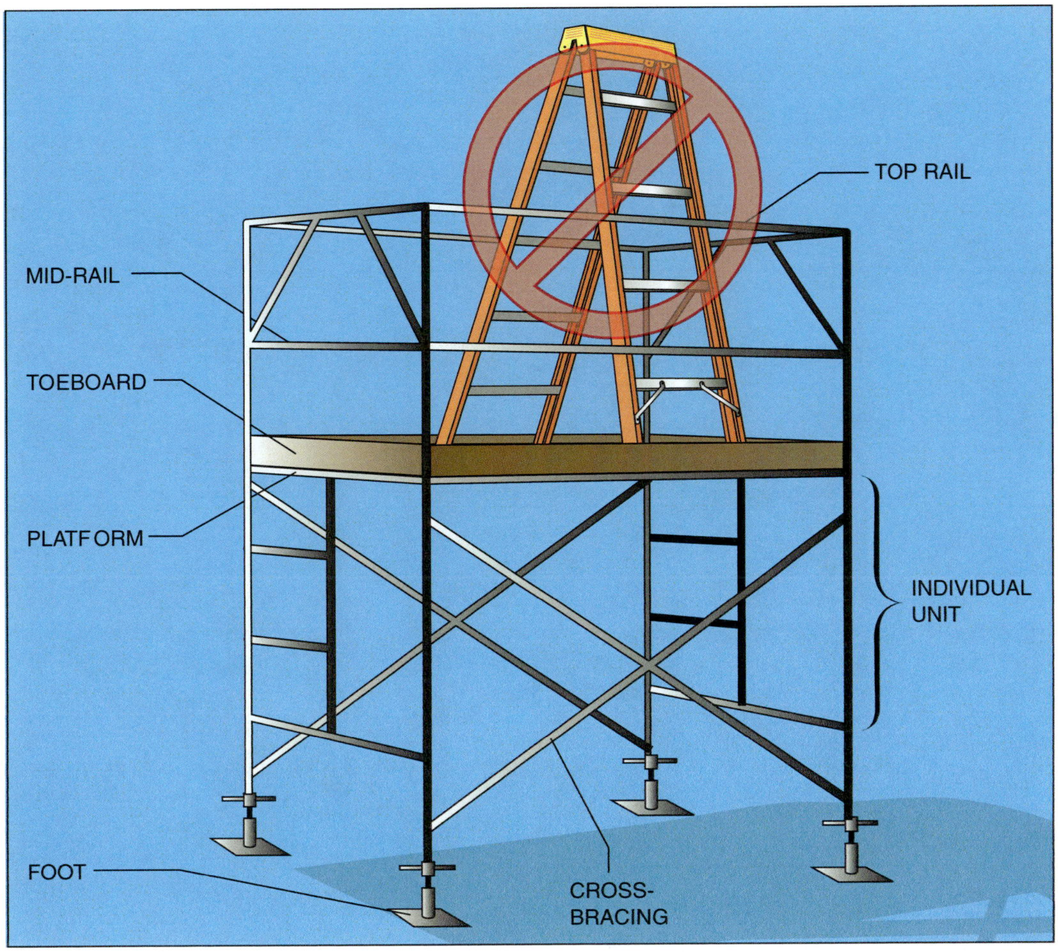

Figure 36 ◆ Typical manufactured scaffold.

7.2.3 Inspecting Scaffolds

Any scaffold that is assembled on the job site should be tagged. These tags indicate whether the scaffold meets OSHA standards and is safe to use. Three colors of tags are used: green, yellow, and red (see *Figure 38*).

A green tag means the scaffold meets all OSHA standards and is safe to use.

A yellow tag means the scaffold does not meet all OSHA standards. An example is a scaffold on which a railing cannot be installed because of equipment interference. To use a yellow-tagged scaffold, you must wear a safety harness attached to a lanyard. You may have to take other safety measures as well.

A red tag means a scaffold is being put up or taken down. Never use a red-tagged scaffold.

Don't rely on the tags alone. Inspect all scaffolds before you use them. Check for bent, broken, or badly rusted tubes. Check for loose joints where the tubes are connected. Any of these problems must be corrected before the scaffold is used.

Make sure you know the weight limit of any scaffold you will be using. Compare this weight limit to the total weight of the people, tools, equipment, and material you expect to put on the scaffold. Scaffold weight limits must never be exceeded.

If a scaffold is more than 10 feet high, check to see that it is equipped with **top rails, mid-rails,** and **toeboards.** All connections must be pinned. That means they must have a piece of metal inserted through a hole to prevent connections from slipping. **Cross-bracing** must be used. A handrail is not the same as cross-bracing. The working area must be completely **planked.**

If it is possible for people to walk under a scaffold, the space between the toeboard and the top rail must be screened. This prevents objects from falling off the work platform and injuring those below.

When you examine a rolling scaffold, check the condition of the wheels and brakes. Be sure the brakes are working properly and can stop the scaffold from moving while work is in progress. Be sure all brakes are locked before you use the scaffold.

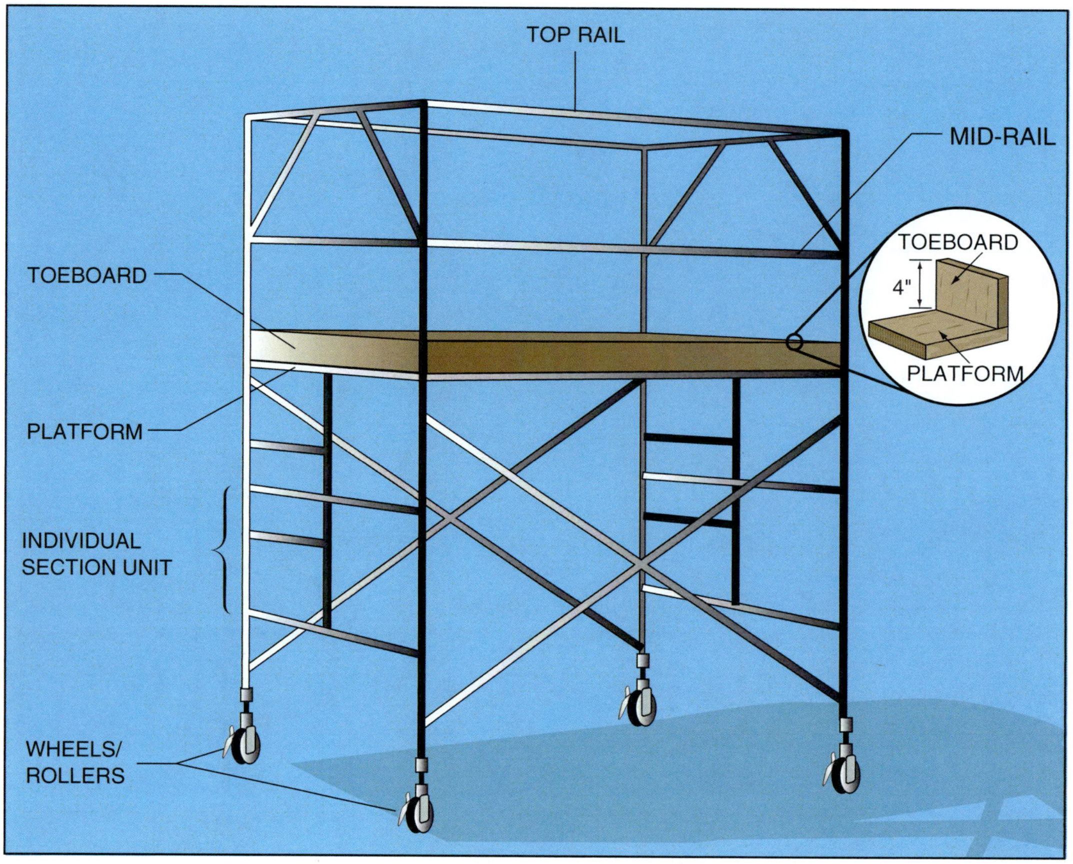

101F37.EPS

Figure 37 ◆ Typical rolling scaffold.

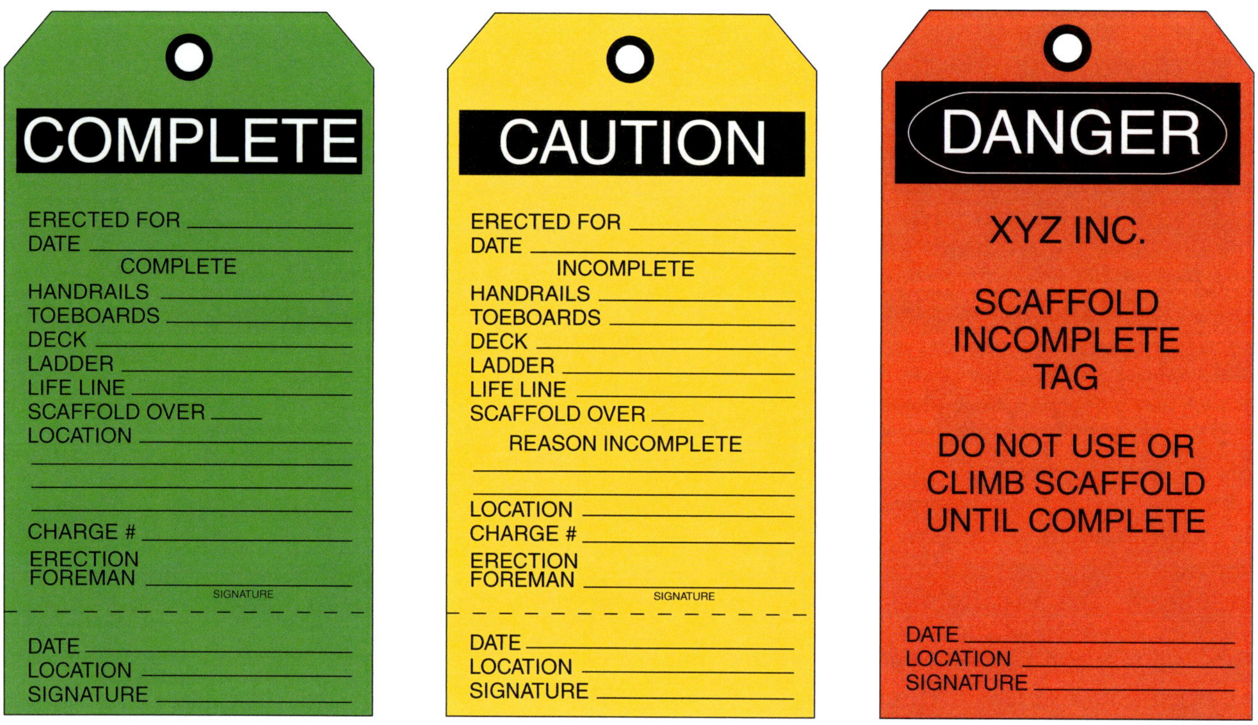

101F38.EPS

Figure 38 ◆ Typical scaffold tags.

7.2.4 Using Scaffolds

Be sure that a competent person inspects the scaffold before you use it.

There should be firm footing under each leg of a scaffold before you put any weight on it. If you are working on loose or soft soil, you can put matting under the scaffold's legs or wheels.

WARNING!
Keep scaffolds a safe distance from power lines in accordance with OSHA guidelines. Refer to *OSHA 1926.*

WARNING!
Falls from scaffolding and falls from ladders each account for 17 percent of deaths among construction workers. Prevent accidents by following OSHA and company guidelines.

When you move a rolling scaffold, always follow these steps:

Step 1 Get off the scaffold.

Step 2 Unlock the wheel brakes.

Step 3 Move the scaffold.

Step 4 Re-lock the wheel brakes.

Step 5 Get back on the scaffold.

WARNING!
Never unlock the wheel brakes of a rolling scaffold while anyone is on it. People on a moving scaffold can lose their balance and fall.

Review Questions

Section 7.0.0

1. Never use a(n) __________ ladder anywhere near electrical current.

a. fiberglass
b. aluminum
c. wooden
d. straight

2. If you lean a straight ladder against the top of a 16-foot wall, the base of the ladder should be __________ feet from the base of the wall.

a. 3
b. 4
c. 5
d. 6

3. With a one-person stepladder, it is safe to __________.

a. stand on the top step
b. climb the back of it
c. stand on the rear braces
d. lock the spreaders in the fully open position

4. Never use a scaffold with a(n) __________ tag.

a. blue
b. red
c. orange
d. yellow

5. A scaffold must be equipped with top rails, mid-rails, and toeboards if it is more than __________ feet high.

a. 5
b. 7
c. 10
d. 12

8.0.0 ◆ HAZARD COMMUNICATION STANDARD

OSHA has a rule that affects every worker in the construction industry. It is called the **Hazard Communication Standard (HazCom).** You may have heard it called the "Right to Know" requirement. It requires all contractors to educate their employees about the hazardous chemicals they may be exposed to on the job site. Employees must be taught how to work safely around these materials.

Many people think that there are very few hazardous chemicals on construction job sites. That isn't true. In the OSHA standard, the term hazardous chemical applies to paint, concrete, and even wood dust, as well as other substances.

8.1.0 Material Safety Data Sheets

A material safety data sheet (MSDS) must accompany every shipment of a hazardous substance and must be available to you on the job site. Use the MSDS to manage, use, and dispose of hazardous materials safely. *Figure 39* shows part of a typical MSDS.

The information on an MSDS includes the following:

- The identity of the substance
- Exposure limits
- Physical and chemical characteristics of the substance
- The kind of hazard the substance presents
- Precautions for safe handling and use
- The reactivity of the substance
- Specific control measures
- Emergency first-aid procedures
- Manufacturer contact for more information

8.2.0 Your Responsibilities Under HazCom

You have the following responsibilities under HazCom:

- Know where MSDSs are on your job site.
- Report any hazards you spot on the job site to your supervisor.
- Know the physical and health hazards of any hazardous materials on your job site, and know and practice the precautions needed to protect yourself from these hazards.
- Know what to do in an emergency.
- Know the location and content of your employer's written hazard communication program.

The final responsibility for your safety rests with you. Your employer must provide you with information about hazards, but you must know this information and follow safety rules.

Material Safety Data Sheet May be used to comply with OSHA's Hazard Communication Standard, 29 CFR 1910 1200. Standard must be consulted for specific requirements.	U.S. Department of Labor Occupational Safety and Health Administration (Non-Mandatory Form) Form Approved OMB No. 1218-0072
IDENTITY (as Used on Label and List)	Note: Blank spaces are not permitted. If any item is not applicable or no information is available, the space must be marked to indicate that.

Section I

Manufacturer's name	Emergency Telephone Number
Address (Number, Street, City, State and ZIP Code)	Telephone Number for Information
	Date Prepared
	Signature of Preparer (optional)

Section II—Hazardous Ingredients/Identity Information

Hazardous Components (Specific Chemical Identity, Common Name(s))	OSHA PEL	ACGIH TLV	Other Limits Recommended	% (optional)

Section III—Physical/Chemical Characteristics

Boiling Point		Specific Gravity (H_2O = 1)	
Vapor Pressure (mm Hg)		Melting Point	
Vapor Density (AIR = 1)		Evaporation Rate (Butyl Acetate = 1)	
Solubility in Water			
Appearance and Odor			

Section IV—Fire and Explosion Hazard Data

Flash Point (Method Used)	Flammable Limits	LEL	UEL
Extinguishing Media			
Special Fire Fighting Procedures			
Unusual Fire and Explosion Hazards			

(Reproduce locally) OSHA 174 Sept. 1985

101F39A.EPS

Figure 39 ◆ Typical MSDS. (1 of 2)

Section V—Reactivity Data

Stability	Unstable		Conditions to Avoid
	Stable		
Incompatibility (Materials to Avoid)			
Hazardous Decomposition or Byproducts			
Hazardous Polymerization	May Occur		Conditions to Avoid
	Will Not Occur		

Section VI—Health Hazard Data

Route(s) of Entry	Inhalation?	Skin?	Ingestion?
Health Hazards (Acute and Chronic)			
Carcinogenicity	NTP?	IARC Monographs?	OSHA Regulated?
Signs and Symptoms of Exposure			
Medical Conditions Generally Aggravated by Exposure			
Emergency and First Aid Procedures			

Section VII—Precautions for Safe Handling and Use

Steps to Be Taken in Case Material Is Released or Spilled

Waste Disposal Method

Precautions to Be Taken in Handling and Storing

Other Precautions

Section VII—Control Measures

Respiratory Protection (Specify Type)		
Ventilation	Local Exhaust	Special
	Mechanical (General)	Other
Protective Gloves		Eye Protection
Other Protective Clothing or Equipment		
Work/Hygienic Practices		

101F39B.EPS

Figure 39 ◆ Typical MSDS. (2 of 2)

Review Questions

Section 8.0.0

1. OSHA's Hazard Communication Standard (HazCom) rule requires all contractors to __________ on-site hazardous chemicals.
   a. store
   b. clean up
   c. remove all
   d. educate employees about

2. HazCom classifies all paint, concrete, and wood dust as __________ materials.
   a. hazardous
   b. common
   c. inexpensive
   d. nonhazardous

3. The information on an MSDS includes __________.
   a. cost and availability
   b. emergency first-aid procedures
   c. point of origin
   d. warranty limitations

4. Under HazCom, if you spot a hazard on your job site you must __________.
   a. report it to your supervisor
   b. leave immediately
   c. notify your co-workers
   d. clean it up

5. Although your employer must provide you with information about hazardous chemicals, the final responsibility for your safety rests with __________.
   a. your immediate supervisor
   b. your site foreman
   c. you
   d. your co-workers

9.0.0 ◆ FIRE SAFETY

Fire is always a hazard on construction job sites. Many of the materials used in construction are flammable. In addition, welding, grinding, and many other construction activities create heat or sparks that can cause a fire. Fire safety involves two elements: fire prevention and fire fighting.

9.1.0 How Fires Start

For a fire to start, three things are needed in the same place at the same time: fuel, heat, and oxygen. If one of these three is missing, a fire will not start.

Fuel is anything that will combine with oxygen and heat to burn. Oxygen is always present in the air. When pure oxygen is present, such as near a leaking oxygen hose or fitting, material that would not normally be considered fuel (including some metals) will burn.

Heat is anything that will raise a fuel's temperature to the **flash point.** The flash point is the temperature at which a fuel gives off enough gases (vapors) to burn. The flash points of many fuels are quite low—room temperature or less. When the burning gases raise the temperature of a fuel to the point at which it ignites, the fuel itself will burn—and keep burning—even if the original source of heat is removed.

What is needed for a fire to start can be shown as a fire triangle (see *Figure 40*). If one element of the triangle is missing, a fire cannot start. If a fire has started, removing any one element from the triangle will put it out.

Research has added a fourth side to the fire triangle concept, resulting in the development of a new model called the Fire Tetrahedron. The fourth element involved in the combustion process is referred to as the "chemical chain reaction." Specific chemical chain reactions between fuel and oxygen molecules are essential to sustaining a fire once it has begun.

9.2.0 Fire Prevention

The best way to ensure fire safety is to prevent a fire from starting. The best way to prevent a fire is to make sure that fuel, oxygen, and heat are never present in the same place at the same time. Here are some basic safety guidelines for fire prevention:

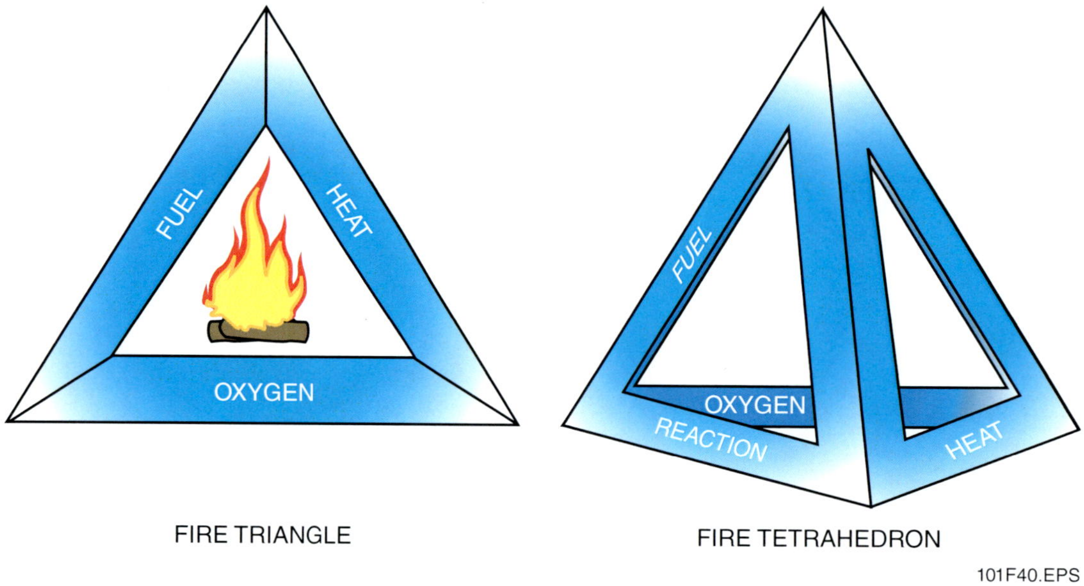

Figure 40 ◆ Basic fire requirements.

- Always work in a well-ventilated area, especially when you are using flammable materials such as shellac, lacquer, paint stripper, or construction adhesives.
- Never smoke or light matches when you are working with or near flammable materials.
- Keep oily rags in approved, self-closing metal containers.
- Store combustible materials only in approved containers.

9.2.1 Flammable and Combustible Liquids

Liquids can be flammable or combustible. Flammable liquids have a flash point below 100°F. Combustible liquids have a flash point at or above 100°F. Fire can be prevented by doing the following things:

- *Removing the fuel* – Liquid does not burn. What burns are the gases (vapors) given off as the liquid evaporates. Keeping liquids in an approved, sealed container prevents evaporation. If there is no evaporation, there is no fuel to burn.
- *Removing the heat* – If the liquid is stored or used away from a heat source, it will not be able to ignite.
- *Removing the oxygen* – The vapor from a liquid will not burn if oxygen is not present. Keeping safety containers tightly sealed prevents oxygen from coming into contact with the fuel.

ON-SITE

Prevention and Preparation Are the Keys to Fire Safety

Any fire in the workplace can cause serious injury or property damage. When chemicals are involved, the risks are even greater. Prevention is the key to eliminating the hazards of fire where you work. Preparation is the key to controlling any fires that do start. Take the following precautions to make sure you are safe from fire in your workplace:

- Keep work areas clean and clutter-free.
- Know how to handle and store chemicals.
- Know what you are expected to do in case of a fire emergency.
- Call for professional help immediately. Don't let a fire get out of control.
- Know what chemicals you work with. You might have to tell firefighters at a chemical fire what kinds of hazardous substances are involved.
- Make sure you are familiar with your company's emergency action plan for fires.
- Use caution when using power tools near flammable substances.

9.2.2 Flammable Gases

Flammable gases used on construction sites include acetylene, hydrogen, ethane, and propane (liquid propane gas, or LPG). To save space, these gases are compressed so that a large amount is stored in a small cylinder or bottle. As long as the gas is kept in the cylinder, oxygen cannot get to it and start a fire. The cylinders should be stored away from sources of heat.

If oxygen is allowed to escape and mix with a flammable gas, the resulting mixture will explode under certain conditions.

WARNING!

Never use grease or oil on the fittings of oxygen bottles and hoses. Never allow greasy or oily rags to come near any part of an oxygen system. Oil and pressurized oxygen form a very dangerous mixture that can ignite at low temperatures.

9.2.3 Ordinary Combustibles

The term ordinary combustibles means paper, wood, cloth, and similar fuels. The easiest way to prevent fire in ordinary combustibles is to keep a neat, clean work area. If there are no scraps of paper, cloth, or wood lying around, there will be no fuel for starting a fire. So establish and maintain good housekeeping habits. Use approved storage cabinets and containers for all waste and other ordinary combustibles.

9.3.0 Firefighting

You are not expected to be an expert firefighter. But you may have to deal with a fire to protect your safety and the safety of others. You need to know the locations of firefighting equipment on your job site. You also need to know which equipment to use on different types of fires. However, only qualified personnel are authorized to fight fires.

Most companies tell new employees where fire extinguishers are kept. If you have not been told, be sure to ask. Also ask how to report fires. The telephone number of the nearest fire department should be clearly posted in your work area. If your company has a company fire brigade, learn how to contact them. Learn your company's fire safety procedures. Know what kind of extinguisher to use for different kinds of fires and how to use them. Make sure all extinguishers are fully charged. Never remove the tag from an extinguisher—it shows the date the extinguisher was last serviced and inspected (see *Figure 41*).

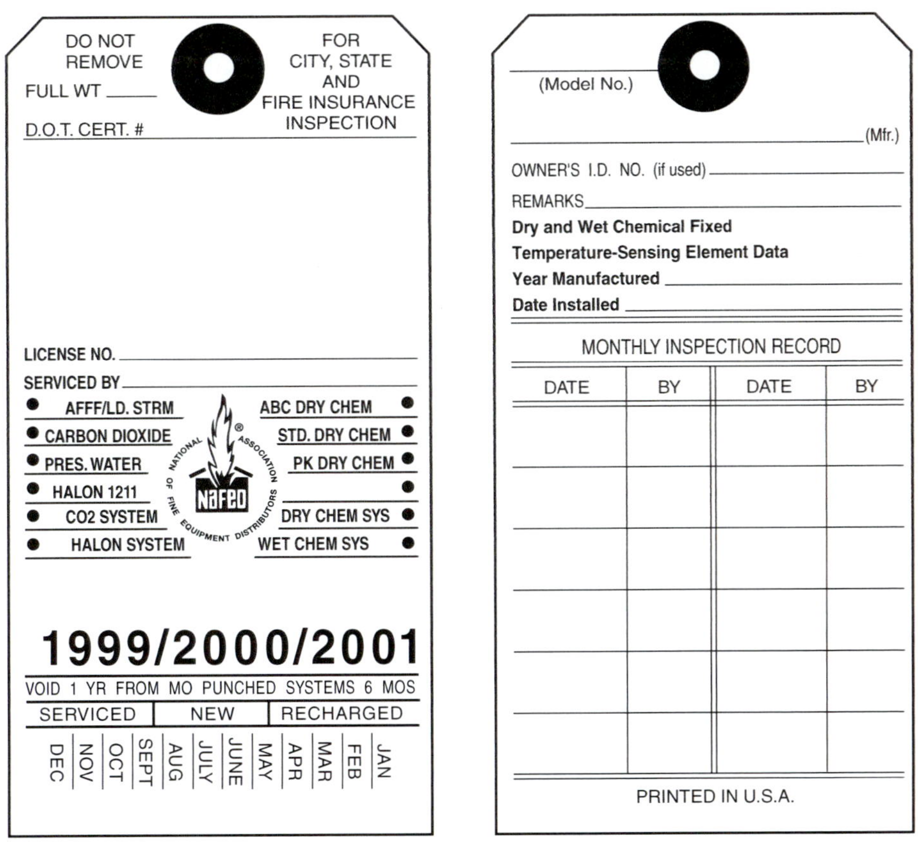

101F41.EPS

Figure 41 ◆ Fire extinguisher tag.

9.3.1 Classes of Fires

Four classes of fuels can be involved in fires (see *Table 5*). You've already learned about liquids, gases, and ordinary combustibles. Another fuel is metal. (You will learn about preventing electrical fires in another module.) Each class of fuel requires a different method of firefighting and a different type of extinguisher.

The label on a fire extinguisher clearly shows the class of fire on which it can be used (see *Figure 42*).

When you check the extinguishers in your work area, you will see that some are rated for more than one class of fire. You can use an extinguisher that has the three codes A, B, and C on it to fight a class A, B, or C fire. But remember, if the extinguisher has only one code letter, do not use it on any other class of fire, even in an emergency. You could make the fire worse and put yourself in great danger.

Table 5 Classes of Fires

Class	Materials and Proper Fire Extinguisher
Class A fires 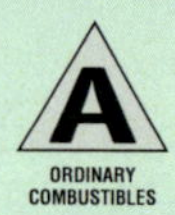 101SA02A.EPS  101SA02B.EPS	These fires involve ordinary combustibles such as wood or paper. Class A fires are fought by cooling the fuel. Class A fire extinguishers contain water. Using a Class A extinguisher on any other type of fire can be very dangerous.
Class B fires  101SA02C.EPS  101SA02D.EPS	These fires involve grease, liquids, or gases. Class B extinguishers contain carbon dioxide (CO_2) or another material that smothers fires by removing oxygen from the fire.
Class C fires  101SA02E.EPS  101SA02F.EPS	These fires are near or involve energized electrical equipment. Class C extinguishers are designed to protect the firefighter from electrical shock. Class C extinguishers smother fires.
Class D fires  101SA02G.EPS	These fires involve metals. Class D extinguishers contain a powder that either forms a crust around the burning metal or gives off gases that prevent oxygen from reaching the fire. Some metals will keep burning even though they have been coated with powder from a Class D extinguisher. The best way to fight these fires is to keep using the extinguisher so the fire will not spread to other fuels.

ON-SITE

How to Use a Fire Extinguisher

Step 1 Hold the extinguisher upright.

Step 2 Pull the pin, breaking the plastic seal.

Step 3 Stand back 8 to 10 feet from the fire. Standing any closer may cause burning objects to scatter, spreading the fire.

Step 4 Aim at the base of the fire.

Step 5 Keep the extinguisher upright. Squeeze the handles together to discharge. Sweep from side to side.

Step 6 Move closer as the fire is extinguished (watch for scattering burning material).

Step 7 When the fire is out, watch for reignition.

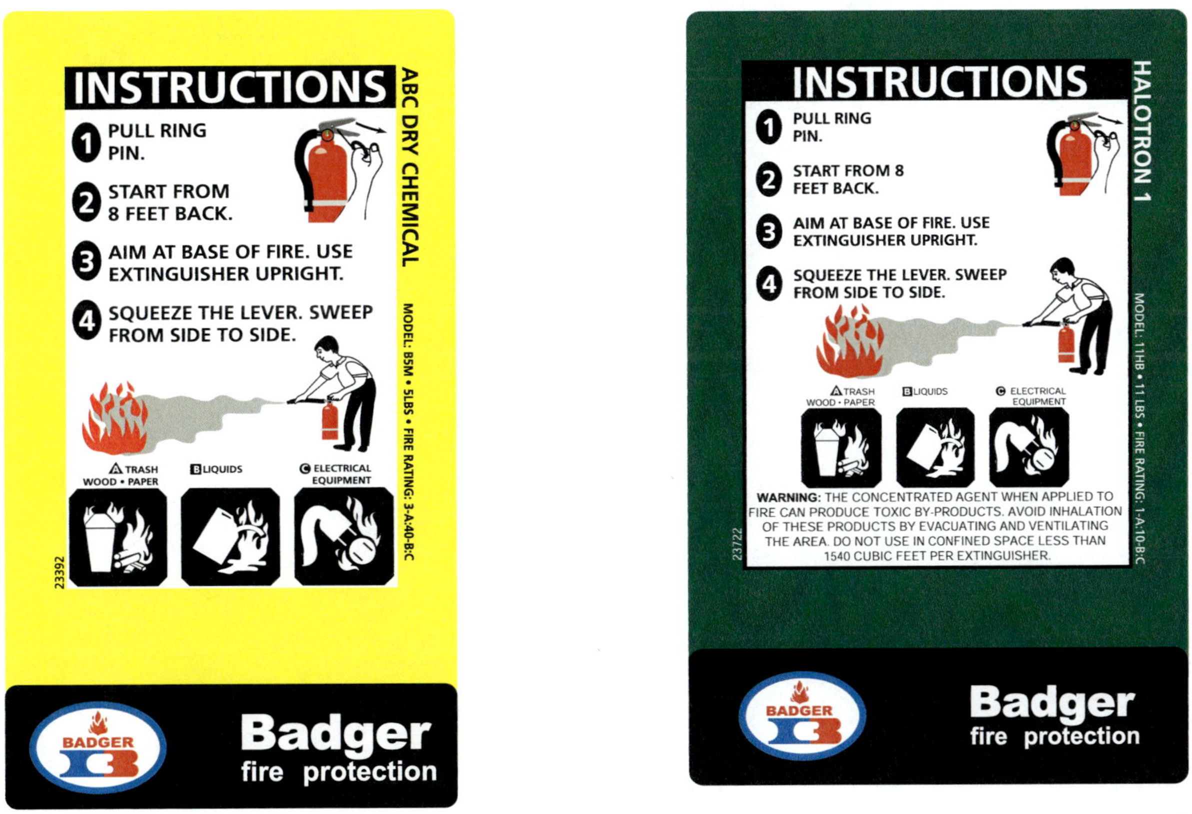

Figure 42 ◆ Typical fire extinguisher labels. (1 of 2)

101F42C.EPS

(C)

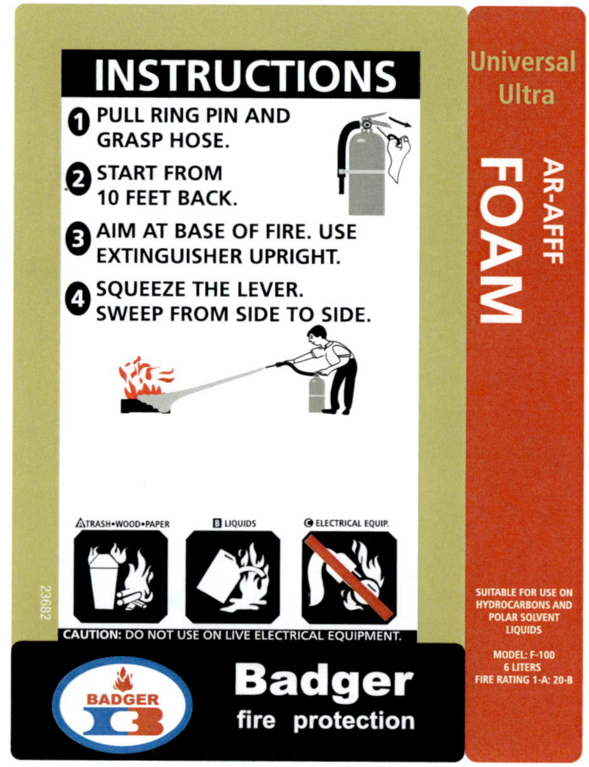

101F42D.EPS

(D)

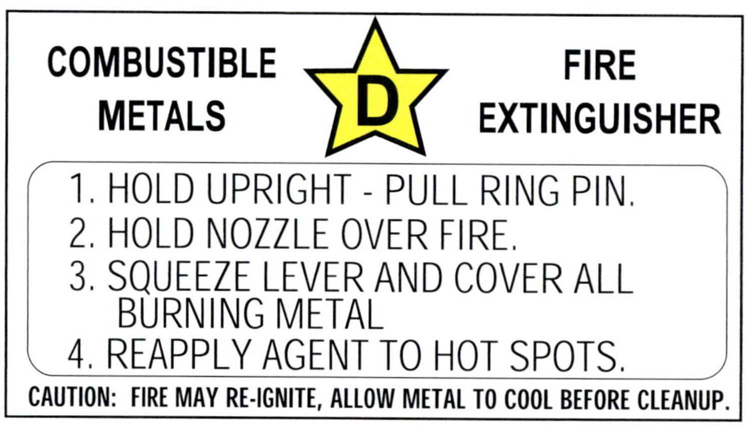

101F42E.EPS

(E)

Figure 42 ◆ Typical fire extinguisher labels. (2 of 2)

Review Questions

Section 9.0.0

1. __________ must be present in the same place at the same time for a fire to occur.

a. Oxygen, carbon dioxide, and heat
b. Oxygen, heat, and fuel
c. Hydrogen, oxygen, and wood
d. Grease, liquid, and heat

2. __________ gas is flammable.

a. Acetylene
b. Carbon dioxide
c. Peroxide
d. Neon

3. A Class D fire involves __________.

a. grease
b. wood
c. electrical equipment
d. metal

4. Fire extinguishers that contain water for fighting fires involving ordinary combustibles are __________ extinguishers.

a. Class A
b. Class B
c. Class C
d. Class D

5. For a grease fire, use a __________ extinguisher.

a. Class A
b. Class B
c. Class C
d. Class D

10.0.0 ◆ ELECTRICAL SAFETY

Some construction workers think that electrical safety matters only to electricians. But on many jobs, no matter what your trade, you will use or work around electrical equipment. Extension cords, power tools, portable lights, and many other pieces of equipment use electricity. If you don't use this equipment safely and properly, the result could be death for you or a co-worker.

Electricity can be described as the flow of electrons through a conductor. This flow of electrons is called electrical current. Some materials—such as silver, copper, steel, and aluminum—are excellent conductors. This means that electrical current flows easily through them. The human body, especially when it is wet, is also a good conductor.

To create an electrical current, a path must be provided in a circular route, or a circuit. If the circuit is interrupted, the electrical current will complete its circular route by flowing along the path of least resistance. This means that it will flow into and through any conductor that is touching it. If it cannot complete its circuit, the electrical current will go to ground. This means that it will find the path of least resistance that allows it to flow as directly as possible into the earth. All of this takes place almost instantly.

If the human body comes in contact with an electrically energized conductor and is in contact with the ground at the same time, the human body becomes the path of least resistance for the electricity. This means the electricity flows through the body in less than the blink of an eye. You can't see that it's about to happen; it just happens. That's why safety precautions are so important when working with and around electrical currents. When a person's body conducts electrical current and the amount of that current is high enough, the person can be electrocuted (killed by electric shock). *Table 6* shows the effects of different amounts of electrical current on the human body and lists some common tools that operate using those currents.

NOTE

Electric shocks or burns are a major cause of accidents in the construction industry. According to the Bureau of Labor Statistics, electrocution is the fourth leading cause of death among construction workers.

WARNING!

All work on electrical equipment should be done with circuits de-energized and cleared or grounded. All conductors, buses, and connections should be considered energized unless proven otherwise.

Table 6 Effects of Electrical Current on the Human Body

Current	Common Item/Tool	Reaction to Current
0.001 amps	Watch battery	Faint tingle
0.005 amps	9-volt battery	Slight shock
0.006–0.025 amps (women) 0.009–0.030 amps (men)	Christmas tree light bulb	Painful shock. Muscular control is lost.
0.050–0.9 amps	Small electric radio	Extreme pain. Breathing stops; severe muscular contractions occur. Death may result.
1.0–9.9 amps	Jigsaw (4 amps); Sawsall® or Port-a-Band® saw (6 amps); portable drill (3–8 amps)	Ventricular fibrillation and nerve damage occur. Death may result.
10 amps and above	ShopVac® (15-gallon); circular saw	Heart stops beating; severe burns occur. Death may result.

WARNING!
Less than 1 amp of electrical current can kill. Always take precautions when working around electricity.

Here's an example. A craftsperson is operating a portable power drill while standing on damp ground. The power cord inside the drill has become frayed, and the electric wire inside the cord touches the metal drill frame. Three amps of current pass from the wire through the frame, then through the craftsperson's body and into the ground. *Table 6* shows that this craftsperson will probably die.

A good method of protection from accidental electrocution is the use of a ground fault circuit interrupter (GFCI). The GFCI is a fast-acting circuit breaker that senses small (as little as approximately 5 milliamps) imbalances in the circuit caused by current leakage to ground. In as little as 1/40 of a second, the GFCI will interrupt the power.

Not all electrical accidents result in death. There are different types of electrical accidents. Any of the following can happen:

- Burns
- Electric shock
- Explosions
- Falls caused by electric shock
- Fires

10.1.0 Basic Electrical Safety Guidelines

OSHA and your company have specific policies and procedures to keep the workplace safe from electrical hazards. You can do many things to reduce the chance of an electrical accident. If you ever have any questions about electrical safety on the job site, ask your supervisor. Here are the basic job-site electrical safety guidelines:

- Use three-wire extension cords and protect them from damage. Never fasten them with staples, hang them from nails, or suspend them from wires. Never use damaged cords.
- Make sure that panels, switches, outlets, and plugs are grounded.
- Never use bare electrical wire.
- Never use metal ladders near any source of electricity.
- Never wear a metal hard hat.
- Always inspect electrical power tools before you use them.
- Never operate any piece of electrical equipment that has a danger tag or lockout device attached to it.
- Use three-wire cords for portable power tools and make sure they are properly connected (see *Figure 43*). The three-wire system is one of the most common safety grounding systems used to protect you from accidental electrical shock. The third wire is connected to a ground. If the insulation in a tool fails, the current will pass to ground through the third wire—not through your body.

NOTE
It is becoming more common to use double-insulated tools because they are safer than relying on a three-wire cord alone.

- Never use worn or frayed cables (see *Figure 44*).
- Make sure all light bulbs have protective guards to prevent accidental contact (see *Figure 45*).

Electrical Cord Safety

ON SITE

Electrical cords are frequently seen on construction sites, yet they are often overlooked. Use the following safety guidelines to ensure your safety and the safety of other workers.

- Every electrical cord should have an Underwriters Laboratory (UL) label attached to it. Check the UL label for specific wattage. Do not plug more than the specified number of watts into an electrical cord.
- A cord set not marked for outdoor use is to be used indoors only. Check the UL label on the cord for an outdoor marking.
- Do not remove, bend, or modify any metal prongs or pins of an electrical cord.
- Extension cords used with portable tools and equipment must be three-wire type and designated for hard or extra-hard use. Check the UL label for the cord's use designation.
- Avoid overheating an electrical cord. Make sure the cord is uncoiled, and that it does not run under any covering materials, such as tarps, insulation rolls, or lumber.
- Do not run a cord through doorways or through holes in ceilings, walls, and floors, which might pinch the cord. Also, check to see that there are no sharp corners along the cord's path. Any of these situations will lead to cord damage.
- Extension cords are a tripping hazard. They should never be left unattended and should always be put away when not in use.

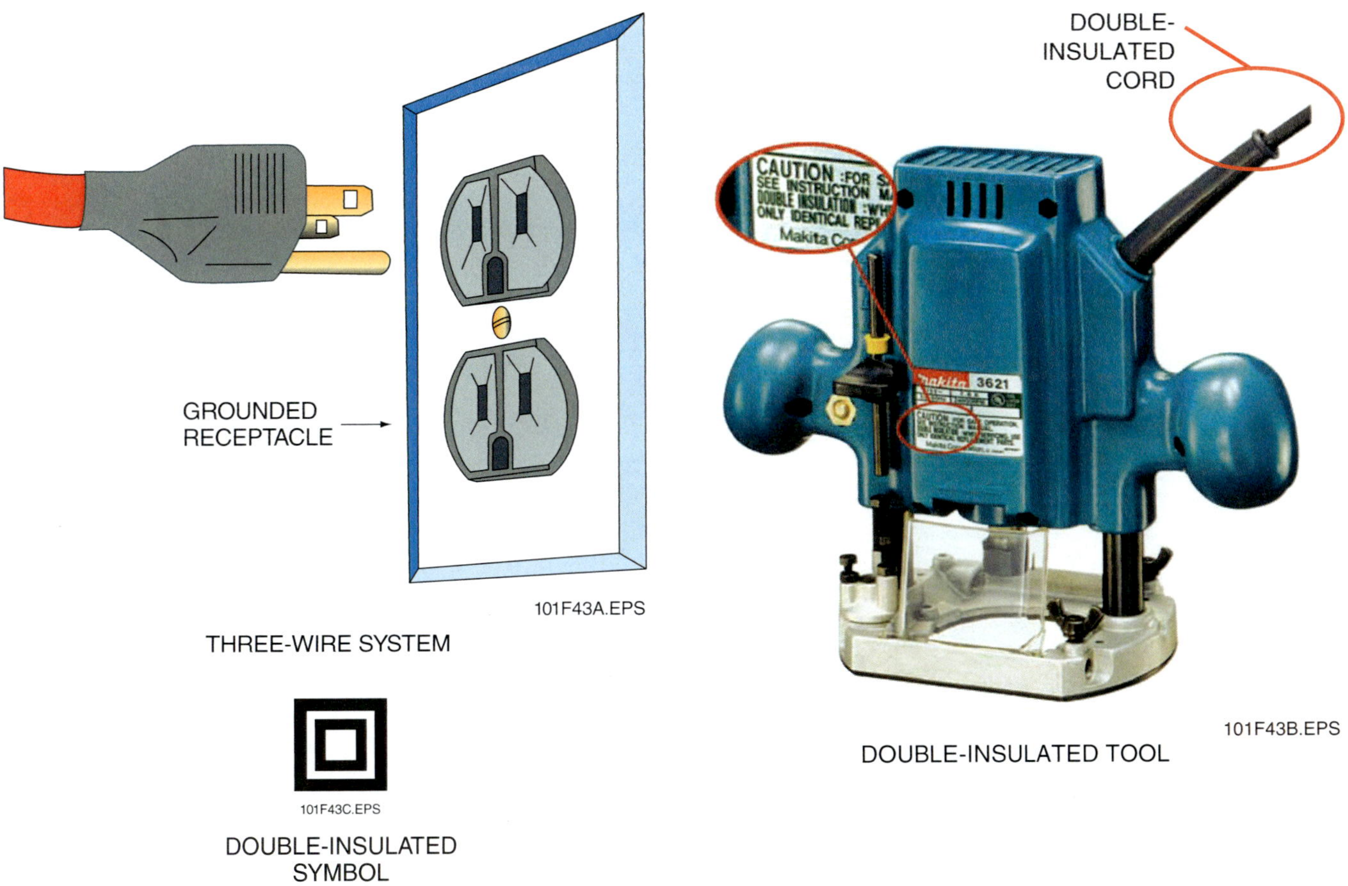

Figure 43 ◆ Three-wire system and double-insulated tool.

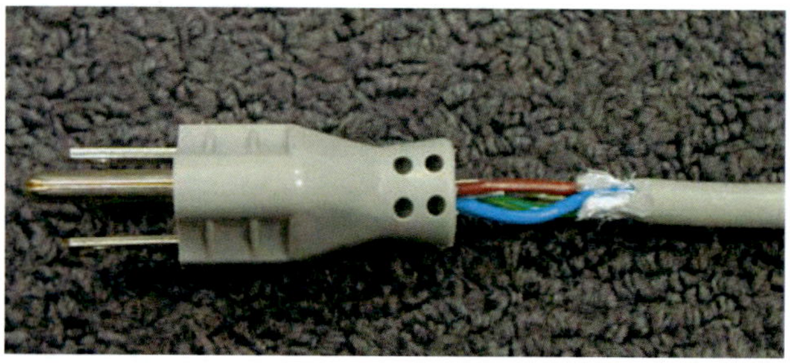

101F44A.EPS

101F44B.EPS

Figure 44 ◆ Never use damaged cords.

- Do not hang temporary lights by their power cords unless they are specifically designed for this use.
- Check the cable and ground prong. Check for cuts in the cords and make sure the cords are clean of grease.
- Use only approved **concealed receptacles** for plugs. If different voltages or types of current are used in the same area, the receptacles should be designed so that the plugs are not interchangeable.
- Any repairs to cords must be performed by a qualified person.
- Use a GFCI (see *Figure 46*) or an assured grounding program (see section 10.2.0) with every tool.
- Always make sure all tools are grounded before use.

CAUTION

All tools used in construction must be ground-fault protected. This helps ensure the safety of workers.

101F45.EPS

Figure 45 ◆ Work light with protective guard.

101F46A.EPS

Figure 46 ◆ GFCI.

10.2.0 Working Near Energized Electrical Equipment

No matter what your trade, your job may include working near exposed electrical equipment or conductors. This is one example of proximity work. Often, **electrical distribution panels, switch enclosures,** and other equipment must be left open during construction. This leaves the wires and components in them exposed. Some or all of the wires and components may be energized. Working near exposed electrical equipment can be safe, but only if you keep a safe working distance.

Regulations and company policies tell you the minimum safe working distances from exposed conductors. The safe working distance ranges from a few inches to several feet, depending on the voltage. The higher the voltage, the greater the safe working distance.

You must learn the safe working distance for each situation. Make sure you never get any part of your body or any tool you are using closer to exposed conductors than that distance. You can get information on safe working distances from your instructor, your supervisor, company safety policies, and regulatory documents.

10.3.0 If Someone Is Shocked

If you are there when someone gets an electrical shock, you can save a life by taking immediate action. Here's what to do:

Step 1 Immediately disconnect the circuit.

Step 2 If you can't disconnect the circuit, do not try to separate the victim from the circuit. If you touch a person who is being electrocuted, the current will flow through you, too.

WARNING!

Do not touch the victim or the electrical source with your hand, foot, or any other part of your body or with any object or material. You could become another victim.

Step 3 Once the circuit is disconnected, give first aid and call an ambulance. If you cannot disconnect the circuit, call an ambulance.

Assured Equipment Grounding Program

An assured equipment grounding program is an alternative to GCFIs. It covers all cord sets, receptacles that are not a part of the permanent wiring of the building or structure, and equipment connected by cord and plug that are available for use or used by employees. The requirements that the program must meet are stated in OSHA's *29 CFR 1926.404*, but employers may provide additional tests or procedures.

OSHA requires two tests. One is a continuity test to ensure that the equipment grounding conductor is electrically continuous. It must be performed on all cord sets, receptacles that are not part of the permanent wiring of the building or structure, and cord- and plug-connected equipment that is required to be grounded. This test may be performed using a simple continuity tester, such as a lamp and battery, a bell and battery, an ohmmeter, or a receptacle tester.

The other test must be performed on receptacles and plugs to ensure that the equipment grounding conductor is connected to its proper terminal. This test can be performed with the same equipment used in the first test.

These tests are required before first use, after any repairs, after damage is suspected to have occurred, and at three-month intervals. Cord sets and receptacles that are essentially fixed and not exposed to damage must be tested at regular intervals of no longer than six months. Any equipment that fails to pass the required tests must not be made available to or used by employees.

Review Questions

Section 10.0.0

1. Observing proper safety precautions when working with and around electrical current is important because the human body ________.
   a. resists the electricity's path
   b. can conduct electrical current
   c. won't conduct electricity
   d. doesn't offer electricity a circular route

2. The ________ system is one of the most common safety grounding systems used with portable power tools.
   a. distribution wire
   b. rubber cord
   c. three-wire cord
   d. insulation plug

3. The minimum safe working distance from exposed electrical conductors ________.
   a. depends on the voltage
   b. is at least 1 foot
   c. is a few inches
   d. is unlimited

4. If someone is being electrically shocked, the first thing you should try to do is ________.
   a. use a metal pole to separate the victim from the circuit
   b. disconnect the circuit
   c. give first aid
   d. call your supervisor

5. If someone is being electrically shocked and you cannot disconnect the circuit, the first thing you should do is ________.
   a. call an ambulance
   b. give first aid
   c. pull the victim away with your hands
   d. use a pole to separate the victim from the circuit

Summary

Notes

Key Terms Quiz

Fill in the blank with the correct key term that you learned from your study of this module.

1. __________ is a formal procedure for taking equipment out of service and ensuring it cannot be operated until a qualified person has returned it to service.
2. __________ is the process of joining metal parts by fusion.
3. Because __________ scrap materials catch fire and burn easily, remove them regularly from the work area.
4. The cutting process results in oxides that mix with molten iron and produce __________.
5. A(n) __________ is any man-made cut, cavity, trench, or depression in an earth surface, formed by removing earth.
6. A(n) __________ identifies unsanitary, hazardous, or dangerous working conditions and has the authority to correct or eliminate them.
7. A(n) __________ is large enough to work in but has limited means of entry or exit; sometimes a permit is required to work in it.
8. Store __________ liquids in safety cans to avoid the risk of fire.
9. If a scaffold is more than 10 feet high, __________—or pieces of wood or metal placed diagonally from the bottom of one rail to the top of another rail that add support to a structure—must be used.
10. The __________ houses the circuits that distribute electricity throughout a structure.
11. A(n) __________ is basically two straight ladders that are connected so the length of the ladder can be changed.
12. Wear __________ to protect your eyes from the _____, which is the sudden, bright light that occurs when you start up a welding operation.
13. To save lives, prevent injuries, and protect the health of America's workers, __________ publishes rules and regulations that employees and employers must follow.
14. If fuel is improperly mixed, it can cause a(n) __________.
15. An opening in a wall or floor is a safety hazard and must be either covered or __________.
16. Even a brief exposure to the ultraviolet light from arc welding can damage your eyes, causing a(n) __________.
17. __________ is an OSHA rule requiring all contractors to educate their employees about the hazardous chemicals they may be exposed to on the job site.
18. The temperature at which a fuel gives off enough gases to burn is called the __________.
19. A(n) __________ is the conducting connection between electrical equipment or an electrical circuit and the earth.
20. When climbing a ladder or scaffold, use a tagline or __________ to pull up your tools.
21. SCBA is an example of a(n) __________, or an assembly of machines used to do a particular job.
22. If a work area is __________, that means it has pieces of material at least 2 inches thick and 6 inches wide used as flooring, decking, or scaffolding.
23. When doing work more than 6 feet above the ground, you must wear a safety harness with a(n) __________ that is attached to a strong anchor point.
24. A good __________ helps prevent or correct conditions that can cause accidents.
25. To prevent a cave-in, follow OSHA regulations for __________ up a trench.
26. Refer to the __________ to learn about how to handle hazardous substances.
27. Overloading, which means exceeding the __________ of a ladder, can cause ladder failure.

28. If you are operating a vehicle on a job site and cannot see to your rear, get a(n) _________ to direct you.

29. Before working in a(n) _________, you must be trained, obtain written authorization, and take the necessary precautions.

30. A(n) _________ is a narrow excavation made below the surface of the ground that is generally deeper than it is wide.

31. The _________ on a scaffold is placed halfway between the toeboard and the top rail.

32. _________ for welding includes a faceshield, ear plugs, and gloves.

33. A(n) _________ has proven his or her extensive knowledge, training, and experience and has successfully demonstrated the ability to solve problems relating to the work.

34. A(n) _________ provides clean air for breathing.

35. Manufactured and rolling are the two basic types of _________.

36. When doing _________, you must be careful not to come into contact with the nearby hazard.

37. A(n) _________ is a self-supporting ladder made of two sections hinged at the top.

38. Use only approved _________ for plugs.

39. A(n) _________ is nonadjustable and consists of two rails, rungs between the rails, and safety feet on the bottom of the rails.

40. A vertical barrier called a(n) _________ is used at floor level on scaffolds to prevent materials from falling.

41. A(n) _________ tells you which way the wind is blowing.

42. A horizontal board called a(n) _________ is used at top-level on all open sides of scaffolding and platforms.

43. Never look at the _________ caused by welding without proper eye protection.

44. An excavating machine called a(n) _________ is used to dig trenches, especially for pipeline and cables.

45. When you are welding and other workers are in the area, set up a protective screen called a(n) _________.

46. A(n) _________ houses electrical switches used to regulate and distribute electricity in a building.

47. _________ is the waste material from welding operations.

Key Terms

Apparatus
Arc
Arc welding
Combustible
Competent person
Concealed receptacle
Confined space
Cross-bracing
Dross
Electrical distribution panel
Excavation
Extension ladder
Flammable
Flash
Flashback
Flash burn
Flash goggles
Flash point
Ground
Guarded
Hand line
Hazard Communication Standard (HazCom)
Lanyard
Lockout/tagout
Management system
Material safety data sheet (MSDS)
Maximum intended load
Mid-rail
Occupational Safety and Health Administration (OSHA)
Permit-required confined space
Personal protective equipment (PPE)
Planked
Proximity work
Qualified person
Respirator
Scaffold
Shoring
Signaler
Slag
Stepladder
Straight ladder
Switch enclosure
Toeboard
Top rail
Trench
Trencher
Welding shield
Wind sock

Profile in Success

Doug Garcia

Chairman, Industrial Training Department
Manatee Technical Institute
West Bradenton, Florida

Doug grew up in Fall River, Massachusetts. After completing high school, he began his career as a union construction laborer. He decided to become a welder and worked in the power plant maintenance industry before turning to shipbuilding. Doug built ships at General Dynamics and submarines at Electric Boat. He worked his way up from shipfitter to lead man, foreman, and ultimately ship superintendent, responsible for the delivery of a ship from the keel on up. Doug became a welding instructor at Diman Regional Vocational Technical High School and then Manatee Technical Institute.

Doug has a BA in vocational education from Fitchburg State College and is a U.S. Department of Labor–approved general industry and construction outreach trainer. He is also an American Welding Society (AWS)–certified welding educator and a certified welding inspector, and holds a Level 2 certification from the American Society for Nondestructive Testing.

What aspect of the construction industry appeals to you most?

What I like the most is the personal satisfaction of seeing the finished product. When you look at a completed project, you are really seeing the result of a collaborative effort between a broad spectrum of trades. In the finished product, you can see how they all come together to make a whole.

How did you decide to become an educator?

Over the course of my career, I have really come to appreciate and understand the need for safety training. It bothers me most to see a great career cut short by a needless accident. Safety is one of the most important concerns in the construction industry. As a whole, the industry gives safety the attention it needs, but ultimately the responsibility still falls on the individual. No matter how good a safety program is, it is only as good as the person who's applying it. The company can only do so much; the employee has to finish the job. So I wanted to help people understand and apply safety on the job.

What do you think it takes to be a success in your trade?

I think that it takes a solid work ethic and pride in one's craftsmanship. You must also feel, or want to feel, a sense of satisfaction from seeing a completed project.

What are some of the things you do in your job?

I have a staff of 17 teachers, 2 of whom are approved construction safety instructors and 2 of whom are general industrial safety instructors. I review the existing curriculum and undertake curriculum development tasks. I try to ensure that the curriculum meets the needs of industry and that we remain current with the changes in construction technology and materials. I assist with job placement and maintain contact with area employers. We provide on-site training for local area employers and help them to develop and implement procedures that are required by the U.S. Occupational Safety and Health Administration (OSHA).

I teach safety instruction to all students in both the construction and general industry tracks at Manatee. I'm also a mentor for new vocational teachers. And I also coordinate the purchase of equipment and supplies. It's a full-time job, and then some!

What do you like most about your job?

I really enjoy seeing former students become successful. Many of them return to the program to serve on advisory committees. They come back to advise the staff and faculty on current trends in their craft area, and they get to make recommendations for the curricula and the purchase of equipment. That way, I get to work with them as professional colleagues and not just as students.

What would you say to someone entering the trades today?

In my classes I stress that no matter what your skill level is, if you are not doing your job safely, then you are jeopardizing all your hard work. Even the most talented individuals can't perform their jobs to the best of their ability if they have been injured or hurt.

I have personally witnessed two construction fatalities. The people most affected are usually the families. Your value as a craft worker is the ability to perform for your employer and to provide for your family. If you are injured or incapable of performing your craft, you are not only losing your earning potential and your potential for advancement, but you are also depriving the industry of a trained craft worker and reducing your family's quality of life.

So you have to ask yourself every day, "How would an accident affect my family?" When I tell my students that, it really gets them thinking. The industry has inherent dangers; there's no need for a careless worker to add to them.

Trade Terms Introduced in This Module

Apparatus: An assembly of machines used together to do a particular job.

Arc: The flow of electrical current through a gas such as air from one pole to another pole.

Arc welding: The joining of metal parts by fusion, in which the necessary heat is produced by means of an electric arc.

Combustible: Capable of easily igniting and rapidly burning; used to describe a fuel with a flash point at or above 100°F.

Competent person: A person who can identify working conditions or surroundings that are unsanitary, hazardous, or dangerous to employees and who has authorization to correct or eliminate these conditions promptly.

Concealed receptacle: The electrical outlet that is placed inside the structural elements of a building, such as inside the walls. The face of the receptacle is flush with the finished wall surface and covered with a plate.

Confined space: A work area large enough for a person to work, but arranged in such a way that an employee must physically enter the space to perform work. A confined space has a limited or restricted means of entry and exit. It is not designed for continuous work. Tanks, vessels, silos, pits, vaults, and hoppers are examples of confined spaces. See also *permit-required confined space.*

Cross-bracing: Braces (metal or wood) placed diagonally from the bottom of one rail to the top of another rail that add support to a structure.

Dross: Waste material resulting from cutting using a thermal process.

Electrical distribution panel: Part of the electrical distribution system that brings electricity from the street source (power poles and transformers) through the service lines to the electrical meter mounted on the outside of the building and to the panel inside the building. The panel houses the circuits that distribute electricity throughout the structure.

Excavation: Any man-made cut, cavity, trench, or depression in an earth surface, formed by removing earth. It can be made for anything from basements to highways. See also *trench.*

Extension ladder: A ladder made of two straight ladders that are connected so that the overall length can be adjusted.

Flammable: Capable of easily igniting and rapidly burning; used to describe a fuel with a flash point below 100°F.

Flash: A sudden bright light associated with starting up a welding torch.

Flashback: A welding flame that flares up and chars the hose at or near the torch connection. It is caused by improperly mixed fuel.

Flash burn: The damage that can be done to eyes after even brief exposure to ultraviolet light from arc welding. A flash burn requires medical attention.

Flash goggles: Eye protective equipment worn during welding operations.

Flash point: The temperature at which fuel gives off enough gases (vapors) to burn.

Ground: The conducting connection between electrical equipment or an electrical circuit and the earth.

Guarded: Enclosed, fenced, covered, or otherwise protected by barriers, rails, covers, or platforms to prevent dangerous contact.

Hand line: A line attached to a tool or object so a worker can pull it up after climbing a ladder or scaffold.

Hazard Communication Standard (HazCom): The Occupational Safety and Health Administration standard that requires contractors to educate employees about hazardous chemicals on the job site and how to work with them safely.

Lanyard: A short section of rope or strap, one end of which is attached to a worker's safety harness and the other to a strong anchor point above the work area.

Lockout/tagout: A formal procedure for taking equipment out of service and ensuring that it cannot be operated until a qualified person has removed the lockout or tagout device (such as a lock or warning tag).

Management system: The organization of a company's management, including reporting procedures, supervisory responsibility, and administration.

Material safety data sheet (MSDS): A document that must accompany any hazardous substance. The MSDS identifies the substance and gives the exposure limits, the physical and chemical characteristics, the kind of hazard it presents, precautions for safe handling and use, and specific control measures.

Maximum intended load: The total weight of all people, equipment, tools, materials, and loads that a ladder can hold at one time.

Mid-rail: Mid-level, horizontal board required on all open sides of scaffolding and platforms that are more than 14 inches from the face of the structure and more than 10 feet above the ground. It is placed halfway between the toeboard and the top rail.

Occupational Safety and Health Administration (OSHA): An agency of the U.S. Department of Labor. Also refers to the Occupational Safety and Health Act of 1970, a law that applies to more than more than 111 million workers and 7 million job sites in the country.

Permit-required confined space: A confined space that has been evaluated and found to have actual or potential hazards, such as a toxic atmosphere or other serious safety or health hazard. Workers need written authorization to enter a permit-required confined space. See also *confined space.*

Personal protective equipment (PPE): Equipment or clothing designed to prevent or reduce injuries.

Planked: Having pieces of material 2 or more inches thick and 6 or more inches wide used as flooring, decking, or scaffolding.

Proximity work: Work done near a hazard but not actually in contact with it.

Qualified person: A person who, by possession of a recognized degree, certificate, or professional standing, or by extensive knowledge, training, and experience, has demonstrated the ability to solve or prevent problems relating to a certain subject, work, or project.

Respirator: A device that provides clean, filtered air for breathing, no matter what is in the surrounding air.

Scaffold: An elevated platform for workers and materials.

Shoring: Using pieces of timber, usually in a diagonal position, to hold a wall in place temporarily.

Signaler: A person who is responsible for directing a vehicle when the driver's vision is blocked in any way.

Slag: Waste material from welding operations.

Stepladder: A self-supporting ladder consisting of two elements hinged at the top.

Straight ladder: A nonadjustable ladder.

Switch enclosure: A box that houses electrical switches used to regulate and distribute electricity in a building.

Toeboard: A vertical barrier at floor level attached along exposed edges of a platform, runway, or ramp to prevent materials and people from falling.

Top rail: A top-level, horizontal board required on all open sides of scaffolding and platforms that are more than 14 inches from the face of the structure and more than 10 feet above the ground.

Trench: A narrow excavation made below the surface of the ground that is generally deeper than it is wide, with a maximum width of 15 feet. See also *excavation.*

Trencher: An excavating machine used to dig trenches, especially for pipeline and cables.

Welding shield: (1) A protective screen set up around a welding operation designed to safeguard workers not directly involved in that operation. (2) A shield that provides eye and face protection for welders by either connecting to helmet-like headgear or attaching directly to a hard hat; also called a welding helmet.

Wind sock: A cloth cone open at both ends mounted in a high place to show which direction the wind is blowing.

Additional Resources

This module is intended to present thorough resources for task training. The following reference works are suggested for further study. These are optional materials for continued education rather than for task training.

Construction Back Safety. Videocassette. 10 minutes. Coastal Training Technologies Corp. Virginia Beach, VA.

Construction Confined Space Entry. Videocassette. 10 minutes. Coastal Training Technologies Corp. Virginia Beach, VA.

Construction Electrical Safety. Videocassette. 10 minutes. Coastal Training Technologies Corp. Virginia Beach, VA.

Construction Fall Protection: Get Arrested! Videocassette. 11 minutes. Coastal Training Technologies Corp. Virginia Beach, VA.

Construction Lockout/Tagout. Videocassette. 10 minutes. Coastal Training Technologies Corp. Virginia Beach, VA.

Construction Safety, 1996. Jimmie Hinze. Englewood Cliffs, NJ: Prentice Hall.

Construction Safety Council Home Page, http://buildsafe.org/home.htm.

Construction Safety Manual, 1998. Dave Heberle. New York: McGraw-Hill.

Construction Stairways & Ladders. Videocassette. 10 minutes. Coastal Training Technologies Corp. Virginia Beach, VA.

Construction Welding Safety. Videocassette. 10 minutes. Coastal Training Technologies Corp. Virginia Beach, VA.

Field Safety, 2003. NCCER. Upper Saddle River, NJ: Prentice Hall.

Handbook of OSHA Construction Safety and Health, 1999. James V. Eidson et al. Boca Raton, FL: Lewis Publishers, Inc.

HazCom For Construction. Videocassette. 11 minutes. Coastal Training Technologies Corp. Virginia Beach, VA.

NAHB-OSHA Jobsite Safety Handbook, 1999. Washington, DC: Home Builder Press. Available online at www.osha.gov.

Occupational Safety and Health Standards for the Construction Industry. Washington, DC: Occupational Safety and Health Administration, U.S. Department of Labor, U.S. Government Printing Office.

Safety Orientation, 2003. NCCER. Upper Saddle River, NJ: Prentice Hall.

Safety Technology, 2003. NCCER. Upper Saddle River, NJ: Prentice Hall.

United States Department of Labor, Occupational Safety and Health Administration Home Page, http://www.osha.gov.

NCCER CURRICULA — USER UPDATE

NCCER makes every effort to keep its textbooks up-to-date and free of technical errors. We appreciate your help in this process. If you find an error, a typographical mistake, or an inaccuracy in NCCER's curricula, please fill out this form (or a photocopy), or complete the online form at **www.nccer.org/olf**. Be sure to include the exact module ID number, page number, a detailed description, and your recommended correction. Your input will be brought to the attention of the Authoring Team. Thank you for your assistance.

Instructors – If you have an idea for improving this textbook, or have found that additional materials were necessary to teach this module effectively, please let us know so that we may present your suggestions to the Authoring Team.

NCCER Product Development and Revision

13614 Progress Blvd., Alachua, FL 32615

Email: curriculum@nccer.org

Online: www.nccer.org/olf

❑ Trainee Guide ❑ Lesson Plans ❑ Exam ❑ PowerPoints Other ______________________

Craft / Level: Copyright Date:

Module ID Number / Title:

Section Number(s):

Description:

Recommended Correction:

Your Name:

Address:

Email: Phone:

Basic Rigging

00106-04

Excellence in Construction Winner—
Institutional, $25–99 Million

Skanska USA Building, Inc. built the second U.S. courthouse in the country designed to meet the General Service Administration's blast-resistant criteria. The Jacksonville, Florida, $84-million, 14-story courthouse encompasses 457,416 square feet and has a four-story rotunda.

00106-04

Basic Rigging

Topics to be presented in this module include:

Overview

Rigging is a common activity on a construction site. It involves transporting materials and equipment from one location to another. Rigging can be dangerous because of the weight, size, or awkwardness of the load. Elements such as training, planning, and using proper procedures will increase the success of a rigging operation.

As with all jobs on a site, rigging operations require specialized hardware. The hardware used in rigging operations works together as a system. Each piece is connected to another. For example, slings are used as links between the load and the lifting device, and hitches are used to arrange the sling to hold the load. The sling and the hitch, therefore, are a connected part of the rigging system.

The ability to control a load is a critical part of rigging operations. Items such as rated capacity, sling alignment, and hardware attachments should be checked before an operation begins. Also, the use of hand signals is critical to the success of the operation. Without proper direction, the operator could lose control of the load and cause property damage or injuries.

Rigging operations affect everyone on site, even if there is no direct involvement in the operation. Always be aware of the rigging operations that are taking place on your work site.

Objectives

When you have completed this module, you will be able to do the following:

1. Identify and describe the use of slings and common rigging hardware.
2. Describe basic inspection techniques and rejection criteria used for slings and hardware.
3. Describe basic hitch configurations and their proper connections.
4. Describe basic load-handling safety practices.
5. Demonstrate proper use of American National Standards Institute (ANSI) hand signals.

Key Trade Terms

ANSI hand signals
Block and tackle
Bridle
Bull ring
Core
Cribbing
Eyebolt
Grommet sling
Hitch
Hoist
Lifting clamp
Load
Load control
Load stress
Master link
One rope lay
Pad eye
Plane
Rated capacity
Rejection criteria
Rigging hook
Shackle
Sheave
Side pull
Sling
Sling angle
Sling legs
Sling reach
Sling stress
Spliced
Strand
Stress
Tag line
Tattle-tail
Threaded shank
Warning-yarn
Wire rope

Required Trainee Materials

1. Appropriate personal protective equipment
2. Sharpened pencils and paper

Prerequisites

Before you begin this module, it is recommended that you successfully complete the following: *Core Curriculum: Introductory Craft Skills*, Modules 00101-04 through 00105-04. Modules 00106-04 through 00108-04 are electives. To receive a successful completion, you must take Module 00106-04 or Modules 00107-04 and 00108-04.

This course map shows all of the modules in *Core Curriculum: Introductory Craft Skills*. The suggested training order begins at the bottom and proceeds up. Skill levels increase as you advance on the course map. The local Training Program Sponsor may adjust the training order.

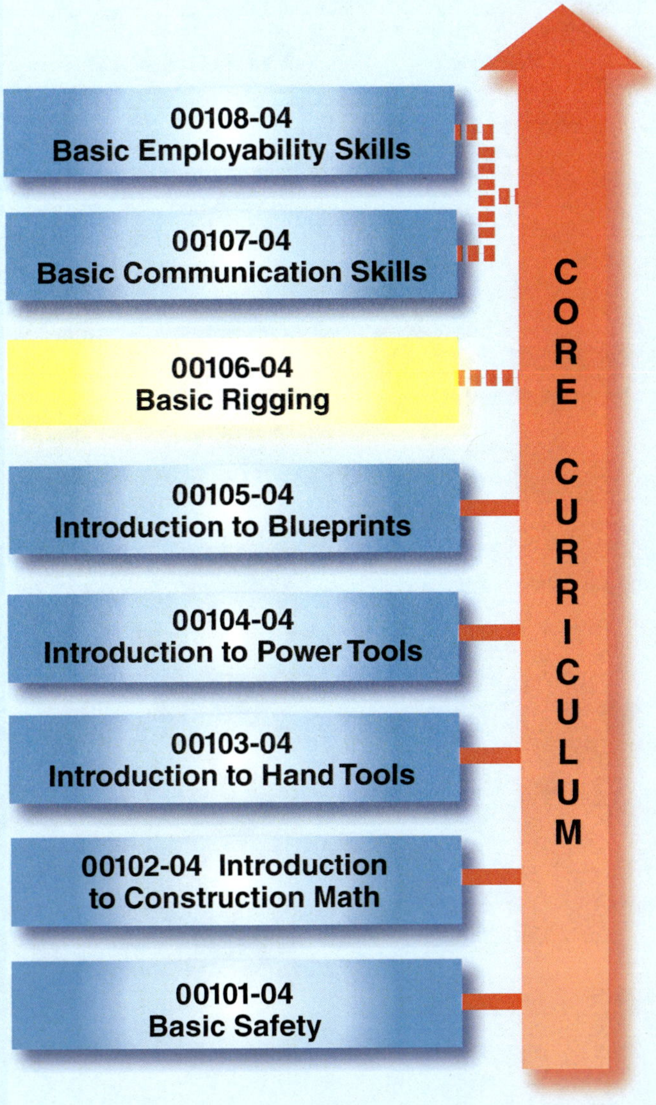

106CMAP.EPS

1.0.0 ◆ INTRODUCTION

Rigging is the planned movement of material and equipment from one location to another using **slings, hoists,** or other types of equipment. Some rigging operations use a loader to move materials around on a job site. Other operations require cranes to lift such **loads.** There are two basic types of cranes: overhead (*Figure 1*) and mobile (*Figure 2*).

> **WARNING!**
> Although this module introduces you to the criteria that are used to make rigging decisions, you will not be making the decisions yourself at this stage in your training. If you ever have any questions about whether a synthetic sling is defective or unsafe, ask your instructor or your immediate supervisor. Always refer to the manufacturer's instructions for the sling rejection criteria for that particular sling.

ON-SITE

CONSTRUCTION

World's Largest Mobile Crane

The world's largest crane was built by Mammoet in the Netherlands. The crane is called the Platform Twinring. There are actually three of them in the fleet. The other two are newer and referred to as PTCs. PTC stands for Platform Twinring Containerised. These two can be disassembled into 20- and 40-ft. shipping containers, and every piece of the crane can be lifted with a normal container crane on any dock anywhere in the world. The fittings are already molded into the ends so the PTCs can go from any container crane onto any truck trailer that normally handles shipping containers. These giant cranes have a maximum load capacity of 1,600 tons.

106SA01.TIF

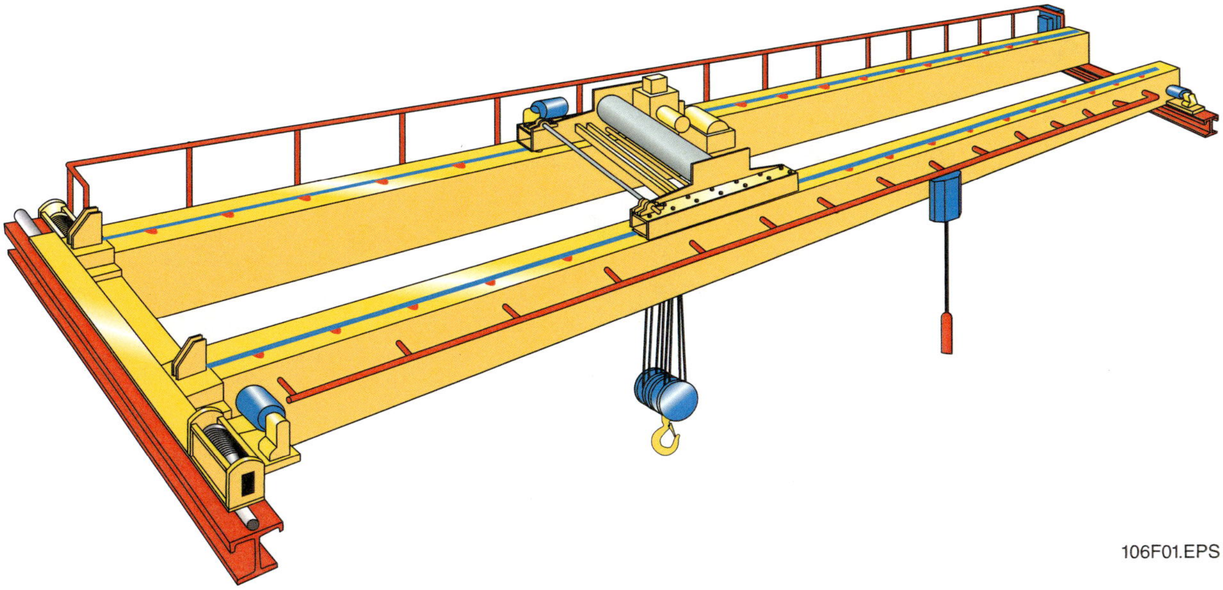

106F01.EPS

Figure 1 ◆ Overhead crane.

106F02A.EPS

106F02B.EPS

Figure 2 ◆ Mobile cranes.

This module provides basic information on

- Slings
- **Hitches**
- Rigging hardware
- The principles and practices behind safe and efficient rigging operations

At this early stage in your career, you will not be asked to perform rigging tasks without supervision. You must not attempt any rigging operations on your own. All rigging operations must be done under the supervision of a competent person. Even so, it is important that you under-

Regulations and Site Procedures

The information in this module is intended as a general guide. The techniques shown here are not the only methods you can use to perform a lift. Many techniques can be safely used to rig and lift different loads (the total amount being lifted).

Some of the techniques for certain kinds of rigging and lifting are spelled out in requirements issued by federal government agencies. Some will be provided at the job site, where you will see written site procedures that address any special conditions that affect lifting procedures on that site. If you have questions about any of these procedures, ask the supervisor at the site.

stand the fundamentals of the following aspects of rigging:

- The types of slings and equipment you may eventually be using
- How to determine whether your equipment is fit for use
- The proper use of the most common types of slings and hardware
- How to select appropriate rigging equipment and how factors such as **load stress** affect that selection

Core-level rigging is designed to introduce you to the basic principles of rigging. These basic principles are true whether you are rigging a single piece of pipe, a bundle of lumber, or an 850-ton steam generator.

Rigging operations can be extremely complicated and dangerous. Do not experiment with rigging operations, and never attempt a lift on your own, without the supervision of an officially recognized qualified person. A lift may appear simple while it is in progress. That is because the people performing the lift know exactly what they are doing. There is no room for guesswork!

No matter whether rigging operations involve a simple vertical lift, a powered hoist, or a highly complicated apparatus, only qualified persons may perform them without supervision.

2.0.0 ◆ SLINGS

During a rigging operation, the load being moved must be connected to the device, such as a crane, that is doing the moving. The connector—the link between the load and the lifting device—is often a sling made of synthetic, chain, or **wire rope** materials. In this section, you will learn about three types of slings:

- Synthetic slings
- Alloy steel chain slings
- Wire rope slings

All slings are required to have identification tags. An identification tag must be securely attached to each sling and clearly marked with the information required for that type of sling. For all three types of slings, that information will include the manufacturer's name or trademark and the **rated capacity** of the type of hitch used with that sling. The rated capacity is the maximum load weight that the sling was designed to carry. An example of a rated capacities chart for synthetic web slings appears in the Appendix. Synthetic, alloy steel chain, and wire rope slings are covered in this module.

The following are the tagging requirements for synthetic slings:

- Manufacturer's name or trademark
- Manufacturer's code or stock number (unique for each sling)
- Rated capacities for the types of hitches used
- Type of synthetic material used in the manufacture of the sling

The following are the tagging requirements for alloy steel chain slings:

- Manufacturer's name or trademark
- Manufactured grade of steel
- Link size (diameter)
- Rated load and the angle on which the rating is based
- New *2003 ASME Standard* requires clarification as to the angle upon which the capacity is based
- **Sling reach**
- Number of **sling legs**

The following are the tagging recommendations for wire rope slings:

- Manufacturer's name or trademark
- Rated capacity in a vertical hitch (other hitches optional)
- Size of wire rope (diameter)
- Manufacturer's code or stock number (unique for each sling)

DID YOU KNOW?

Sling Identification Tags

The identification tags that are required on all slings must be securely attached to the slings and clearly marked with the information required for each type of sling.

WARNING!

All slings and hardware have a rated capacity, also called lifting capacity, working capacity, working load limit (WLL), or safe working load (SWL). Under no circumstances should you ever exceed the rated capacity! Overloading may result in catastrophic failure. Rated capacity is defined as the maximum load weight a sling or piece of hardware or equipment can hold or lift.

2.1.0 Synthetic Slings

Synthetic slings are widely used to lift loads, especially easily damaged ones. In this section, you will learn about two types of synthetic slings: synthetic web slings and round slings.

2.1.1 Synthetic Web Sling Design and Characteristics

Synthetic web slings provide several advantages over other types of slings:

- They are soft and wider than wire rope or chain slings. Therefore, they do not scratch or damage machined or delicate surfaces (*Figure 3*).
- They do not rust or corrode and therefore will not stain the loads they are lifting.
- They are lightweight, making them easier to handle than wire rope or chain slings. Most synthetic slings weigh less than half as much as a wire rope that has the same rated capacity. Some new synthetic fiber slings weigh one-tenth as much as wire rope.
- They are flexible. They mold themselves to the shape of the load (*Figure 4*).
- They are very elastic, and they stretch under a load much more than wire rope does. This stretching allows synthetic slings to absorb shocks and to cushion the load.
- Loads suspended in synthetic web slings are less likely to twist than those in wire rope or chain slings.

106F03.EPS

Figure 3 ◆ Surface protection.

106F04.EPS

Figure 4 ◆ Synthetic web sling shaping.

Synthetic web slings should not be exposed to temperatures above 180°F. They are also susceptible to cuts, abrasions, and other wear-and-tear damage. To prevent damage to synthetic web slings, riggers use protective pads (*Figure 5*).

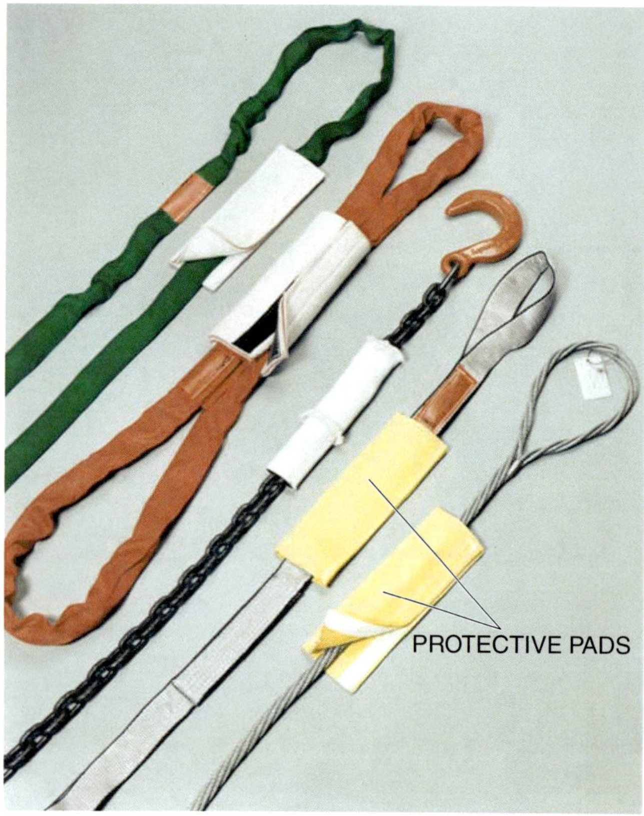

Figure 5 ◆ Protective pads.

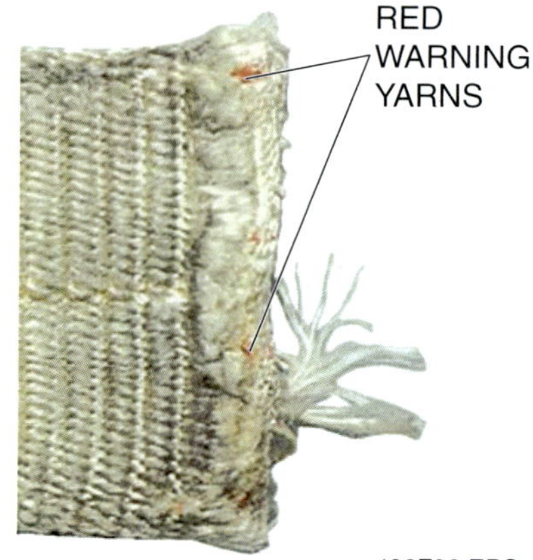

Figure 6 ◆ Synthetic web sling warning-yarns.

CAUTION

If the sling does not come with protective pads, use other kinds of softeners of sufficient strength or thickness to protect the sling. Pieces of old sling, fire hose, canvas, or rubber can be used.

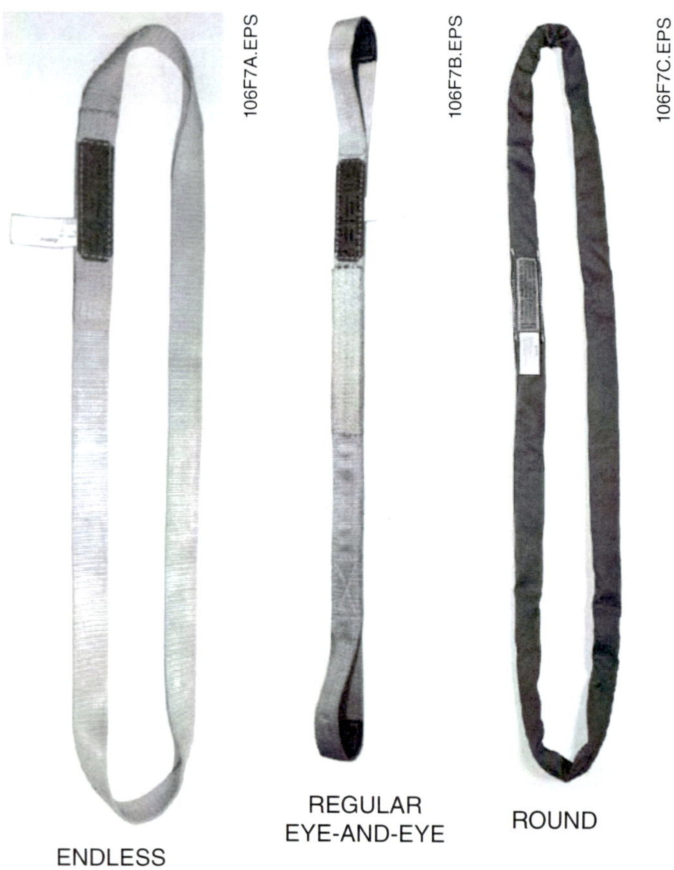

Figure 7 ◆ Synthetic web slings.

Most synthetic web slings are manufactured with red core **warning-yarns.** These are used to let the rigger know whether the sling has suffered too much damage to be used. When the yarns are exposed, the synthetic web sling should not be used (*Figure 6*). Red core yarns should not be used exclusively.

2.1.2 Types of Synthetic Web Slings

Synthetic web slings are available in several designs. The most common are the following (*Figure 7*):

- Endless web slings, which are also called **grommet slings.**
- Round slings, which are endless and made in a continuous circle out of polyester filament yarn. The yarn is then covered by a woven sleeve.
- Synthetic web eye-and-eye slings, which are made by sewing an end of the sling directly to the sling body. Standard eye-and-eye slings have eyes on the same **plane;** twisted eye-and-eye slings have eyes at right angles to each other (*Figure 8*) and are primarily used for choker hitches (which you will learn about later).

Synthetic web eye-and-eye slings are also available with hardware end fittings instead of fabric eyes. The standard end fittings are made of either aluminum or steel. They come in male and female configurations (*Figure 9*).

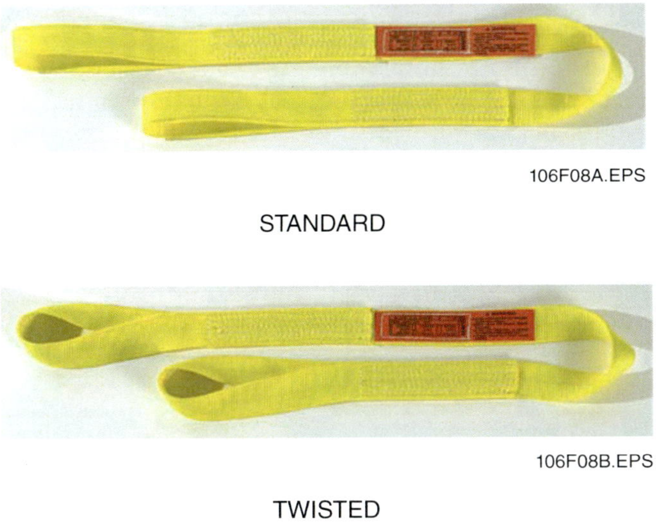

Figure 8 ◆ Eye-and-eye synthetic web slings.

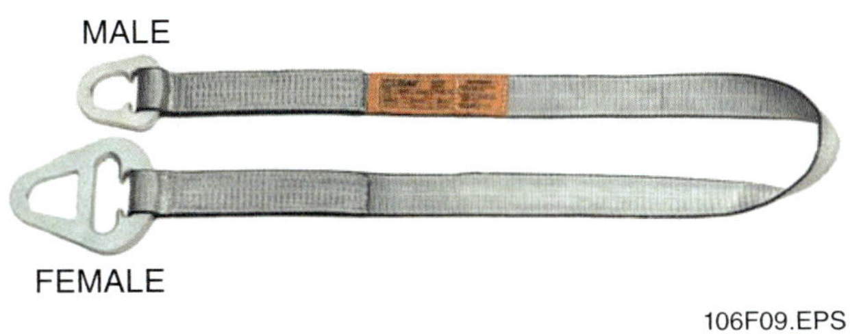

Figure 9 ◆ Synthetic web sling hardware end fittings.

2.1.3 Round Sling Design and Characteristics

Twin-Path® slings are made by wrapping a synthetic yarn around a set of spindles to form a loop. A protective jacket encases the core yarn (*Figure 10*).

Twin-Path® slings are made of a synthetic fiber, such as polyester. The material used to make up the **strands** and the number of wraps in a loop determine the rated capacity of the sling, or how much weight it can handle. Aramid round slings, for example, have a greater rated capacity for their size than the web-type slings do.

Twin-Path® slings are also available in a design with two separate wound loops of strand jacketed together side-by-side (*Figure 11*). This design greatly increases the lifting capacity of the sling.

The jackets of these slings are available in several materials for various purposes, including heat-resistant Nomek®, polyester, and bulked nylon (Covermax™) (*Figure 12*). Twin-Path® slings featuring K-Spec® yarn weigh at least 50 percent less than a polyester round sling of the same size and capacity.

Twin-Path® slings are equipped with **tattle-tail** yarns to help the rigger determine whether the

Figure 10 ◆ Synthetic endless-strand jacketed sling.

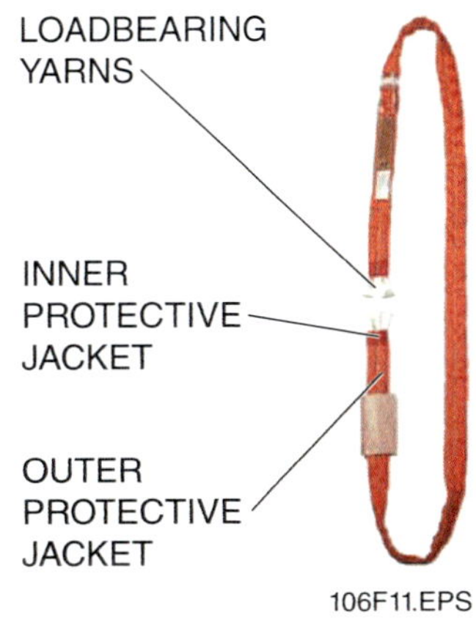

Figure 11 ◆ Twin-Path® sling.

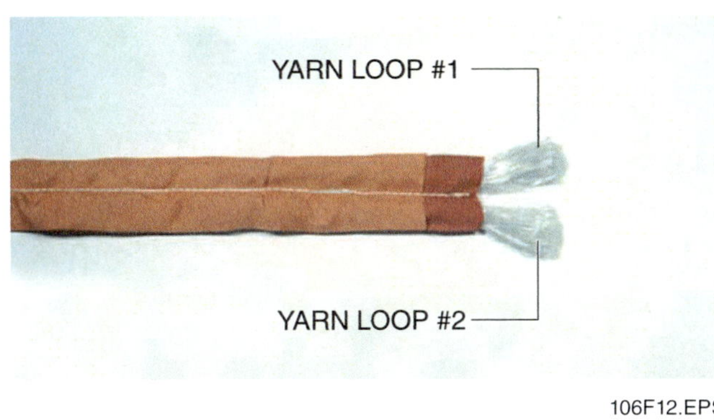

Figure 12 ◆ Twin-Path® sling makeup.

DID YOU KNOW?

Any sling without an identification tag must be removed from service immediately. It may be turned over to a qualified rigging inspector.

sling has become overloaded or stretched beyond a safe limit (*Figure 13*). These slings are also available with a fiber-optic inspection cable running through the strand (*Figure 14*). When you direct a light at one end of the fiber, the other end will light up to show that the strand has not been broken.

The way the sling will be used determines what material is used for the jackets of round slings. Polyester is normally used for light- to medium-duty sling jackets. Covermax™, a high-strength material with much greater resistance to cutting and abrasion, is used to make a sturdier jacket for heavy-duty uses.

2.1.4 Synthetic Sling Inspection

Like all slings, synthetic slings must be inspected before each use to determine whether they are in good condition and can be used. They must be inspected along the entire length of the sling, both visually (looking at them) and manually (feeling them with your hand). If any **rejection criteria** are met, the sling must be removed from service.

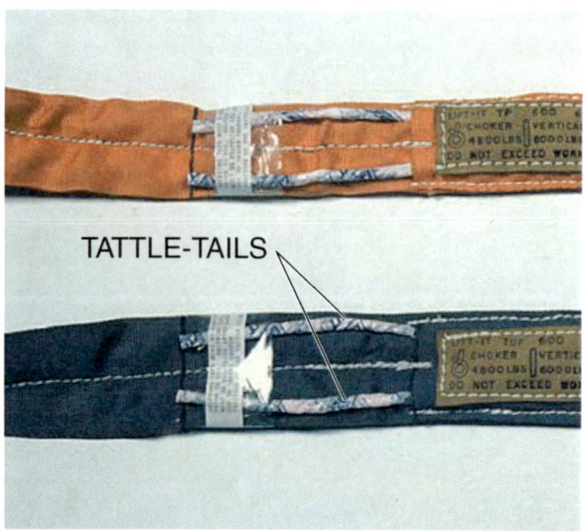

106F13.EPS

Figure 13 ◆ Tattle-tails.

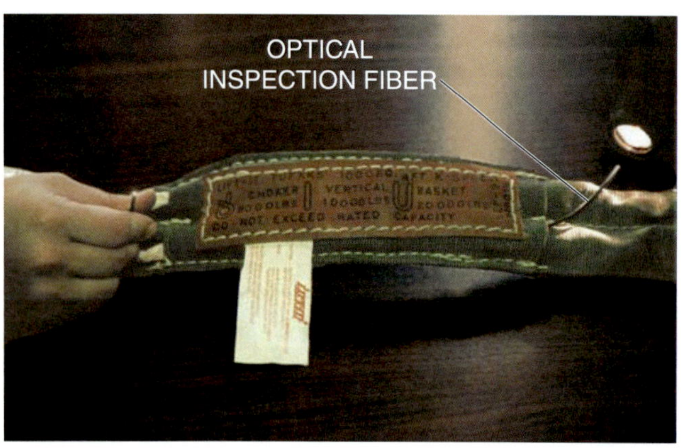

106F14.EPS

Figure 14 ◆ Fiber-optic inspection cable.

2.1.5 Synthetic Sling Rejection Criteria

If any synthetic sling meets any of the rejection criteria presented in this section, it must be removed from service immediately. In addition, the rigger has to exercise sound judgment. Along with looking for any single major problem, the rigger needs to watch for combinations of relatively minor defects in the synthetic sling. Combinations of minor damage may make the sling unsafe to use, even though such defects may not be listed in the rejection criteria.

WARNING!

The rigger must always inspect synthetic slings before using them, every time. The rigger must inspect them by looking at them as well as by feeling them, because sometimes you can feel damage that you cannot see.

If you are helping to inspect synthetic slings, alert your supervisor if you suspect any defects at all, especially if you find any of the following synthetic sling damage rejection criteria (*Figure 15*):

- A missing identification tag or an identification tag that cannot be read. Any synthetic sling without an identification tag must be removed from service immediately.
- Abrasion that has worn through the outer jacket, or that has exposed the loadbearing yarn of the sling, or that has exposed the warning-yarn of a web sling.
- A cut that has severed the outer jacket or exposed the loadbearing yarn (single-layer jacket) of a round sling or that has exposed the warning-yarn of a web sling.
- A tear that has exposed the inner jacket or the loadbearing yarns (single-layer jacket) of a round sling or that has exposed the warning-yarn of a web sling.
- A puncture.
- Broken or worn stitching in the **splice** or stitching of a web sling.
- A knot that cannot be removed by hand in either a web or round sling.
- A snag in the sling that reveals the warning-yarn of a web sling or a snag that tears through the outer jacket or exposes the loadbearing yarns of a round sling.
- Crushing of either a web sling or a round sling. Crushing in a web sling feels like a hollow pocket or depression in the sling. Crushing in a round sling feels like a hard, flat spot underneath the jacket.

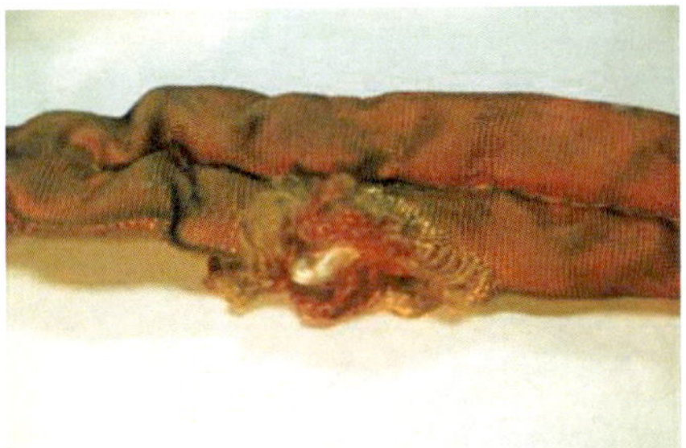

106F15A.EPS

(A)
JACKET AND WEB
ABRASION

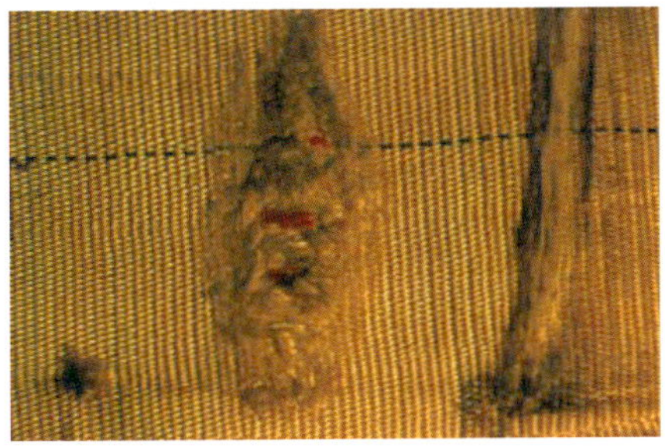

106F15B.EPS

(B)
JACKET AND WEB
SLING ABRASION

106F15C.EPS

(C)
SEVERE ABRASION

106F15D.EPS

(D)
MINOR ABRASION

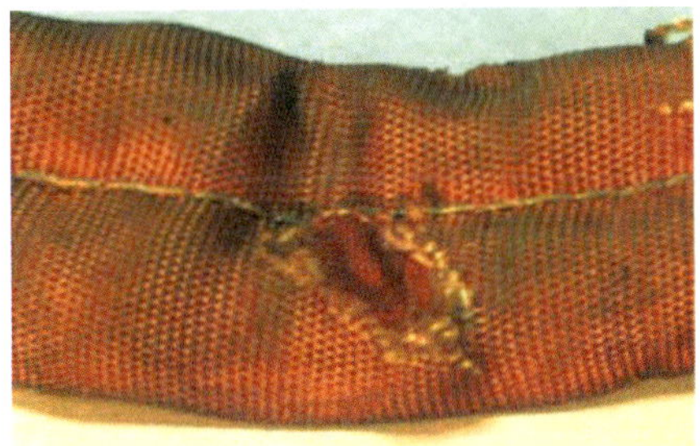

106F15E.EPS

(E)
OUTER JACKET
CUT

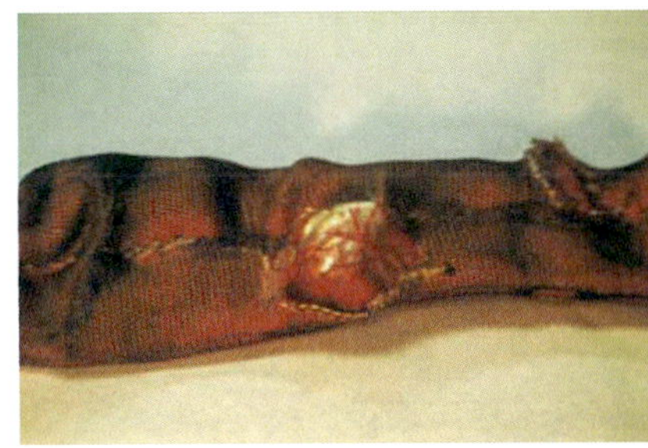

106F15F.EPS

(F)
INNER AND OUTER
JACKETS CUTS

Figure 15 ◆ Sling damage rejection criteria. (1 of 2)

- Damage from overload (called tensile damage or overstretching) in a web or round sling. Tensile damage in a web sling is evident when the weave pattern of the fabric begins to pull apart. Twin-Path® slings have tattle-tails. When the tail has been pulled into the jacket, it indicates that the sling may have been overloaded.
- Chemical damage, including discoloration, burns, and melting of the fabric or jacket.
- Heat damage, ranging from friction burns to melting of the sling material, the loadbearing strands, or the jacket. Friction burns give the webbing material a crusty or slick texture. Heat damage to the jacket of a round sling looks like glazing or charring. In round slings with heat-resistant jackets, you may not be able to see heat damage to the outside of the sling; however, the internal yarns may have been damaged. You can detect that damage by carefully handling the sling and feeling for brittle or fused fibers inside the sling or by flexing and folding the sling and listening for the sound of fused yarn fibers cracking or breaking.
- Ultraviolet (UV) damage. The evidence of UV damage is a bleaching-out of the sling material, which breaks down the synthetic fibers. Web slings with UV damage will give off a powder-like substance when they are flexed and folded. Round slings, especially those made of Cordura nylon and other specially treated synthetic fibers, have a much greater resistance to UV damage. In round slings, UV damage shows up as a roughening of the fabric texture where no other sign of damage, such as abrasion, can be found.
- Loss of flexibility caused by the presence of dirt or other abrasives. You can tell that a sling has lost flexibility when the sling becomes stiff. Abrasive particles embedded in the sling material act like tiny blades that cut apart the internal fibers of the sling every time the sling is stretched, flexed, or wrapped around a load.

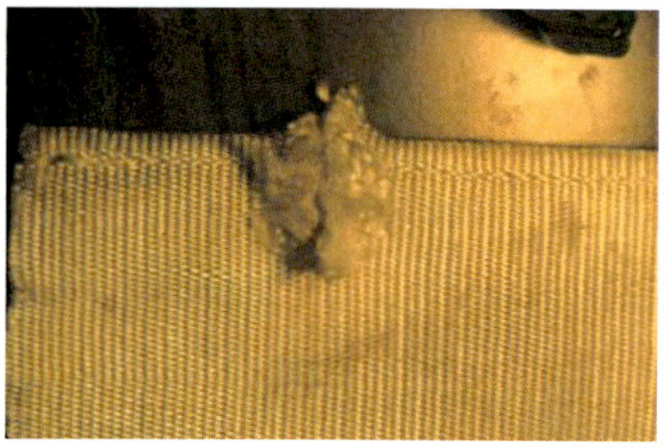

106F15G.EPS

(G)
CUT

106F15H.EPS

(H)
CUT WITH WARNING
THREADS SHOWING

106F15I.EPS

(I)
PUNCTURE

106F15J.EPS

(J)
BROKEN SPLICE
OR STITCHING

106F15K.EPS

(K)
SNAG SHOWING
RED THREAD

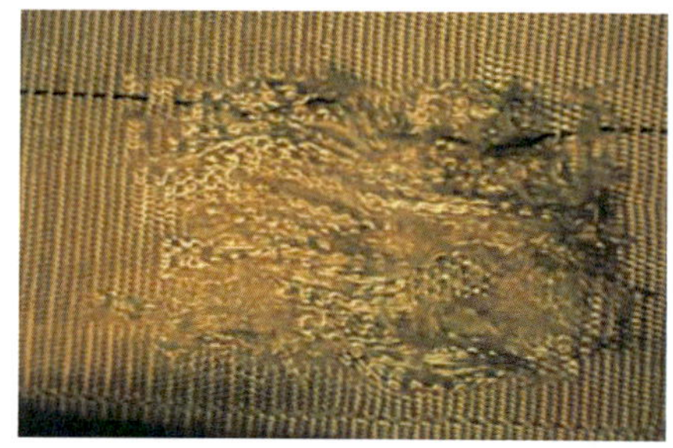

106F15L.EPS

(L)
TENSILE DAMAGE

106F15M.EPS

(M)
TENSILE BREAK

106F15N.EPS

(N)
OVERLOAD DAMAGE
(TATTLE-TAILS PULLED IN)

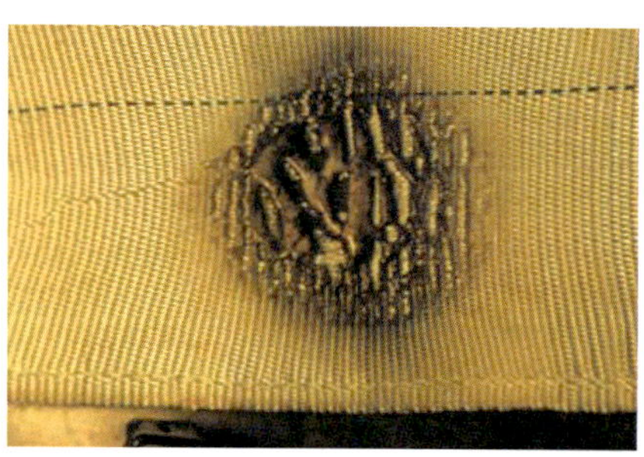

106F15O.EPS

(O)
FRICTION BURN FROM
ABRASION AND HEAT DAMAGE

106F15P.EPS

(P)
SEVERE HEAT DAMAGE

Figure 15 ◆ Sling damage rejection criteria. (2 of 2)

Sling Jackets

If only the outer jacket of a sling is damaged, under the rejection criteria, that sling can be sent back to the manufacturer for a new jacket. It must be tested and certified before it is used again. It is much less costly to do this than to replace the sling. However, slings that are removed because of heat- or tension-related jacket damage cannot be returned for repair and must be disposed of by a qualified person.

Review Questions

Section 2.0.0–2.1.5

1. Identification tags for slings must include the _____.
   a. type of protective pads to use
   b. type of damage sustained during use
   c. color of the tattle-tail
   d. manufacturer's name or trademark

2. A sling that exhibits _____ must be removed from service immediately.
   a. a missing identification tag
   b. any kind of abrasion
   c. flexibility
   d. paint on the sling

3. An advantage of synthetic slings is that they are _____.
   a. narrower than wire rope or chain slings
   b. lightweight and flexible
   c. more rigid than other types of slings
   d. less stretchable than wire rope

4. Synthetic slings must be inspected _____.
   a. the first time they are ever used
   b. before every use
   c. every fifth time they are used
   d. by sight only

5. When a tattle-tail has been pulled into the jacket, the sling has been _____.
   a. chemically damaged
   b. overheated
   c. overloaded
   d. crushed

2.2.0 Alloy Steel Chain Slings

Alloy steel chain, like wire rope (which you will learn about later), can be used in many different rigging operations. Chain slings are often used for lifts in high heat or rugged conditions.

The use of chain slings is discouraged, however. A chain sling weighs much more than a wire rope sling of the same capacity. It is also harder to inspect and can fail without warning. Synthetic slings are preferred. But you will encounter chain slings in the field, so you need to know how to use them.

2.2.1 Alloy Steel Chain Sling Design and Characteristics

Steel chain slings used for overhead lifting must be made of alloy steel. The higher the alloy grade, the safer and more durable the chain is. Alloy steel chains commonly used for most overhead lifting are marked with the number 8, the number 80, the number 800, or the letter A (*Figure 16*).

Steel chain slings have two basic designs with many variations:

- Single- and double-basket slings (*Figure 17*) do not require end-fitting hardware. The chain is attached to the **master link** in a permanent basket hitch or hitches.
- Chain **bridle** slings are available with two to four legs (*Figure 18*). Chain bridle slings require some type of end-fitting hardware—eye hooks, grab hooks, plate hooks, sorting or pipe hooks (*Figure 19*), or links.

Alloy steel chain slings are used for lifts when temperatures are high or where the slings will be subjected to steady and severe abuse. Most alloy steel chain slings can be used in temperatures up to 500°F with little loss in rated capacity.

Even though alloy steel chain slings can withstand extreme temperature ranges and abusive working conditions, these slings can be damaged if loads are dropped on them, if they are wrapped around loads with sharp corners (unless you use softeners), or if they are exposed to intense temperatures.

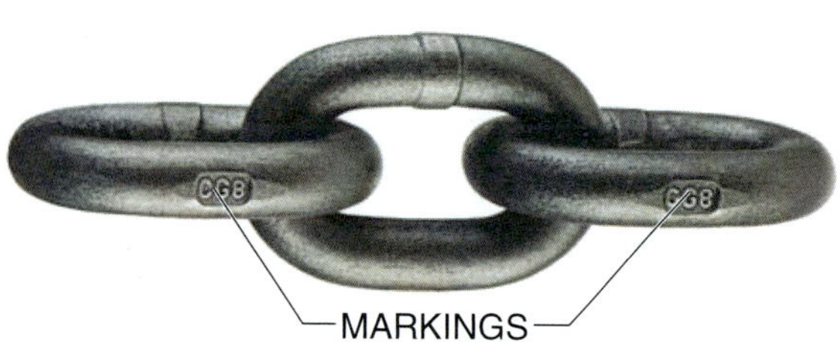

Figure 16 ◆ Markings on alloy steel chain slings.

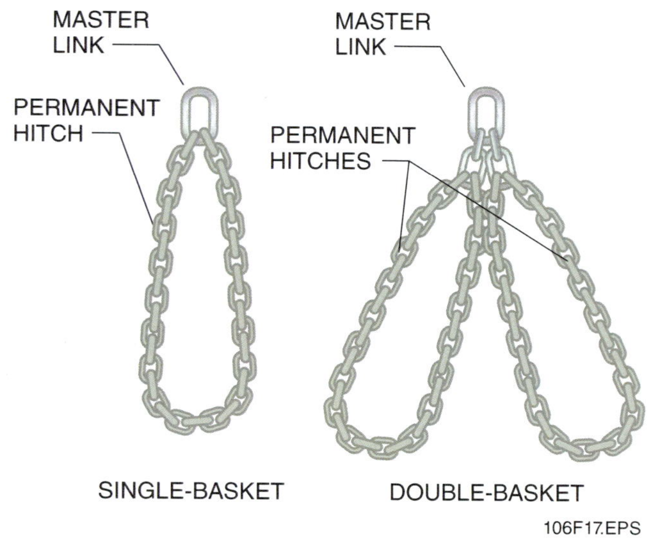

Figure 17 ◆ Chain slings.

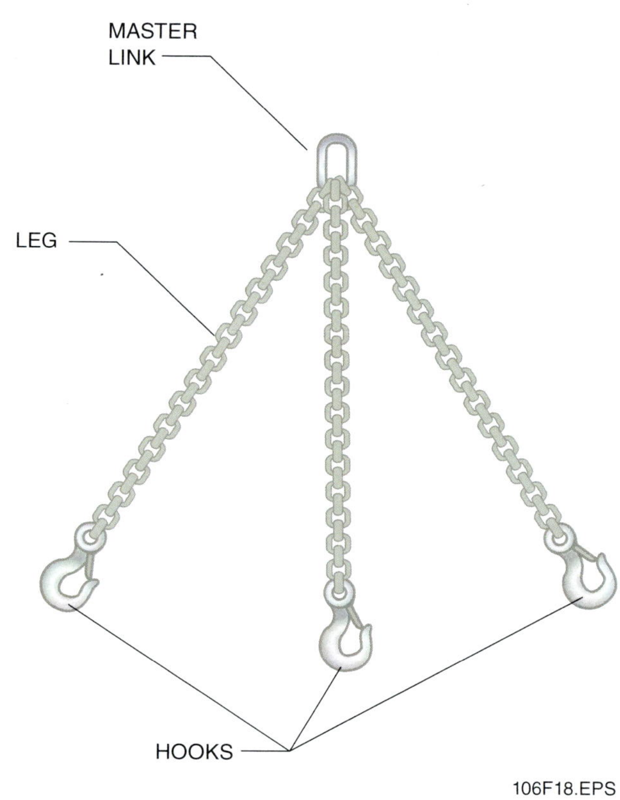

Figure 18 ◆ Three-leg chain bridle.

CAUTION

Never drag alloy steel chain slings across hard surfaces, especially concrete. Friction with abrasive surfaces wears out the chain and weakens the sling.

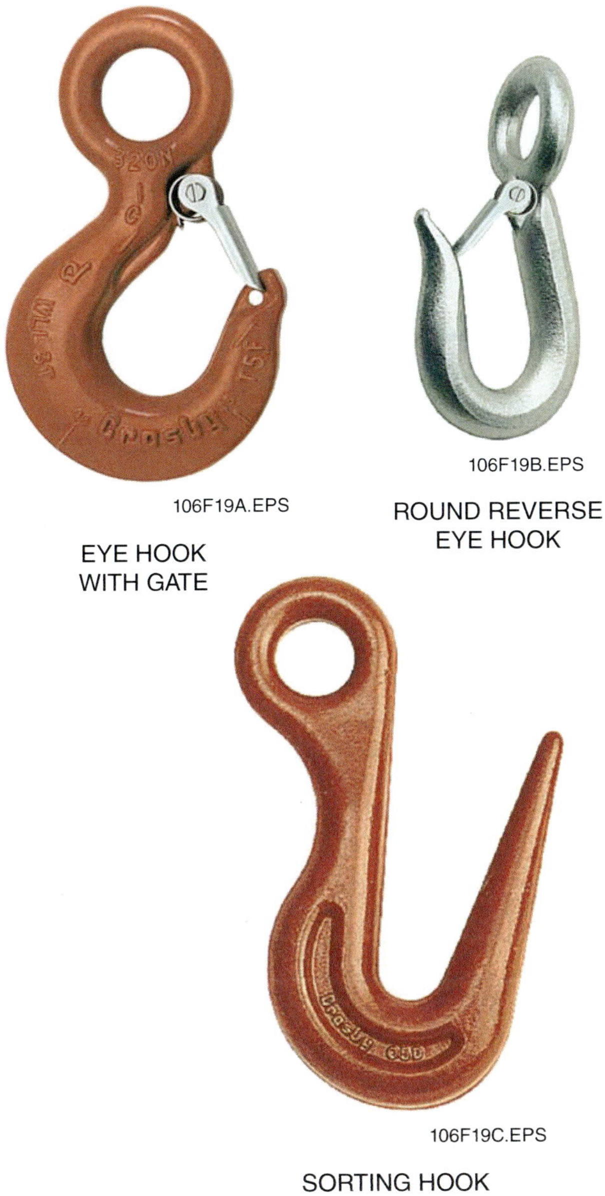

Figure 19 ◆ Eye and sorting hooks.

2.2.2 Alloy Steel Chain Inspection

Before each use, the rigger must inspect the alloy steel chain sling to determine whether it is safe to use. This module introduces you to the criteria involved in making this decision, but at this point in your training you will not make the decision yourself. If you have any questions about whether a sling is defective or unsafe and should not be used, ask your instructor or your immediate supervisor.

Steel chain slings must be visually inspected before each lift. A chain must be removed from service if a deficiency in the chain matches any of the rejection criteria. A chain may also need to be removed from service if something shows up that does not exactly match the rejection criteria.

2.2.3 Alloy Steel Chain Rejection Criteria

An alloy steel chain sling must be removed from service for any of the following defects (*Figure 20*):

- Missing or illegible identification tag
- Cracks
- Heat damage
- Stretched links. Damage is evident when the link grows long and when the barrels—the long sides of the links—start to close up
- Bent links
- Twisted links
- Excessive rust or corrosion, meaning rust or corrosion that cannot be easily removed with a wire brush
- Cuts, chips, or gouges resulting from impact on the chain
- Damaged end fittings, such as hooks, clamps, and other hardware
- Excessive wear at the link-bearing surfaces
- Scraping or abrasion

2.3.0 Wire Rope Slings

Wire rope slings (*Figure 21*) are made of high-strength steel wires formed into strands wrapped around a supporting **core.** They are lighter than chain, can withstand substantial abuse, and are easier to handle than chain slings. They can also withstand relatively high temperatures. However, because wire rope slings can slip, the use of synthetic slings is preferred. Wire rope is still being used, though, so you need to learn about the design and characteristics, applications, inspection, and maintenance of wire rope slings.

2.3.1 Wire Rope Sling Design and Characteristics

There are many types of wire ropes. They all consist of a core that supports the rope, and center strand wires, each with many high-grade steel wires wound around them to form the strands.

Wire rope is designed to operate like a self-adjusting machine. This means a wire rope has moving parts, as a machine does. A wire rope's moving parts are the wires that make up the strands and the core of the rope itself. These moving parts interact with one another by sliding and adjusting. This sliding and adjusting compensates for the ever-changing **stresses** placed upon a working rope. Because of this, a wire rope's rated capacity depends on it being in good condition. This means its wires must be able to move the way they were designed to.

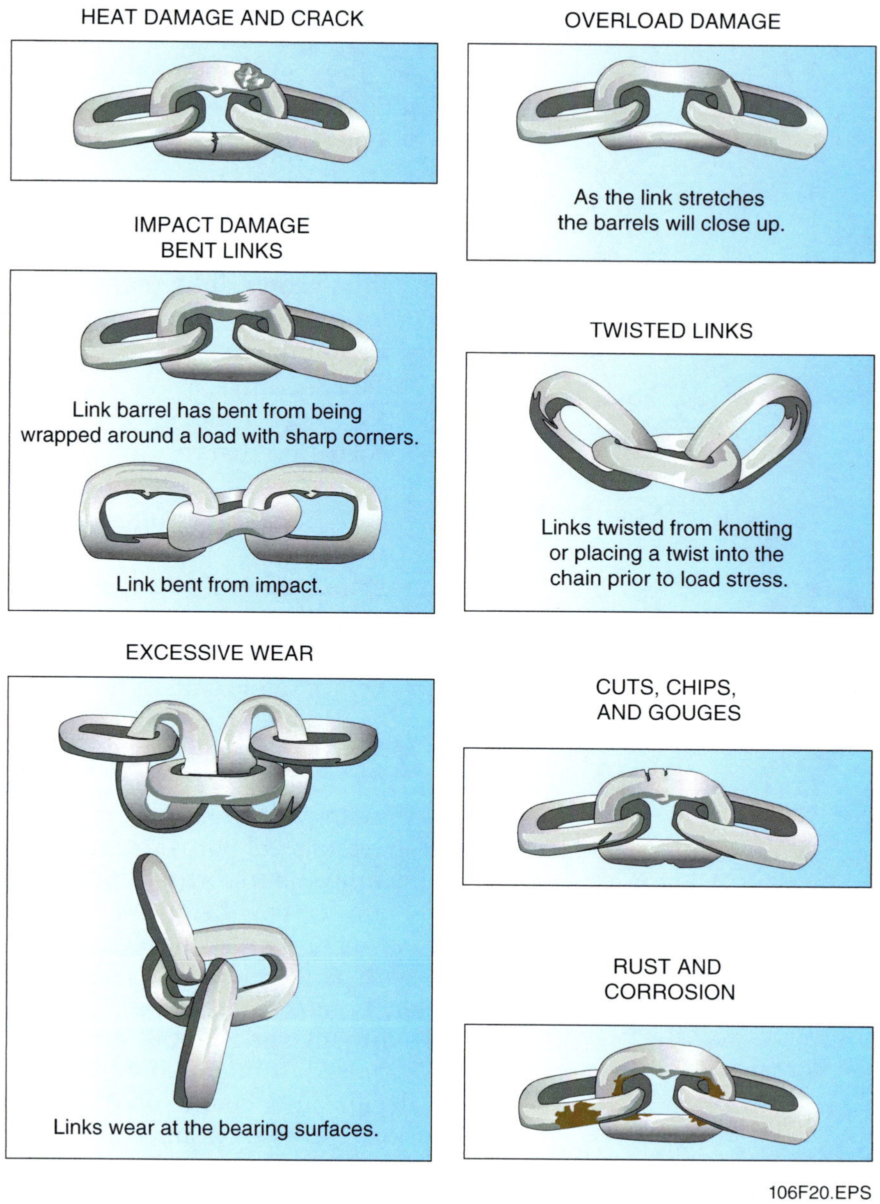

Figure 20 ◆ Damage to chains.

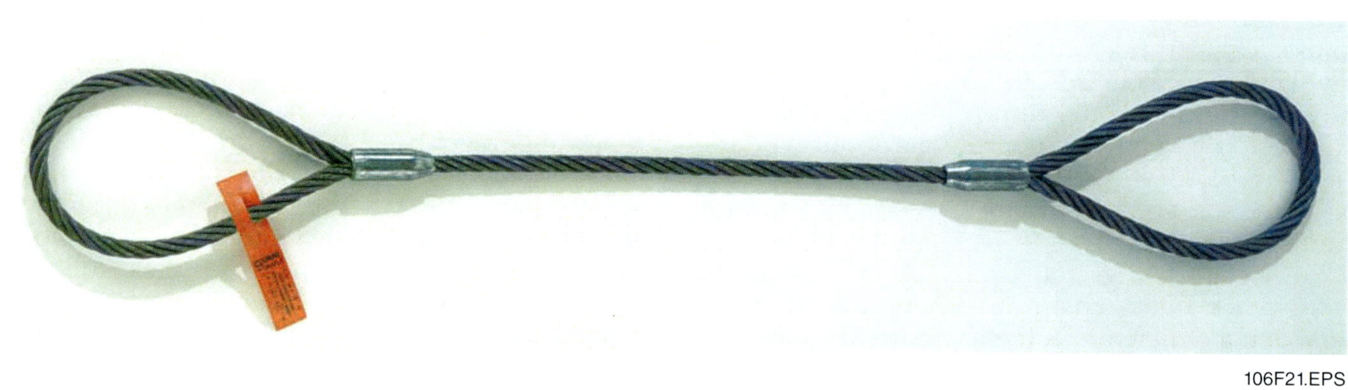

Figure 21 ◆ Wire rope sling.

The three basic components of a wire rope are as follows (*Figure 22*):

- A supporting core
- High-grade steel wires
- Multiple center wires

There are three basic types of supporting cores for wire rope (*Figure 23*): fiber cores, strand cores, and independent wire rope cores. Fiber cores are usually made of synthetic fibers, but they also can be made of natural vegetable fibers, such as sisal. Strand cores are made by using one strand of the same size and type as the rest of the strands of rope. Independent wire rope cores are made of a separate wire rope with its own core and strands. The core rope wires are much smaller and more delicate than the strand wires in the outer rope.

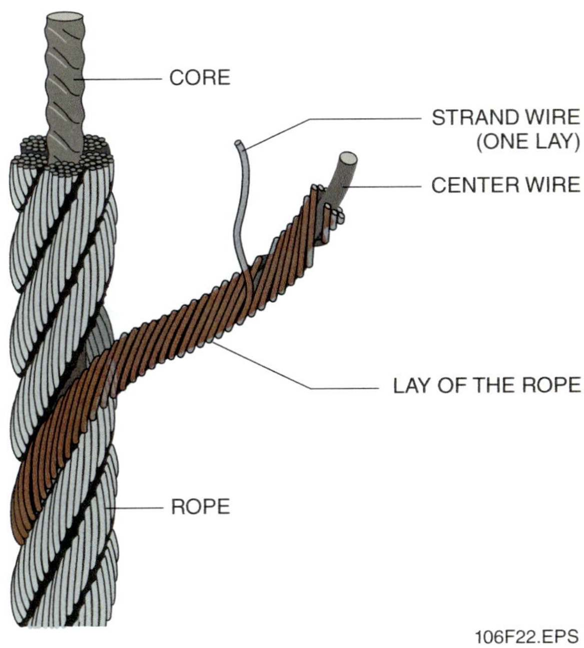

Figure 22 ◆ Wire rope components.

The various materials used to form the supporting core of a wire rope have both desirable characteristics and drawbacks, depending on how they are to be used. Fiber core ropes, for example, may be damaged by heat at relatively low temperatures (180°F to 200°F) as well as by exposure to caustic chemicals.

The function of the core in a wire rope is to support the strands so that the strands keep their original shapes. When the core supports the strands, their high-grade steel wires can slide and adjust against each other. The ability to adjust makes them less likely to be damaged by stress when the rope is bent around **sheaves** or loads, or when it is placed at an angle during rigging.

2.3.2 Wire Rope Sling Inspection

Like other slings, wire rope slings must be inspected before each use. If the wire rope is damaged, it must be removed from service. Always ask your instructor or immediate supervisor if you have any questions about whether a wire rope sling is suitable for use.

2.3.3 Wire Rope Sling Rejection Criteria

Following an inspection, a sling may be rejected based on several common types of damage (*Figure 24*), including broken wires, kinks, birdcaging, crushing, corrosion and rust, and heat damage. At this stage in your training, you need to be able to demonstrate that you understand the rejection criteria for wire ropes. Only a qualified person can actually make the decision to use a wire rope in a rigging operation or to discard it if it is damaged. If you think a wire rope may be damaged, bring it to the attention of your instructor or immediate supervisor. Additional information about possible types of damage is as follows.

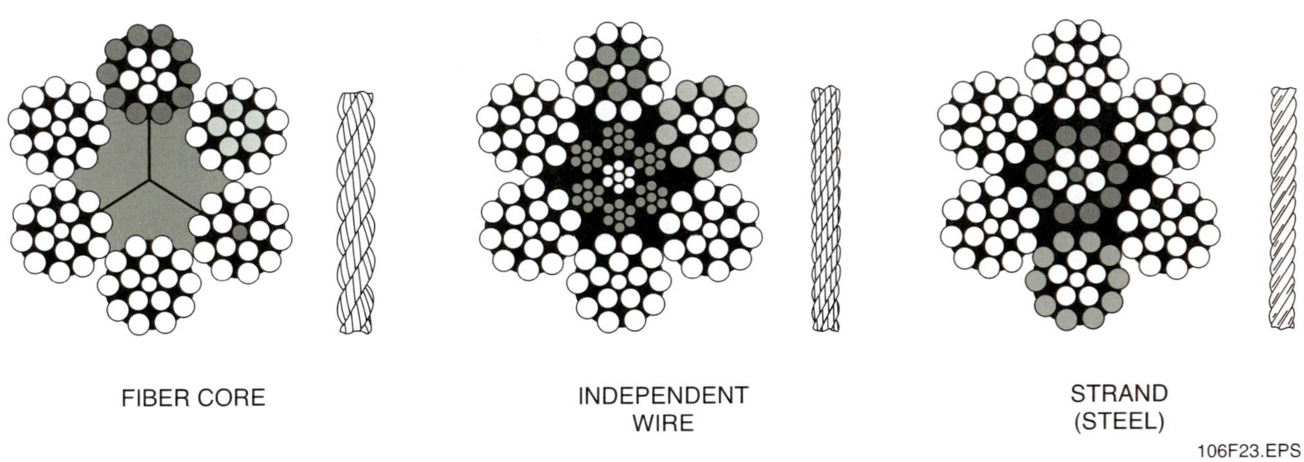

Figure 23 ◆ Wire rope supporting cores.

- *Broken wires* – Broken wires in the strands of a wire rope lessen the material strength of the rope and interfere with the interaction among the rope's moving parts. External broken wires usually mean normal fatigue, but internal or severe external breaks should be investigated closely. Internal or severe breaks in a wire rope mean it has been used improperly.

  Rejection criteria for broken wires consider how many wires are broken in one lay length of rope, or **one rope lay.** One rope lay is a term that defines the lengthwise distance it takes for one strand of wire to make one complete turn around the core (*Figure 25*). Different wire ropes have different one rope lays, so it is important to inspect each wire rope closely when looking for broken wires.
- *Kinks* – Kinking, or distortion of the rope, is a very common type of damage. Kinking can result in serious accidents. Sharp kinks restrict or prevent the movement of wires in the strands at the area of the kink. This means the rope is damaged and must not be used. Ropes with kinks in the form of large, gradual loops in a corkscrew configuration must be removed from service.
- *Birdcaging* – This damage occurs when a load is released too quickly and the strands are pulled or bounced away from the supporting core. The wires in the strands cannot compensate for the change in stress level by adjusting inside the strands. The built-up stress then finds its own release out through the strands.

  Birdcaging usually occurs in an area where already-existing damage prevents the wires from moving to compensate for changes in stress, position, and bending of the rope. Any sign of birdcaging is cause to remove the rope from service immediately.
- *Crushing* – This results from setting a load down on a sling or from hammering or pounding a sling into place. Crushing of the sling prevents the wires from adjusting to changes in stress, changes in position, and bends. A crushed sling usually results in the crushing or breaking of the core wires directly beneath the damaged strands. If crushing occurs, the sling must be removed from service immediately.
- *Corrosion and rust* – Corrosion and rust of wire rope are the result of improper or insufficient lubrication. Corrosion and rust are considered excessive if there is surface scaling or rust that cannot easily be removed with a wire brush, or if they occur inside the rope. If corrosion and rust are excessive, the rope must be removed from service.
- *Heat damage* – Heat damage embrittles wires in the strands and core in the area directly affected by the heat's contact, but also in a surrounding area up to 12" in each direction. Heat damage appears as discoloration and sometimes the actual melting of the wire rope. A wire rope that has been damaged by heat must be removed from service.

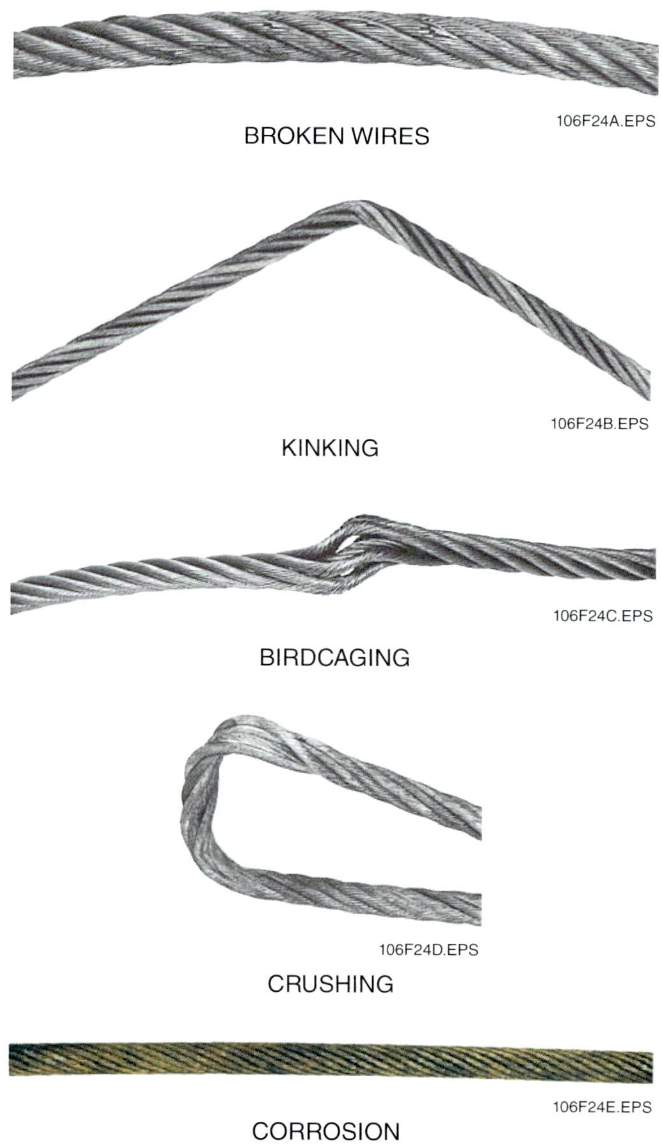

Figure 24 ◆ Common types of wire rope damage.

Figure 25 ◆ One rope lay.

Review Questions

Section 2.2.0–2.3.3

1. The best sling to use in lifts where high temperatures or abusive conditions are expected is _____.
   a. grommet steel chain
   b. round steel chain
   c. alloy steel chain
   d. carbon steel chain

2. A chain sling that _____ must not be used.
   a. exhibits links stretched from overloading
   b. exhibits paint on more than half of the links
   c. is more than one year old
   d. has an identification tag

3. In wire rope, the rope component that supports the strands is the _____.
   a. high-grade steel wire
   b. core
   c. center wire
   d. sling

4. The type of rope core that is susceptible to heat damage at relatively low temperatures is the _____.
   a. fiber core
   b. strand core
   c. independent wire rope core
   d. supporting core

5. The term used to describe the damage done when the strands bounce away from the core is _____.
   a. kinking
   b. corkscrewing
   c. shock-loading
   d. birdcaging

3.0.0 ◆ HITCHES

As you have learned, the link between the load and the lifting device is often a sling made of synthetic, alloy steel chain, or wire rope material. The way the sling is arranged to hold the load is called the rigging configuration or hitch. Hitches can be made using just the sling or by using connecting hardware as well. There are three basic types of hitches:

- Vertical
- Choker
- Basket

One of the most important parts of the rigger's job is making sure that the load is held securely. The type of hitch the rigger uses depends on the type of load to be lifted. Different hitches are used to secure, for example, a load of pipes, a load of concrete slabs, or a load of heavy machinery.

Controlling the movement of the load once the lift is in progress is another extremely important part of the rigger's job. Therefore, the rigger must also consider the intended movement of the load when choosing a hitch. For example, some loads are lifted straight up and then straight down. Other loads are lifted up, turned 180 degrees in midair, and then set down in a completely different place. In this section, you will learn about how each of the three basic types of hitches—vertical, choker, and basket—is used to both secure the load and control its movement.

3.1.0 Vertical Hitch

The single vertical hitch is used to lift a load straight up. It forms a 90-degree angle between the hitch and the load. With this hitch, some type of attachment hardware is needed to connect the sling to the load (*Figure 26*). The single vertical hitch allows the load to rotate freely. If you do not want the load to rotate freely, some method of **load control** must be used, such as a **tag line.** These methods are explained later in this module.

WARNING!

All rigging operations are dangerous, and extreme care must be used at all times. A straight-up-and-down vertical lift is every bit as dangerous as a lift that involves rotating a load in midair and moving it to a different place. Only a qualified person may select the hitch to be used in any rigging operation.

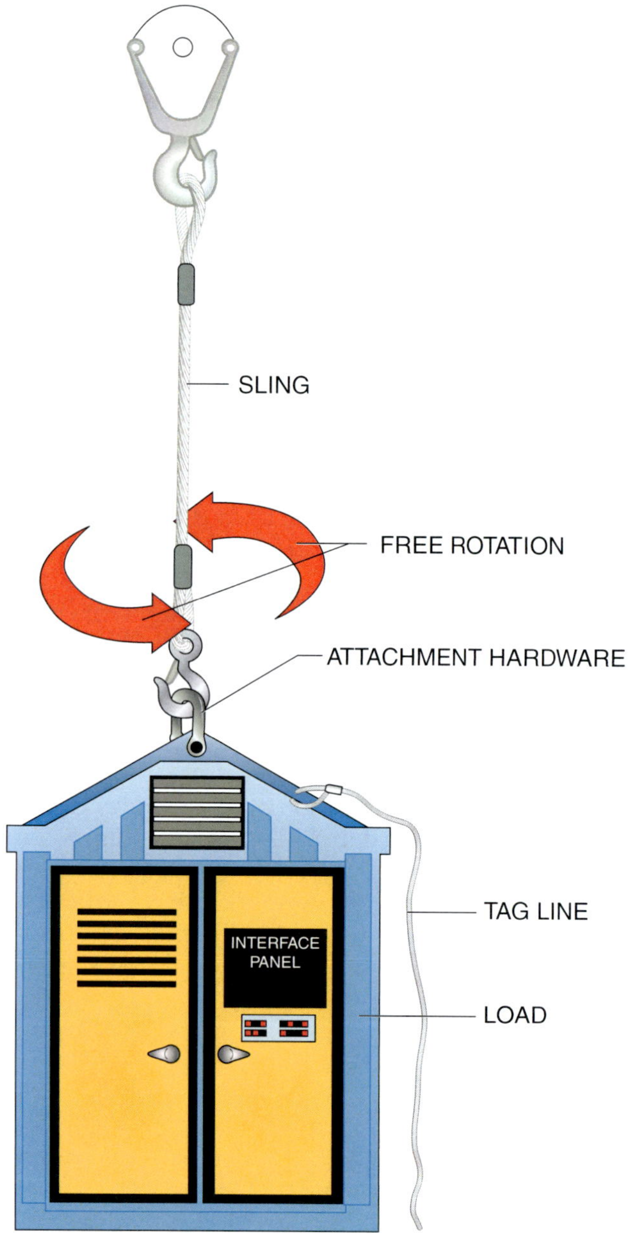

Figure 26 ◆ Single vertical hitch.

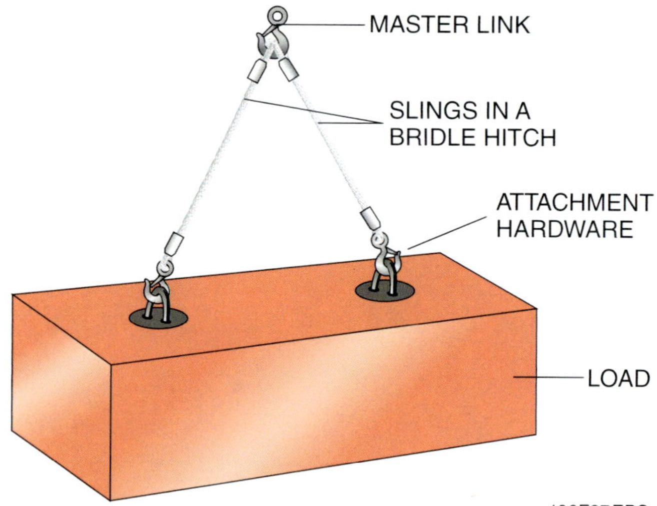

Figure 27 ◆ Bridle hitch.

Figure 28 ◆ Spreader beam.

Another classification of vertical hitch is the bridle hitch (*Figure 27*). The bridle hitch consists of two or more vertical hitches attached to the same hook, master link, or **bull ring.** The bridle hitch allows the slings to be connected to the same load without the use of such devices as a spreader beam, which is a stiff bar used when lifting large objects with a crane hook (*Figure 28*).

The multiple-leg bridle hitch (*Figure 29*) consists of three or four single hitches attached to the same hook, master link, or bull ring. Multiple-leg bridle hitches provide increased stability for the load being lifted. A multiple-leg bridle hitch is always considered to have only two of the legs supporting the majority of the load and the rest of the legs balancing it.

3.2.0 Choker Hitch

The choker hitch is used when a load has no attachment points or when the attachment points are not practical for lifting (*Figure 30*). The choker hitch is made by wrapping the sling around the load and passing one eye of the sling through a **shackle** to form a constricting loop around the load. It is important that the shackle used in a choker hitch be oriented properly, as shown in

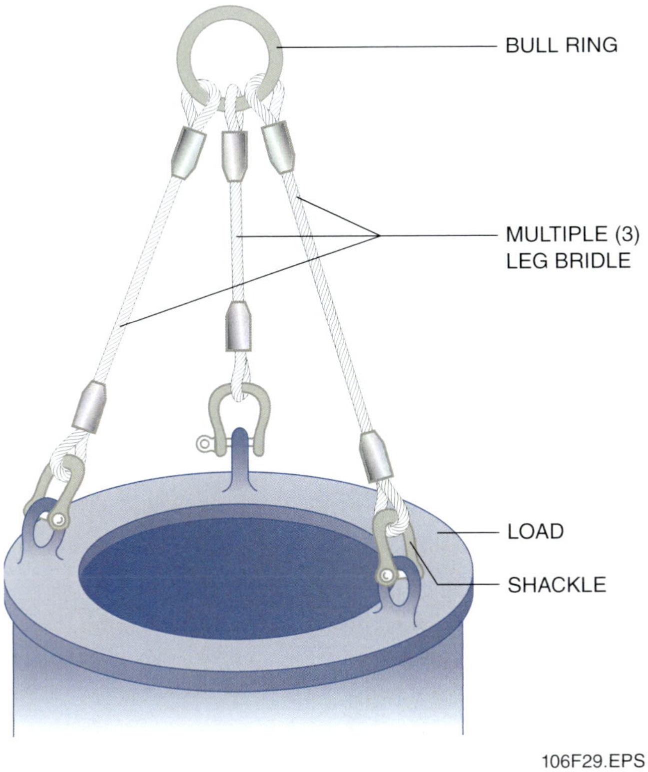

Figure 29 ◆ Multiple-leg bridle hitch.

Figure 30 ◆ Choker hitch.

Figure 30. The choker hitch affects the capacity of the sling, reducing it by a minimum of 25 percent. This reduction must be considered when choosing the proper sling.

The choker hitch does not grip the load securely. It is not recommended for loose bundles of materials because it tends to push loose items up and out of the choker. Many riggers use the choker hitch for bundles, mistakenly believing that forcing the choke down provides a tight grip. In fact, it serves only to drastically increase the stress on the choke leg (*Figure 31*).

Figure 31 ◆ Choker hitch constriction.

When an item more than 12' long is being rigged, the general rule is to use two choker hitches spaced far enough apart to provide the stability needed to transport the load (*Figure 32*).

To lift a bundle of loose items, or to maintain the load in a certain position during transport, a double-wrap choker hitch (*Figure 33*) may be useful. The double-wrap choker hitch is made by wrapping the sling completely around the load, and then wrapping the choke end around again and passing it through the eye like a conventional choker hitch. This enables the load weight to produce a constricting action that binds the load into the middle of the hitch, holding it firmly in place throughout the lift.

Forcing the choke down will drastically increase the stress placed on the sling at the choke point. To gain gripping power, use a double-wrap choker hitch. The double-wrap choker uses the load weight to provide the constricting force, so there is no need to force the sling down into a tighter choke.

A double-wrap choker hitch is ideal for lifting bundles of items, such as pipes and structural steel. It will also keep the load in a certain position, which makes it ideal for equipment installation lifts. Lifting a load longer than 12' requires two of these hitches.

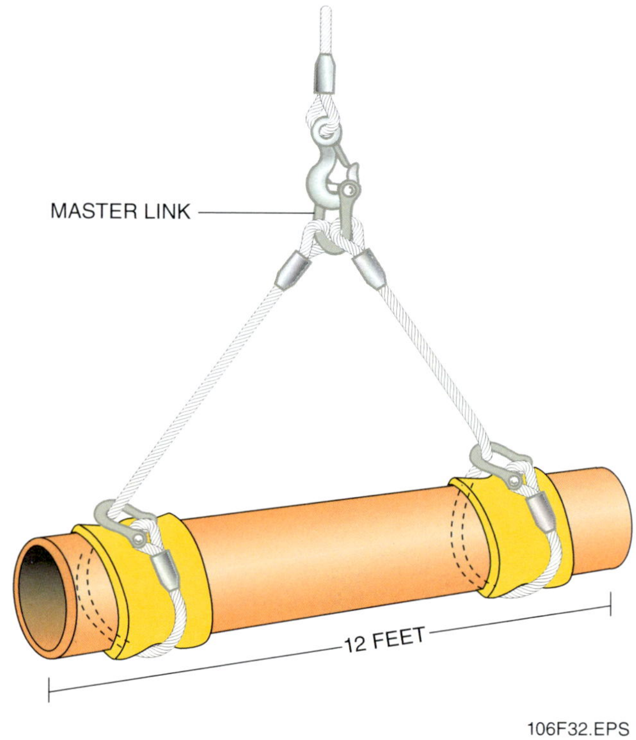

Figure 32 ◆ Double choker hitch.

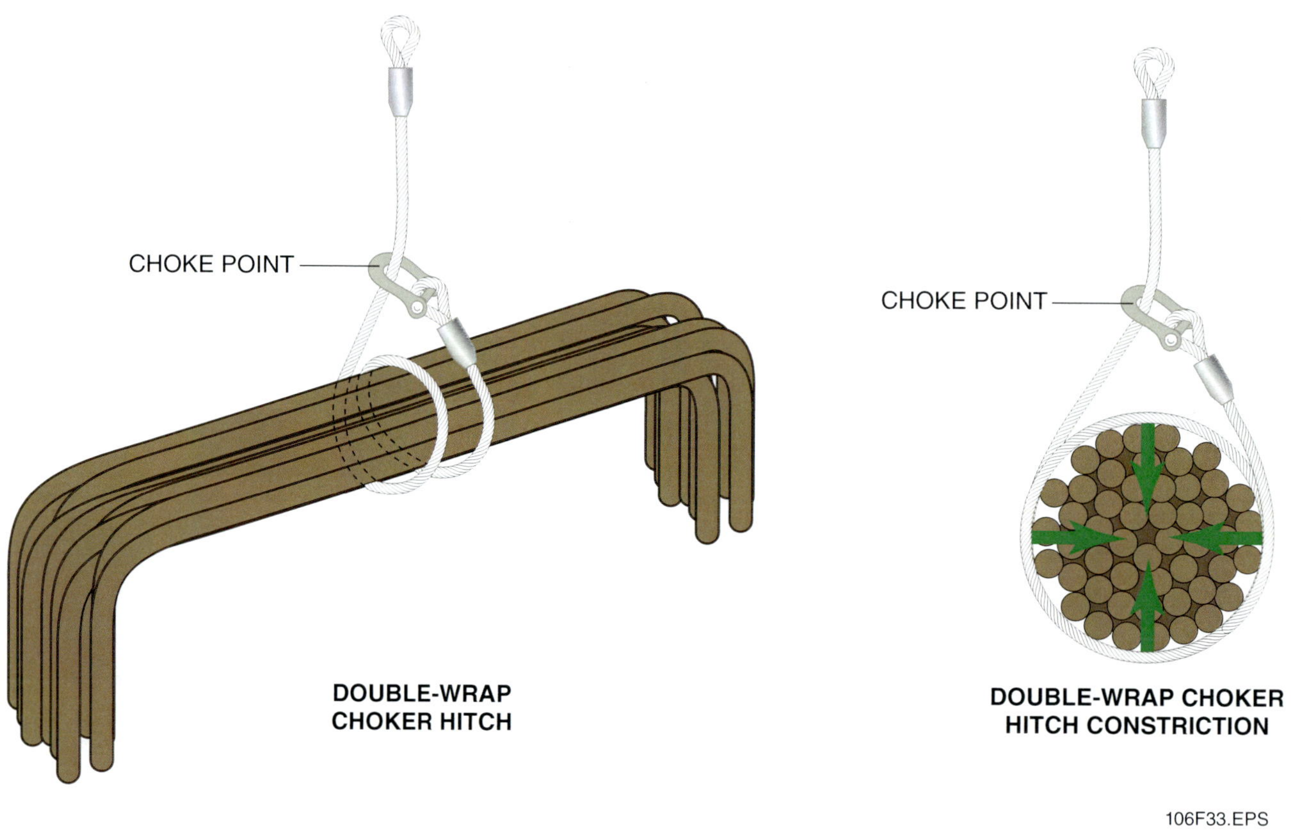

Figure 33 ◆ Double-wrap choker hitch and double-wrap choker hitch constriction.

3.3.0 Basket Hitch

Basket hitches are very versatile and can be used to lift a variety of loads. A basket hitch is formed by passing the sling around the load and placing both eyes in the hook (*Figure 34*). Placing a sling into a basket hitch effectively doubles the capacity of the sling. This is because the basket hitch creates two sling legs from one sling.

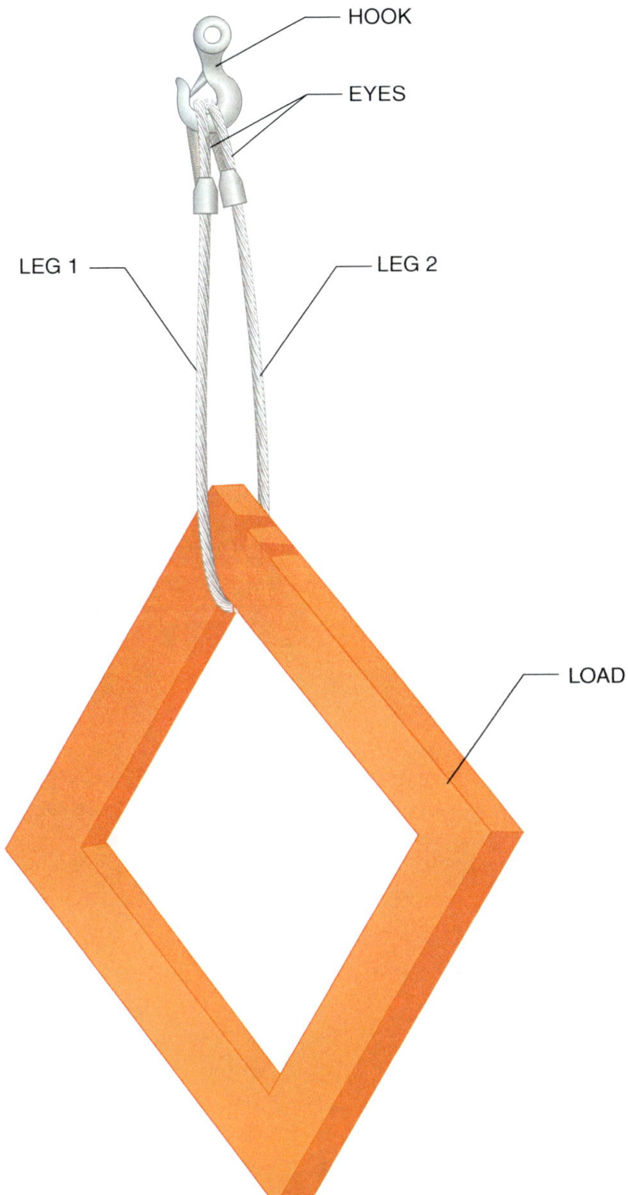

106F34.EPS

Figure 34 ◆ Basket hitch.

CAUTION

A basket hitch should not be used to lift loose materials. Loads placed in a basket hitch should be balanced.

The double-wrap basket hitch combines the constricting power of the double-wrap choker hitch with the capacity advantages of a basket hitch (*Figure 35*). This means it is able to hold a larger load more tightly.

The double-wrap basket hitch requires a considerably longer sling length than a double-wrap choker hitch. If it is necessary to join two or more slings together, the load must be in contact with the sling body only, not with the hardware used to join the slings. The double-wrap basket hitch provides support around the load. Just as with the double-wrap choker hitch, the load weight provides the constricting force for the hitch.

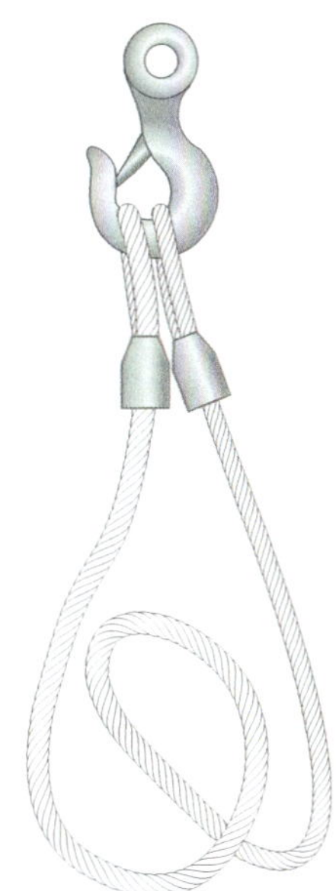

106F35.EPS

Figure 35 ◆ Double-wrap basket hitch.

Load Oscillation

Because only three or four small points (the points where the hitches connect to the load) transfer the entire weight of the load, a load will oscillate (swing back and forth like a pendulum) as it is moved. Because of this load oscillation, at any given moment it is impossible to tell which slings are supporting the weight and which ones are providing stability.

To understand load oscillation, imagine the following example. A glass of water filled to the rim is placed on a table. Four people each place one finger under each corner of the table. They try to lift the table and walk across the room without spilling any of the water. The table under the glass oscillates as the people move the table, and a large amount of the water is probably spilled on the way across the room. This is because only a small amount of surface area of their fingers is in contact with the table, which causes the weight of the table to shift back and forth, or oscillate.

Now imagine that each person is allowed to use both hands on a corner of the table, thereby spreading the contact stress over a larger surface area. Now they could probably make it all the way across the room without spilling much, if any, of the water.

Review Questions

Section 3.0.0

1. The multiple-leg _____ hitch uses three or four single hitches to increase the stability of the load.

a. vertical
b. choker
c. bridle
d. basket

2. The _____ allows slings to be connected to the same load without using a spreader beam.

a. pendulum
b. choker hitch
c. basket hitch
d. bridle hitch

3. A _____ is ideal for lifting bundles of items such as pipes.

a. choker hitch
b. double-wrap choker hitch
c. bull ring
d. vertical hitch

4. The hitch made by wrapping the sling around the load and then passing one eye of the sling through a shackle is called the _____ hitch.

a. basket
b. bridle
c. choker
d. grommet

5. The hitch made by passing the sling around the load and placing both eyes in the hook is called the _____ hitch.

a. basket
b. bridle
c. vertical
d. choker

4.0.0 ◆ RIGGING HARDWARE

Rigging hardware is as crucial as the crane, the slings, or any specially designed lifting frame or hoisting device. If the hardware that connects the slings to either the load or the master link were to fail, the load would drop just as it would if the crane, hoist, or slings were to fail. Hardware failure related to improper attachment, selection, or inspection contributes to a great number of the deaths, serious injuries, and property damages in rigging accidents. The importance of hardware selection, maintenance, inspection, and proper use cannot be stressed enough. The regulations and requirements for rigging hardware are as stringent as those governing cranes and slings.

4.1.0 Shackles

A shackle is an item of rigging hardware used to attach an item to a load or to couple slings together. A shackle can be used to couple the end of a wire rope to eye fittings, hooks, or other connectors. It consists of a U-shaped body and a removable pin.

4.1.1 Shackle Design and Characteristics

Shackles used for overhead lifting should be made from forged steel, not cast steel. Quenched and tempered steel is the preferred material because of its increased toughness, but at a minimum, shackles must be made of drop- or hammer-forged steel.

All shackles must have a stamp that is clearly visible, showing the manufacturer's trademark, the size of the shackle (determined by the diameter of the shackle's body, not by the diameter of the pin), and the rated capacity of the shackle.

4.1.2 Types of Shackles

Shackles are available in two basic classes, identified by their shapes: anchor shackles and chain shackles (*Figure 36*). Both anchor and chain shackles have three basic types of pin designs, each with a unique function, as shown in *Figure 36*: the screw pin shackle, the round pin or straight pin shackle, and the safety shackle. The screw pin shackle design has become the most widely used type in general industry.

4.1.3 Specialty Shackles

Specialty shackles are available for specific applications where a standard shackle would not work well. For example, wide-body shackles (*Figure 37*) are for heavy lifting applications.

Synthetic web sling shackles (*Figure 38*) are designed with a wide throat opening and a wide bow that is contoured to provide a larger, nonslip surface area to accommodate the wider body of synthetic web slings.

4.1.4 Shackle Inspection and Rejection Criteria

Shackles, like any other type of hardware, must be inspected by the rigger before each use to make sure there are no defects that would make the shackle unsafe. Each lift may cause some degree of damage or may further reveal existing damage.

If any of the following conditions exists, a shackle must be removed from use:

- Bends, cracks, or other damage to the shackle body
- Incorrect shackle pin or improperly substituted pin
- Bent, broken, or loose shackle pin
- Damaged threads on threaded shackle pin
- Missing or illegible capacity and size markings

ON-SITE

Using Screw Pin Shackles

One of the nation's leading manufacturers of rigging hardware has judged the former practice of backing the pin off a quarter turn to be unsafe, because it allows the shackle to stretch under the load. The new practice is to engage the pin but not to tighten it fully. If you tighten or torque the pin fully, the load stress from the sling may torque it even more. This will stretch and jam the pin's threads.

Pins should never be swapped in shackles. The threadings of different brands of shackles and of different brands of pins may not engage properly with one another. In addition, there is no easy way to tell how much reserve strength is left in a pin or a shackle. Also, the rated capacities of different shackles and pins may not be compatible. The shackle pin should be placed and secured into the shackle between uses.

106F36A.EPS

SCREW PIN
ANCHOR SHACKLE

106F36B.EPS

SCREW PIN
CHAIN SHACKLE

106F36C.EPS

ROUND PIN
ANCHOR SHACKLE

106F36D.EPS

ROUND PIN
CHAIN SHACKLE

106F36E.EPS

SAFETY
ANCHOR SHACKLE

106F36F.EPS

SAFETY
CHAIN SHACKLE

Figure 36 ◆ Shackles.

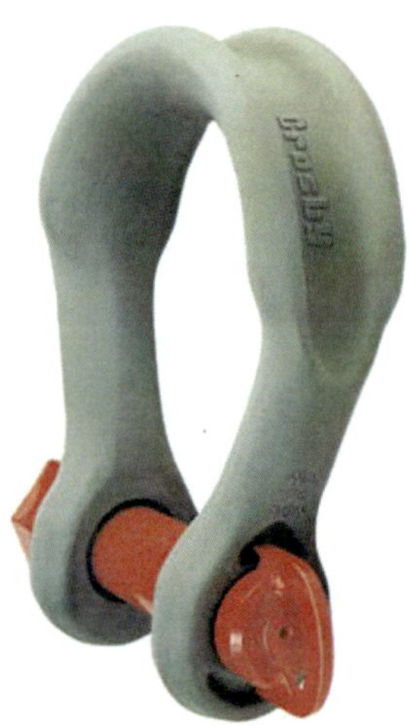

106F37.EPS

Figure 37 ◆ Wide-body shackle.

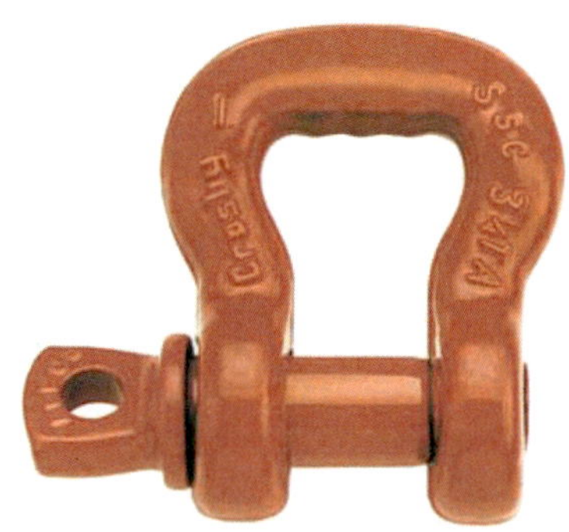

106F38.EPS

Figure 38 ◆ Synthetic web sling shackle.

4.2.0 Eyebolts

An **eyebolt** is an item of rigging hardware with a **threaded shank.** The eyebolt's shank end is attached directly to the load, and the eyebolt's eye end is used to attach a sling to the load.

4.2.1 Eyebolt Design and Characteristics

Eyebolts for overhead lifting should be made of drop- or hammer-forged steel. Eyebolts are available in three basic designs with several variations (*Figure 39*):

- Unshouldered eyebolts, designed for straight vertical pulls only
- Shouldered eyebolts, with a shoulder that is used to help support the eyebolt during pulls that are slightly angular
- Swivel eyebolts, also called hoist rings, designed for angular pulls from 0 to 90 degrees from the horizontal plane of the load

4.2.2 Unshouldered Eyebolts

Unshouldered eyebolts are designed for vertical pulls only, not angular pulls. In the installation of an unshouldered eyebolt, the sling must not pull perpendicular to the eyebolt (*Figure 40*). This will cause the eyebolt to bend and probably break off.

Figure 39 ◆ Eyebolt variations.

4.2.3 Shouldered Eyebolts

Shouldered eyebolts can be used for some angular pulls. Some manufacturers set the limit for angular pulling at 45 degrees from the horizontal plane of the load. Shouldered eyebolts lose capacity as the angle of the pull deviates from the vertical (*Figure 41*). Always check manufacturer ratings for load capacities at various pulling angles and eyebolt sizes.

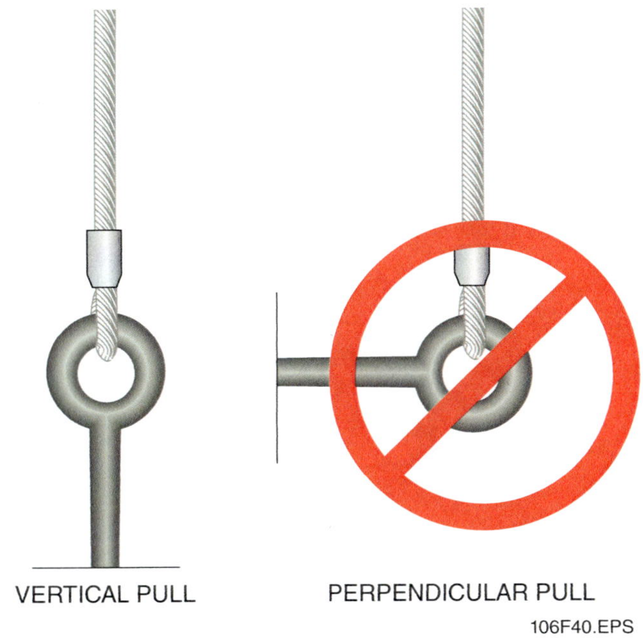

Figure 40 ◆ Installing an unshouldered eyebolt.

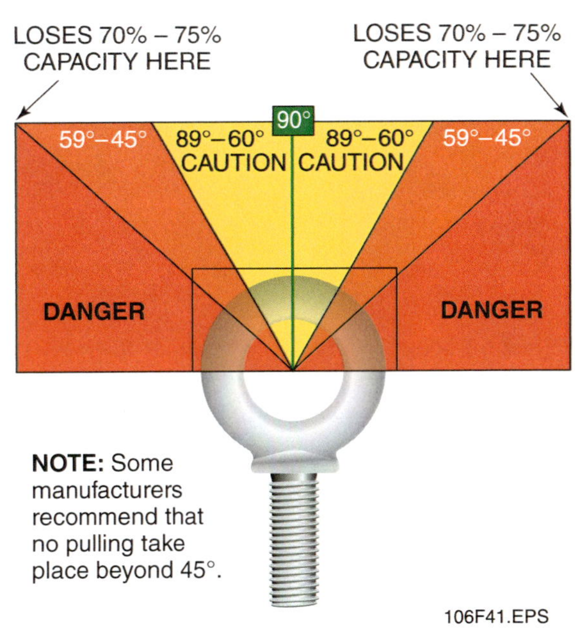

Figure 41 ◆ Effects of angular pull on shouldered eyebolts.

Most of the installation requirements for shouldered eyebolts are similar to the installation requirements for unshouldered eyebolts. However, the limit for angular pulling depends on proper installation. For example, in order for a shouldered eyebolt to support the load weight, it has to be installed so the eyebolt's shoulder is flush with the load surface (*Figure 42*). If the shoulder is not flush with the load surface, the eyebolt acts the same as an unshouldered eyebolt. Shouldered eyebolts must be positioned so the sling is in the same plane as the eyebolt.

Figure 42 ◆ Proper installation of shouldered eyebolts.

4.2.4 Swivel Eyebolts

Swivel eyebolts are specially designed for angular pulls. They are available in through-bolt and machine-bolt styles. A through-bolt swivel eyebolt is designed so that the shank passes completely through the members it connects. A machine-bolt swivel eyebolt has a straight shank with a conventional head, such as a square or hexagonal head.

A swivel eyebolt's seating surface must be flush with the surface of the load or it will not function any better than an unshouldered eyebolt. Because the seating base of the swivel eyebolt is considerably larger than that of the unshouldered or shouldered eyebolt, it often will not install into the same area that a shouldered or unshouldered eyebolt would. Swivel eyebolts are self-aligning; their bases rotate 360 degrees, and the bail swivels 180 degrees. The rated capacity of a swivel eyebolt is based upon its 0- to 90-degree pull in any direction. There is no reduction factor for angular pulls (*Figure 43*).

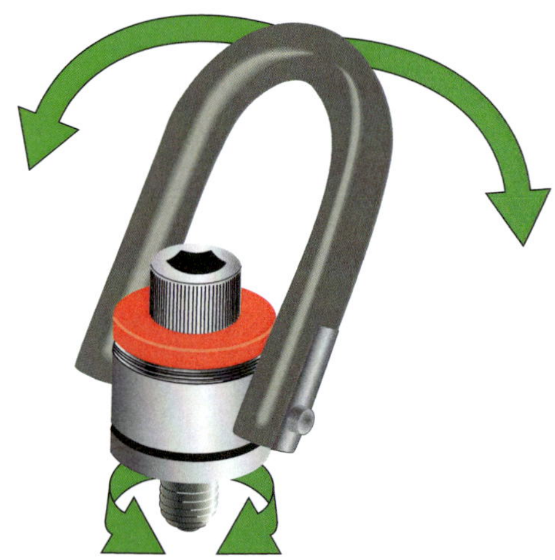

THREAD SIZE (inches)	SAMPLE RATED CAPACITY FROM 0–180 (pounds)
1/2"	2,500
5/8"	4,000
3/4"	5,000
7/8"	8,000
1"	10,000
1 1/2"	24,000
2"	30,000

Bolt specification is Grade 8 alloy socket head cap screw to ASTM A 574.

106F43.EPS

Figure 43 ◆ Effects of angular pull on swivel eyebolts.

4.2.5 Eyebolt Inspection and Rejection Criteria

Eyebolts must be visually inspected before each use, and a qualified rigging inspector must inspect eyebolts once a year. They should be in like-new condition, although minor surface rust, superficial scraping, or minor nicks are permissible. Any degree of defect beyond these may be reason to remove eyebolts from service, even if the defect is not described in the rejection criteria below. Rejection criteria for eyebolts include the following (*Figure 44*):

- Scraping or abrasion that results in a noticeable loss of material
- A bent shank or distorted threads
- Stress cracks
- Rust or corrosion that cannot be easily removed with a wire brush
- Elongation from overload of the eye or bottle-necking of the shank
- Damaged or worn threads
- Deformation or twist from side loading
- Wear

4.3.0 Lifting Clamps

Lifting clamps are used to move loads such as steel plates or concrete panels without the use of slings. Loads are moved one item at a time only.

4.3.1 Lifting Clamp Design and Characteristics

Lifting clamps are designed to bite down on a load and use the jaw tension to secure the load during transport (*Figure 45*). They are available in a wide variety of designs. Lifting clamps must be made of forged steel, and they must be stamped with their rated capacity.

Some clamps use the weight of the load to produce and sustain the clamping pressure. Others use an adjustable cam that is set and tightened to maintain a secure grip on the load.

Lifting clamps are designed to carry one item at a time, regardless of the capacity or jaw dimensions of the clamp or the thickness or weight of the item being lifted. Most clamps have a cam, which is adjustable, and a jaw, which is fixed. The item to be lifted is placed in the jaw of the clamp, which bites down onto the item as load stress is applied or as the adjustable cam is tightened.

In order for the clamp to hold an item securely, the cam and the jaw must bite or grip both sides of a single item. Placing more than one plate or sheet into the clamp prevents both the cam and

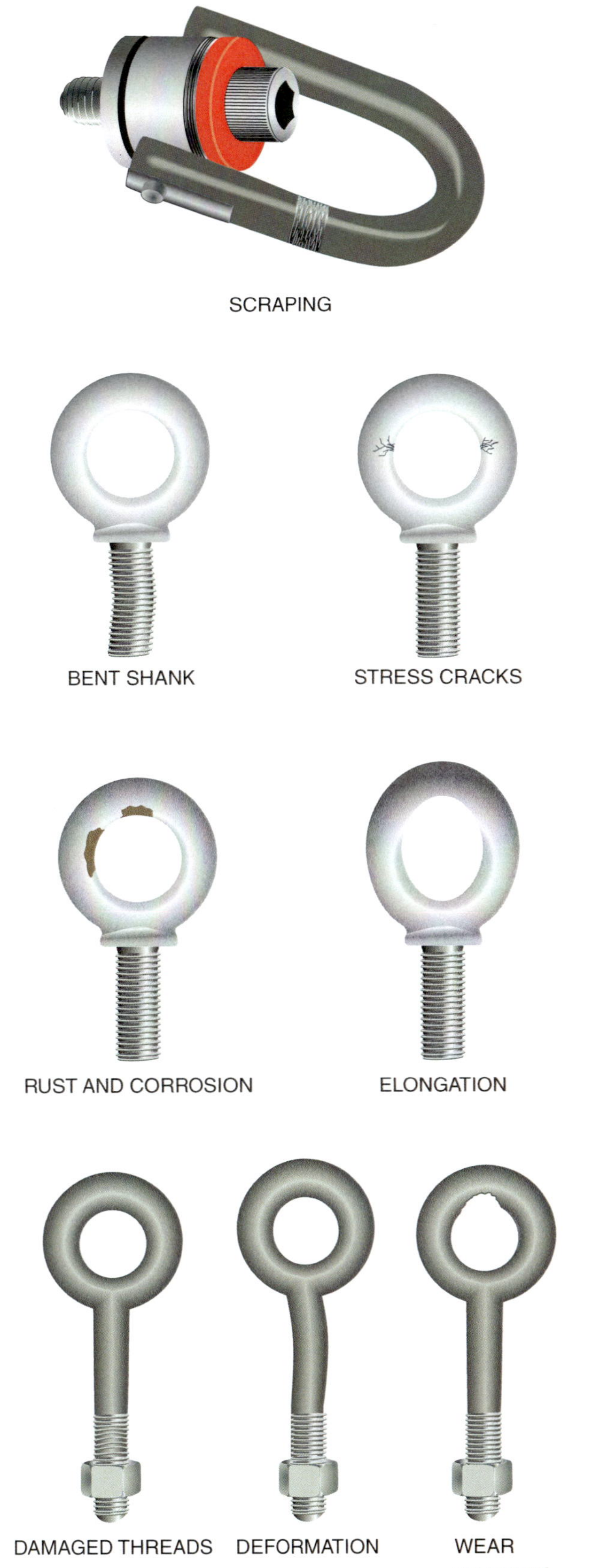

Figure 44 ◆ Rejection criteria for eyebolts.

Figure 45 ◆ Basic nonlocking clamp.

the jaw from securing both sides of the plate or sheet. Clamp designs vary with the application, so it is important to match the design to the intended application. Lifting clamps must function smoothly and adjust with no mechanical difficulty. At least two lifting clamps should always be used when lifting an item. Lifting clamps must be placed to ensure that the load remains balanced. Loads lifted with both non-marring clamps and plate clamps must be lifted slowly and smoothly. These types of clamps are designed to grip the load gently yet securely. Their cams and jaws do not bite down too forcefully into the load, which might damage it. Therefore, you must be very careful to avoid any sudden movements that could jar the load out of these types of clamps.

4.3.2 Types of Lifting Clamps

Lifting clamps must be carefully selected for the specific lifting application, based on their design. There are four types of specialized lifting clamps and several variations in each type (*Figure 46*): linkage-type cam clamps, locking cam clamps, screw-adjusted cam clamps, and non-marring clamps.

Figure 46 ◆ Nonstandard types of lifting clamps.

4.3.3 Lifting Clamp Inspection and Rejection Criteria

Lifting clamps, like any other type of rigging hardware, must be inspected by the rigger before every use to make sure there are no defects that would make the clamp unsafe. Each lift may cause some degree of damage or may further reveal existing damage.

If any of the following conditions exists, a clamp must be removed from use (*Figure 47*):

- Cracks
- Abrasion, wear, or scraping
- Any deformation or other impact damage to the shape that is detectable during a visual examination
- Excessive rust or corrosion, meaning rust or corrosion that cannot be removed easily with a wire brush
- Excessive wear of the teeth
- Heat damage
- Loose or damaged screws or rivets
- Worn springs

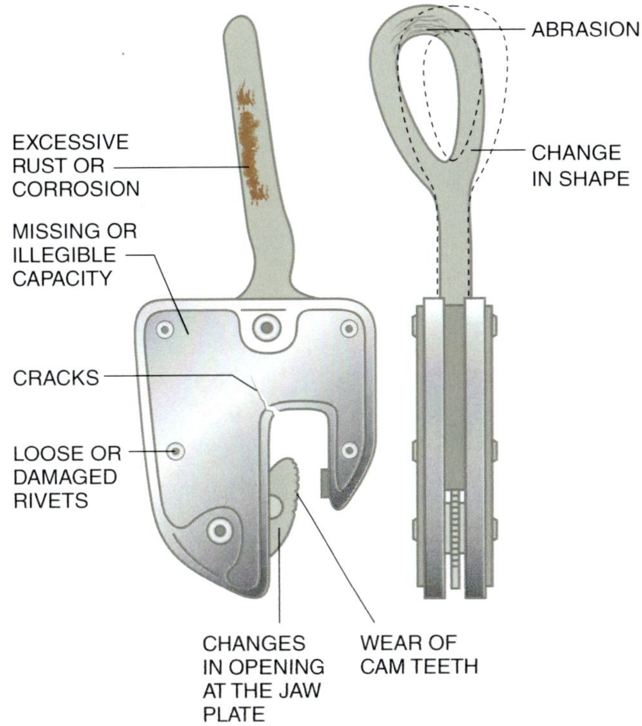

Figure 47 ◆ Rejection criteria for lifting clamps.

4.4.0 Rigging Hooks

A **rigging hook** is an item of rigging hardware used to attach a sling to a load. Although there are many classes of rigging hooks used for hoist hooks and rigging, there are only six basic types of rigging hooks (*Figure 48*):

- Eye hooks are the most common type of end fitting hook. These hooks may or may not have safety latches or gates.
- Sorting hooks, also called pipe hooks, are used to lift pipe sections or containers by inserting the hook into the load, thereby avoiding the need for shackles or additional hardware.
- Reverse eye hooks position the point of the hook perpendicular to the eye.
- Sliding choker hooks are installed onto the sling when it is made. The hooks, which can be positioned anywhere along the sling body, are used to secure the sling eye in a choker hitch. Sliding choker hooks are available for steel chain slings.
- Grab hooks are used on steel chain slings. These hooks fit securely in the chain link, so that choker hitches can be made and chains can be shortened.
- Shortening clutches, a more efficient version of the grab hooks, provide a secure grab of the shortened sling leg with no reduction in the capacity of the chain because the clutch fully supports the links.

106F48A.EPS

EYE HOOK

106F48B.EPS

SORTING HOOK

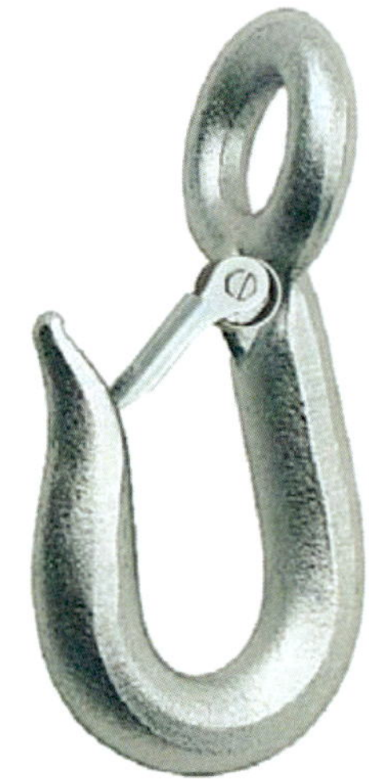

106F48C.EPS

ROUND REVERSE EYE HOOK

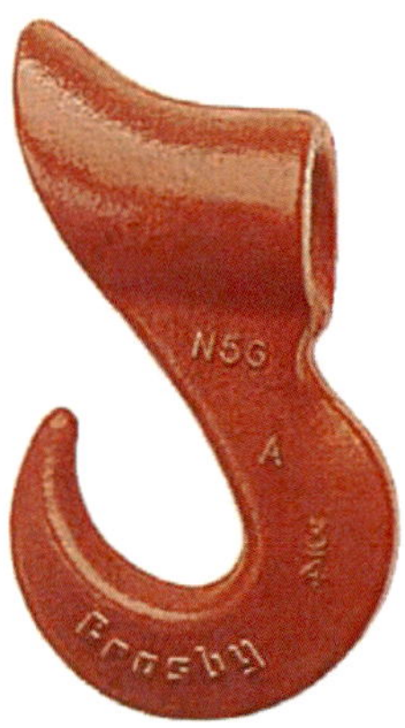

106F48D.EPS

SLIDING CHOKER HOOK

106F48E.EPS

GRAB HOOK

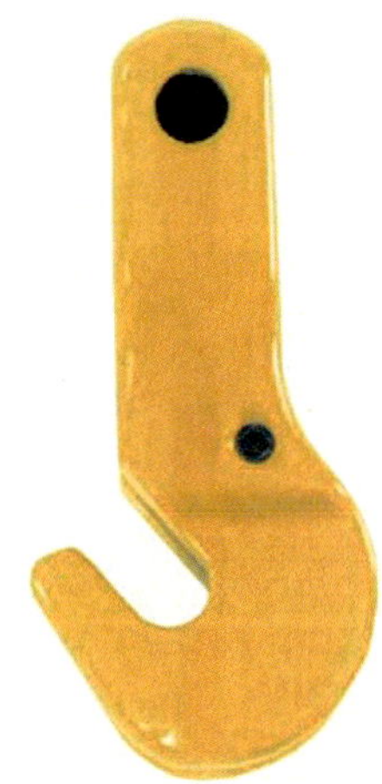

106F48F.EPS

SHORTENING CLUTCH

Figure 48 ◆ Rigging hooks.

4.4.1 Rigging Hook Design and Characteristics

Hooks used for rigging must be made of drop- or hammer-forged steel. All hooks used in rigging operations must meet the characteristics and performance criteria described in this module.

Safety latches or gates are installed in rigging hooks to prevent a sling from coming out of a hook or load when the sling is slackened. If a safety latch is installed in a rigging hook, the latch must be in good working condition. Damaged safety latches can be easily replaced. Report any damage you detect in a safety hook to your instructor or immediate supervisor.

4.4.2 Rigging Hook Inspection and Rejection Criteria

When hooks are installed as end fittings, they must be inspected along with the rest of the sling before each use. Slings with hook-type end fittings need to be removed from service for any of the following defects (*Figure 49*):

- Wear, scraping, or abrasion
- A broken or missing safety latch

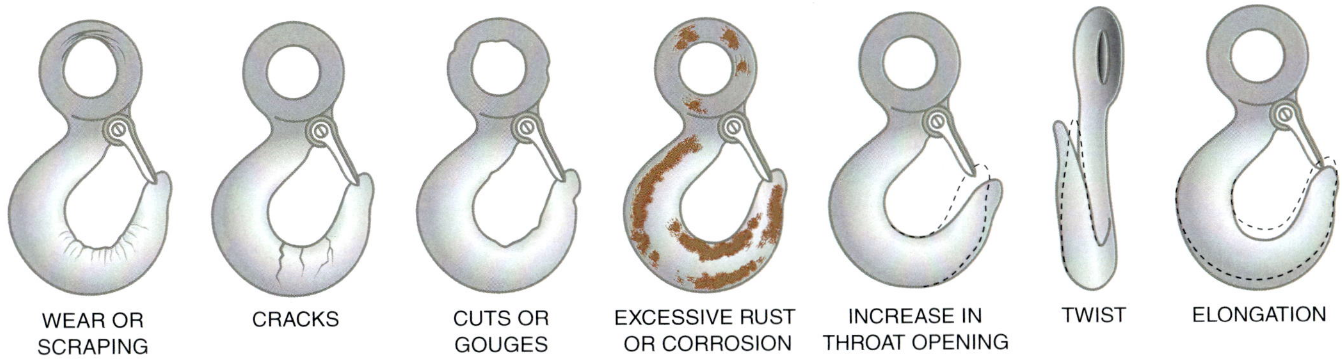

Figure 49 ◆ Rejection criteria for rigging hooks.

- Cracks
- Cuts, gouges, nicks, or chips
- Excessive rust or corrosion, meaning rust or corrosion that cannot be easily removed with a wire brush
- A twist of 10 degrees or more from the unbent plane of the hook
- An increase in the throat opening of the hook of 15 percent or more—easy to detect if the hook is equipped with a safety latch, because the latch will no longer bridge the throat opening
- An increase of 5 percent or more in the shank or overall elongation of the hook

Review Questions

Section 4.0.0

1. The piece of rigging hardware used to couple the end of a wire rope to eye fittings, hooks, or other connections is a(n) _____.

a. eyebolt
b. shackle
c. clamp
d. U-bolt

2. The type of bolt used only for straight vertical pulls is a(n) _____.

a. U-bolt
b. swivel eyebolt
c. unshouldered eyebolt
d. shouldered eyebolt

3. The piece of rigging hardware designed to move loads such as steel plates or concrete panels without the use of slings is called a _____.

a. shackle clamp
b. C-clamp
c. lifting clamp
d. U-bolt clamp

4. Excessive rust or corrosion on an item of rigging hardware is rust or corrosion that cannot easily be removed with _____.

a. a wet towel
b. a chisel
c. your hand
d. a wire brush

5. _____ are installed in rigging hooks to prevent a sling from coming out of a hook when the sling is slackened.

a. Safety latches
b. Locking cam clamps
c. Swivel eyebolts
d. Unshouldered eyebolts

5.0.0 ◆ SLING STRESS

When **sling angle** decreases, **sling stress** increases. This is one of the most important facts you need to know to conduct rigging safely. It is essential that you understand this concept, because not understanding it—or misunderstanding it—could get you or someone else killed.

WARNING!

To rig and move a load safely, the rigger must understand the specific type and amount of stress placed on rigging components such as slings, hooks, and loads. If calculations are not made accurately, the rigging procedure could fail and damage the load and injure or kill the workers.

Here's a good way to understand sling stress. Imagine there are four loads, each weighing 2,000 pounds. If the first load (*Figure 50*) has two slings straight up and down (90 degrees), then each sling is itself holding a weight of 1,000 pounds.

The second load has slings at 60 degrees (*Figure 51*). Each sling still has the original 1,000 pounds of weight pulling down on it. However, because the slings are at 60-degree angles, the **side pull** now adds more stress to the sling. Side pull means the slings are being pulled in, or sideways, by the load's weight as well as down.

The third load (*Figure 52*) also weighs 2,000 pounds. Each sling still has the same 1,000 pounds of weight pulling down on it. But because the sling angle has decreased to 45 degrees, the amount of side pull has increased. This decrease in sling angle adds another 414 pounds of side pull, increasing the total weight the slings must hold.

The fourth load (*Figure 53*) also weighs 2,000 pounds. Each sling still has the same 1,000 pounds of weight pulling down on it. But here the sling angle has decreased to 30 degrees, which causes side pull to increase the sling stress to 1,000 extra pounds. This decrease in sling angle means each sling now has to hold up a total of 2,000 pounds.

As you can see from examples 1 through 4, as the sling angle decreases, the sling stress increases. By the time the sling angle gets to 30 degrees, the sling stress has doubled. Although the total weight of the load has stayed the same in each example, the stress on the sling has increased with each reduction in the sling's angle. Below 45 degrees, sling stress increases dramatically. The greater the sling stress, the greater the effect on the lift's safety.

See the Appendix in the back of this module for more information on calculating sling stress.

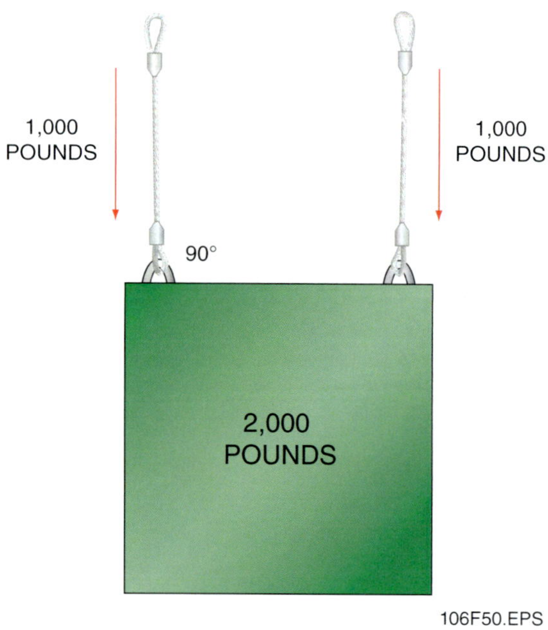

Figure 50 ◆ Sling stress example 1.

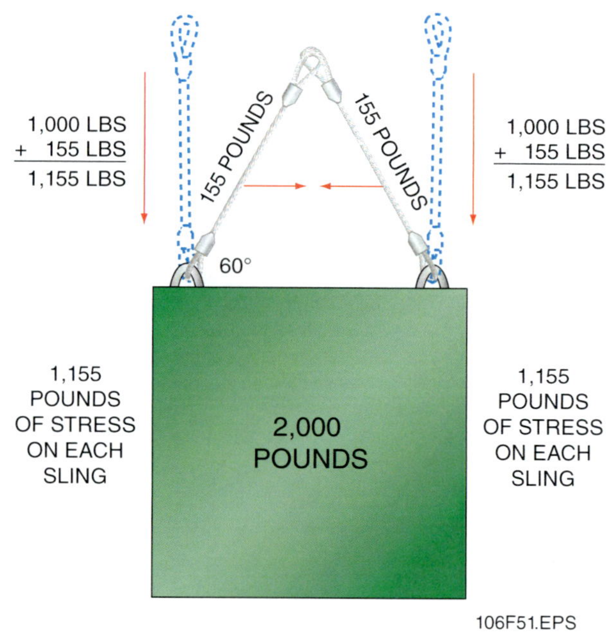

Figure 51 ◆ Sling stress example 2.

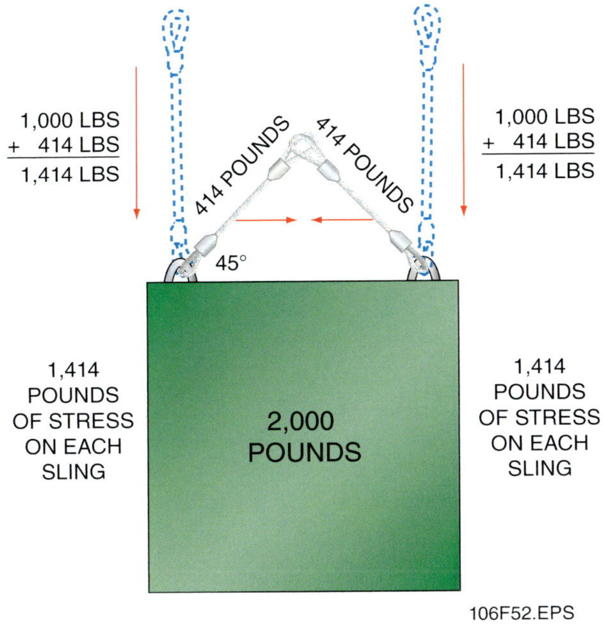

Figure 52 ◆ Sling stress example 3.

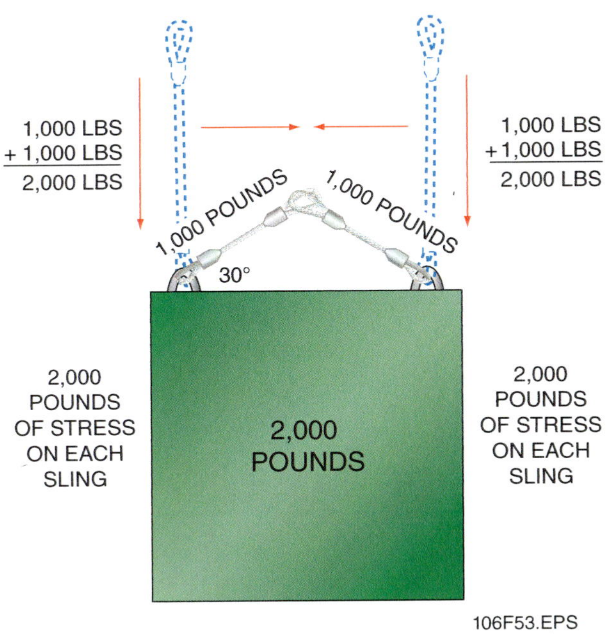

Figure 53 ◆ Sling stress example 4.

Review Questions

Section 5.0.0

1. When sling angle decreases, sling stress _____.

a. stays the same
b. decreases
c. increases
d. becomes zero

2. Side pull means the _____.

a. load is being pulled from its side
b. same as sling angle
c. load stress has doubled
d. slings are being pulled both in and down by the load's weight

3. A decrease in sling angle can add more side pull, increasing the total weight the slings must hold.

a. True
b. False

4. When a sling's angle goes from 90 degrees to 30 degrees, the _____.

a. sling stress doubles
b. sling stress becomes zero
c. sling stress increases 30 pounds
d. amount of side pull decreases

5. The greater the sling stress, _____.

a. the less you have to worry about the lift's safety
b. the greater the effect on the lift's safety
c. the safer the lift
d. the more slings you must use

6.0.0 ◆ HOISTS

A hoist gives you a mechanical advantage for lifting a load, allowing you to move objects that you cannot lift manually. All hoists use a pulley system to transmit power and lift a load. Some hoists are mounted on trolleys and use electricity or compressed air for power. In this section, you will learn about a simple hoisting mechanism called a **block and tackle** and about a more complex hoisting mechanism called a chain hoist.

- *Block and tackle* – A block and tackle is a simple rope-and-pulley system used to lift loads (*Figure 54*). By using fixed pulleys and a wire rope attached to a load, the rigger can raise and lower the load by winding the wire rope around a drum.
- *Chain hoists* – Chain hoists may be operated manually or mechanically. There are three types of chain hoists (*Figure 55*): manual, electric, and pneumatic. Because electric and pneumatic chain hoists use mechanical, not manual, power, they are known as powered chain hoists. All chain hoists use a gear system to lift heavy loads (*Figure 56*). The gearing is coupled to a sprocket that has a chain with a hook attached to it. The load is hooked onto a chain and the gearing turns the sprocket, causing the chain to travel over the sprocket and moving the load. The hoist can be suspended by a hook connected to an appropriate anchorage point, or it can be suspended from a trolley system (*Figure 57*).

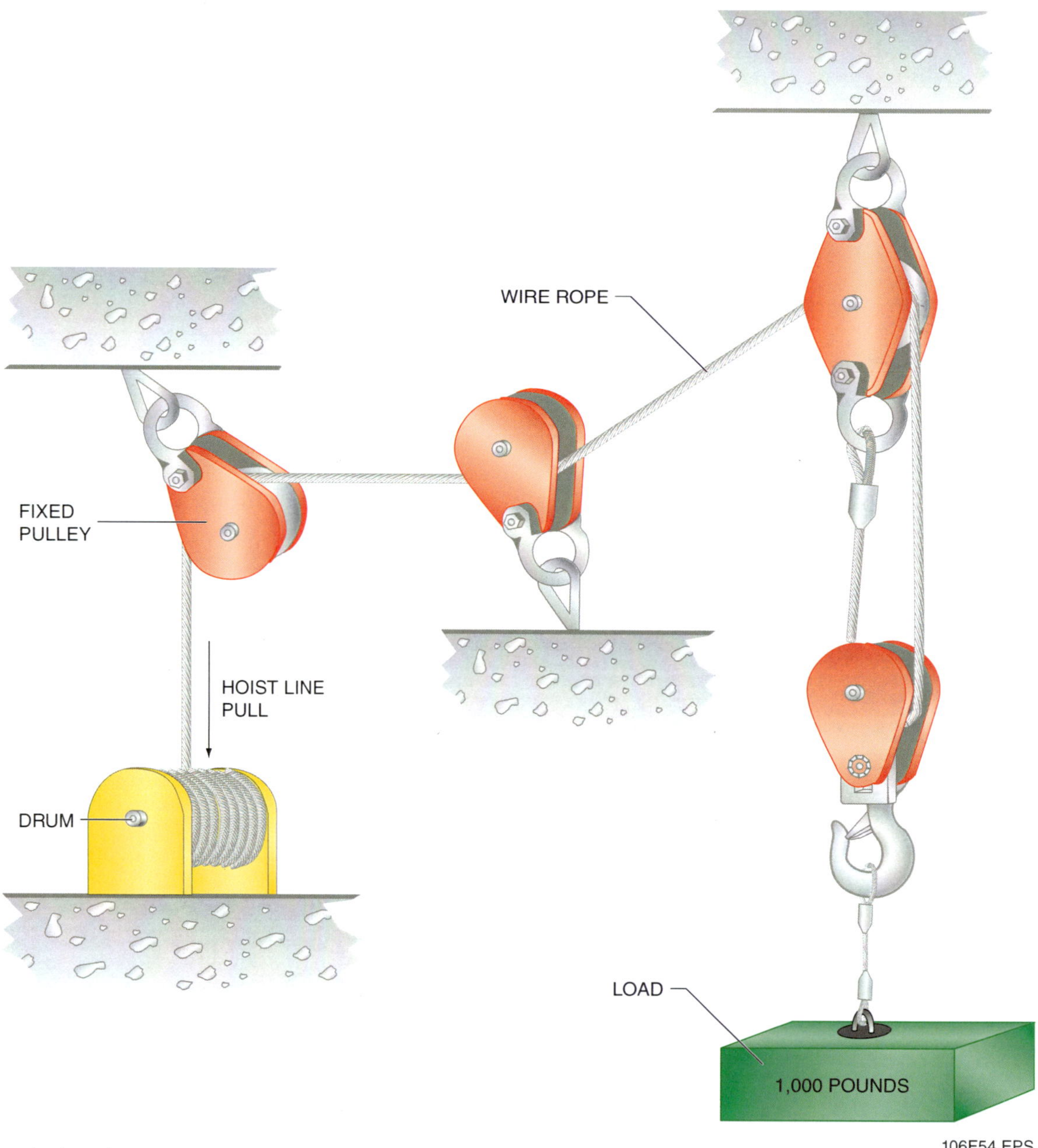

Figure 54 ◆ Block and tackle hoist system.

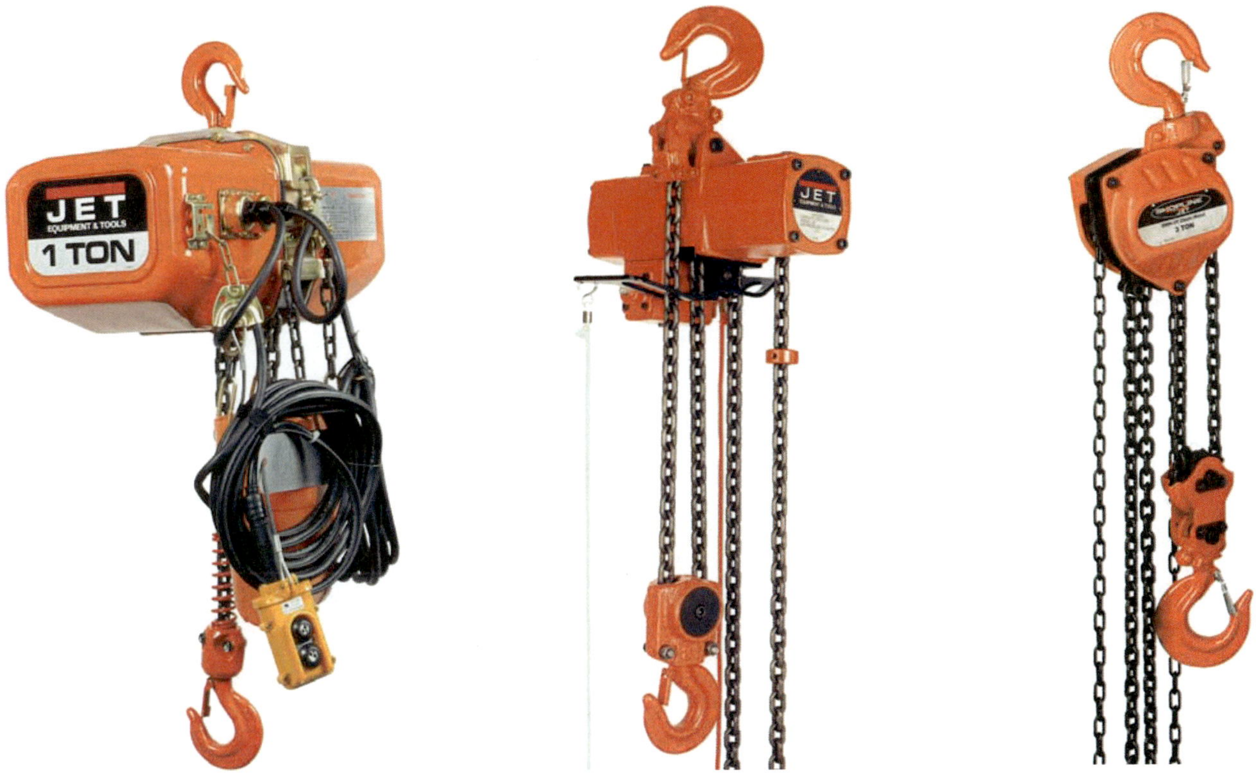

106F55A.EPS

106F55B.EPS

106F55C.EPS

ELECTRIC

PNEUMATIC

MANUAL

Figure 55 ◆ Types of chain hoists.

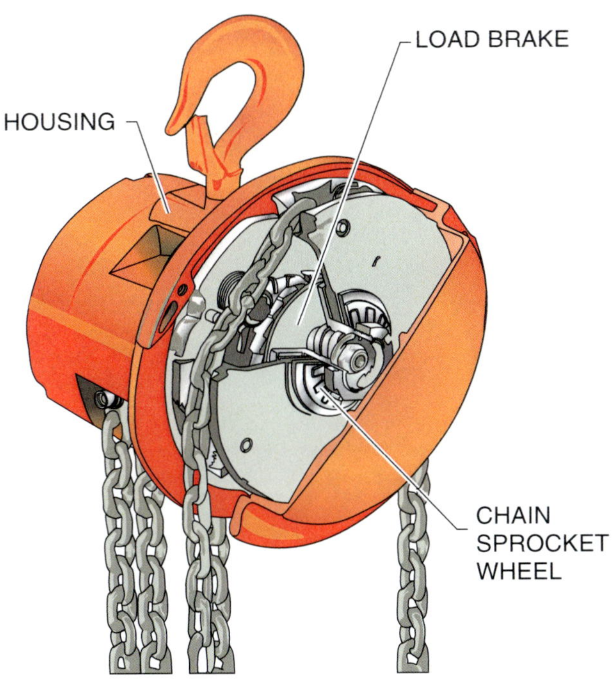

106F56.EPS

Figure 56 ◆ Chain hoist gear system.

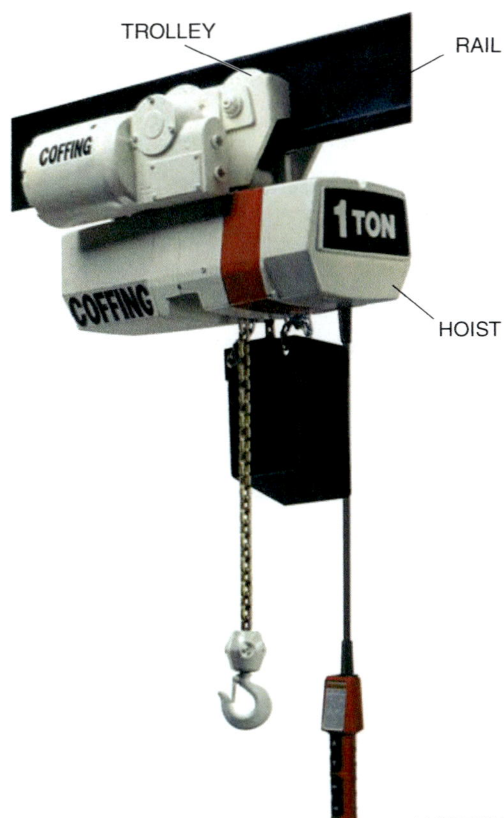

106F57.EPS

Figure 57 ◆ Hoist suspended from a trolley system.

6.1.0 Operation of Chain Hoists

Chain hoists are operated manually (by hand) or by electric or pneumatic power. In this section, you will learn some of the fundamental operating procedures for using both hand chain hoists and powered chain hoists.

- *Hand chain hoist* – To use a hand chain hoist, the rigger suspends the hoist above the load to be lifted, using either the suspension hook or the trolley mount. The rigger then attaches the hook to the load and pulls the hand chain drop to raise the load. The load will either rise or fall, depending on which side of the chain drop is pulled.
- *Powered chain hoist* – To use an electric or a pneumatic powered chain hoist, the rigger positions the chain hoist on the trolley above the load to be lifted, attaches the hook to the load, and uses the control pad to raise the load. Only qualified persons may use powered chain hoists.

6.2.0 Hoist Safety and Maintenance

In addition to the general safety rules you learned in the *Basic Safety* module, there are some specific safety rules for working with hoists.

- Always use the appropriate personal protective equipment when working with and around any lifting operations.
- Make sure that the load is properly balanced and attached correctly to the hoist before you attempt the lift. Unbalanced loads can slide or shift, causing the hoist to fail.
- Keep gears, chains, and ropes clean. Improper maintenance can shorten the working life of chains and ropes.
- Lubricate gears periodically to keep the wheels from freezing up.
- Never perform a lift of any size without proper supervision.

CAUTION

Never use a come-along for vertical overhead lifting. Use a come-along only to move loads horizontally over the ground. Be careful not to confuse a come-along with a ratchet lever hoist. Ratchet lever hoists have both a friction-type holding brake and a ratchet-and-pawl load control brake. Come-alongs have only a spring-load ratchet that holds the pawl in place. If the ratchet and pawl fail, the overhead load falls.

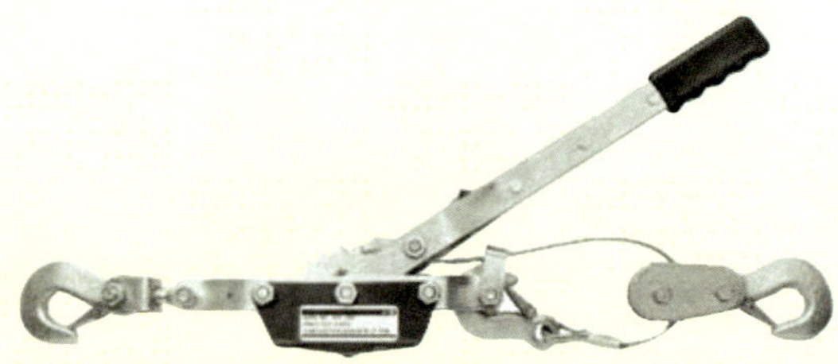

106SA02A.EPS

COME-ALONG

106SA02B.EPS

RATCHET LEVER HOIST

Review Questions

Section 6.0.0

1. All hoists use _____ to transmit power and lift a load.
   a. pneumatic gears
   b. a pulley system
   c. powered chains
   d. trolley mounts

2. The three types of chain hoists are _____.
   a. block, tackle, and pulley
   b. hook, gear, and sprocket
   c. anchor, suspension, and ratchet-and-pawl
   d. manual, electric, and pneumatic

3. To use a hand chain hoist, the rigger _____ to raise and lower the load.
   a. pulls the hand chain drop
   b. winds the steel rope
   c. uses the control panel
   d. ratchets the lever handle

4. Powered chain hoists must be used only by _____.
   a. those who can't lift loads manually
   b. mechanical engineers
   c. qualified persons
   d. rigging trainees

5. Make sure the load is _____ before you attempt the lift.
   a. heavy enough
   b. properly balanced
   c. detached from the hoist
   d. connected to an anchorage point

7.0.0 ◆ RIGGING OPERATIONS AND PRACTICES

In this module, you have learned some of the basics about tools used to perform rigging operations. You have also learned some of the basic rigging practices used to conduct lifts. Now you will learn how to apply this basic knowledge toward participating in a safe and efficient rigging operation.

7.1.0 Rated Capacity

All slings, hardware, and rigging devices are required to be clearly marked with their rated capacity. Paper tags cannot be used. Rated capacity means the same thing as the working load limit (WLL) or a safe working load (SWL). The rated capacity of slings and rigging hardware must be determined to make sure that loads are safely and effectively lifted and transported.

WARNING!
Under no circumstances should you ever exceed the rated capacity! Overloading may result in catastrophic failure.

7.2.0 Sling Attachment

Before attaching slings to a load, riggers must remember several important points. The rigger must select the most appropriate type of sling (synthetic, alloy steel, or wire rope) for the load to be lifted and inspect the sling for damage. Once the type of sling is chosen, the rigger determines the best rigging configuration or hitch (vertical [bridle], choker, or basket) to lift the load. Knowing the type of sling and the best hitch, the rigger then chooses the appropriate rigging hardware to connect the sling to the load. To make the safest selection, the rigger must know the total weight of the load, the angle at which the sling(s) will attach to the load, and the total sling stress applied during the lift. Only then can a sling be rigged safely to a load.

When using slings, follow these safety guidelines:

- Never try to shorten a sling by wrapping it around the hoist hook before attaching the eye to the hook.
- Never try to shorten the legs of a sling by knotting, twisting, or wrapping the slings around one another.
- Never try to shorten a chain sling by bolting or wiring the links together.

WARNING!

Use the correct length sling for each lift. Trying to shorten a sling may cause an accident.

- Make sure all personnel are clear of the load before you take full load strain on the slings.
- Never try to adjust the slings while a strain is being taken on the load.
- Make sure all personnel keep their hands away from the slings and the load during hoisting.
- Use sling softeners whenever possible. Sling softeners protect the sling from abrasion, cuts, heat, and chemical damage. Whether the softeners or pads are manufactured or made in the field, using them regularly will considerably extend the life of the sling. If no standard sleeves are available, use rubber belting or sections of old slings. Softeners prevent kinking in wire rope. Corner buffers are specially made for this purpose, but if they are not available, the rigger can use whatever is handy to protect the rope. Wood, old web slings, and factory-made buffers will provide some protection. Softeners for chain slings protect the links contacting the corners of the load and keep the chain links from scraping or crushing the load itself.

7.3.0 Hardware Attachment

Most rigging hitches require some type of hardware to attach the slings to a piece of equipment or crane. You have learned about shackles, eyebolts, and hooks. This section summarizes the basics of how this equipment is attached to the load.

7.3.1 Shackles

Shackles have several installation requirements to prevent damage or failure during the lift. For instance, the shackle must remain in line with the sling so that the shackle does not become side-loaded and get pulled apart by the load stress. Also, when a screw pin type of shackle is installed, the pin should not be overtightened or it will stretch the threads. Overtightening makes the shackle difficult to remove and may damage the shackle so that it has to be removed from service (*Figure 58*).

Figure 58 ◆ Don't overtighten the pin.

7.3.2 Eyebolts

Eyebolts must be safely and properly installed. Use the following guidelines:

Eyebolts must be installed so that the plane of the eye is in direct line with the plane of the sling when the sling is positioned at an angle other than vertical (*Figure 59*). Otherwise, the eyebolt may fold over under the load stress, which could severely deform the eyebolt or even lead to a fracture or a complete failure of the eye.

Eyebolts with shoulders or swivel bases must be installed so that the shoulder or base is flush with the surface of the load and makes positive contact around the entire circumference of the shoulder or base.

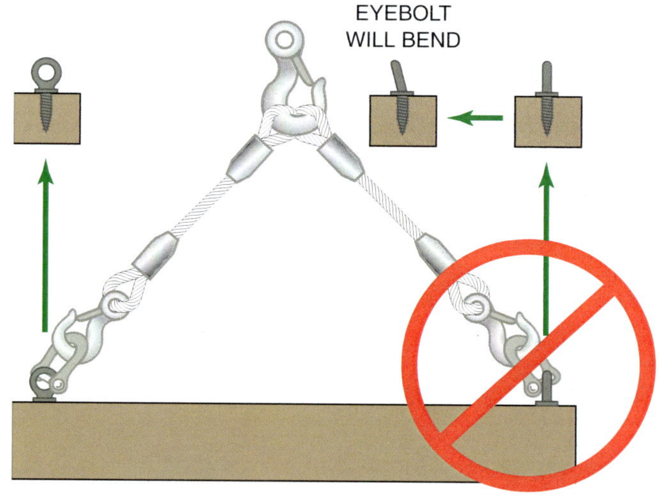

Figure 59 ◆ Eyebolt orientation.

CAUTION

Use a pair of shackles in the eyebolts instead of threading a sling through hardware around a load. Threading the sling through eyebolts as shown and applying stress will cause the eyebolts to pull toward each other and shear off the bolts.

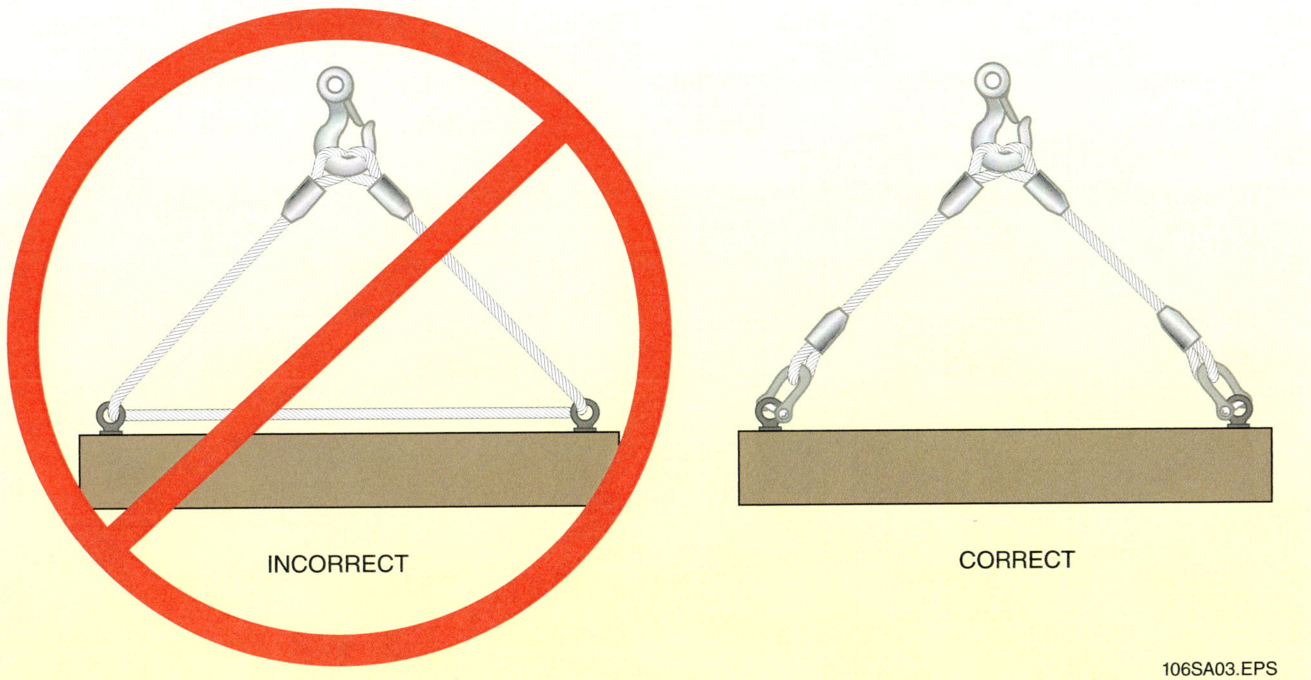

106SA03.EPS

CAUTION

The severe side stress of low sling angles may cause a load to buckle into the center because of side pull. Use extreme caution when attaching a sling to swivel eyebolts, even though the swivel eyebolt does not have restrictions on sling angles.

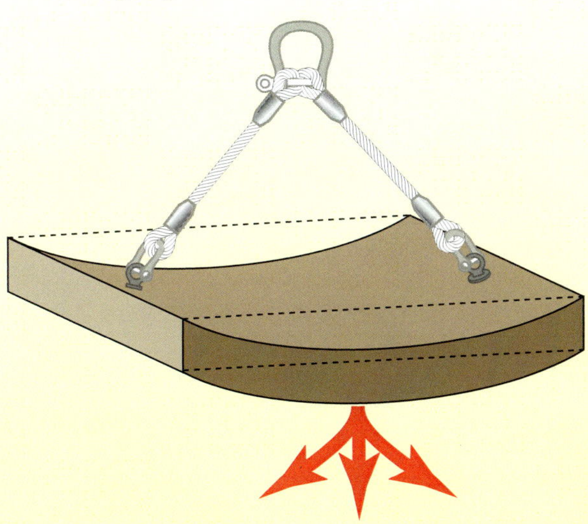

106SA04.EPS

7.3.3 Hooks

Hooks must be carefully attached to the load to prevent binding of the hook. When a hook becomes bound in a **pad eye** or eyebolt, it can become point-loaded. A point-loaded hook will easily stretch open and slip out of the load. When a hook becomes point-loaded, the capacity of the hook decreases drastically the closer the load moves toward the point of the hook (*Figure 60*). Using a shackle instead of placing the hook directly into the pad eye or eyebolt will prevent this from happening.

7.4.0 Load Control

Safe and efficient load control (*Figure 61*) involves communication, the use of physical load control techniques, the safe handling of loads, taking landing zone precautions, and following sling disconnection and removal practices.

7.4.1 American National Standards Institute (ANSI) Hand Signals

Load navigation is accomplished by using common signals—either verbal signals given by radio

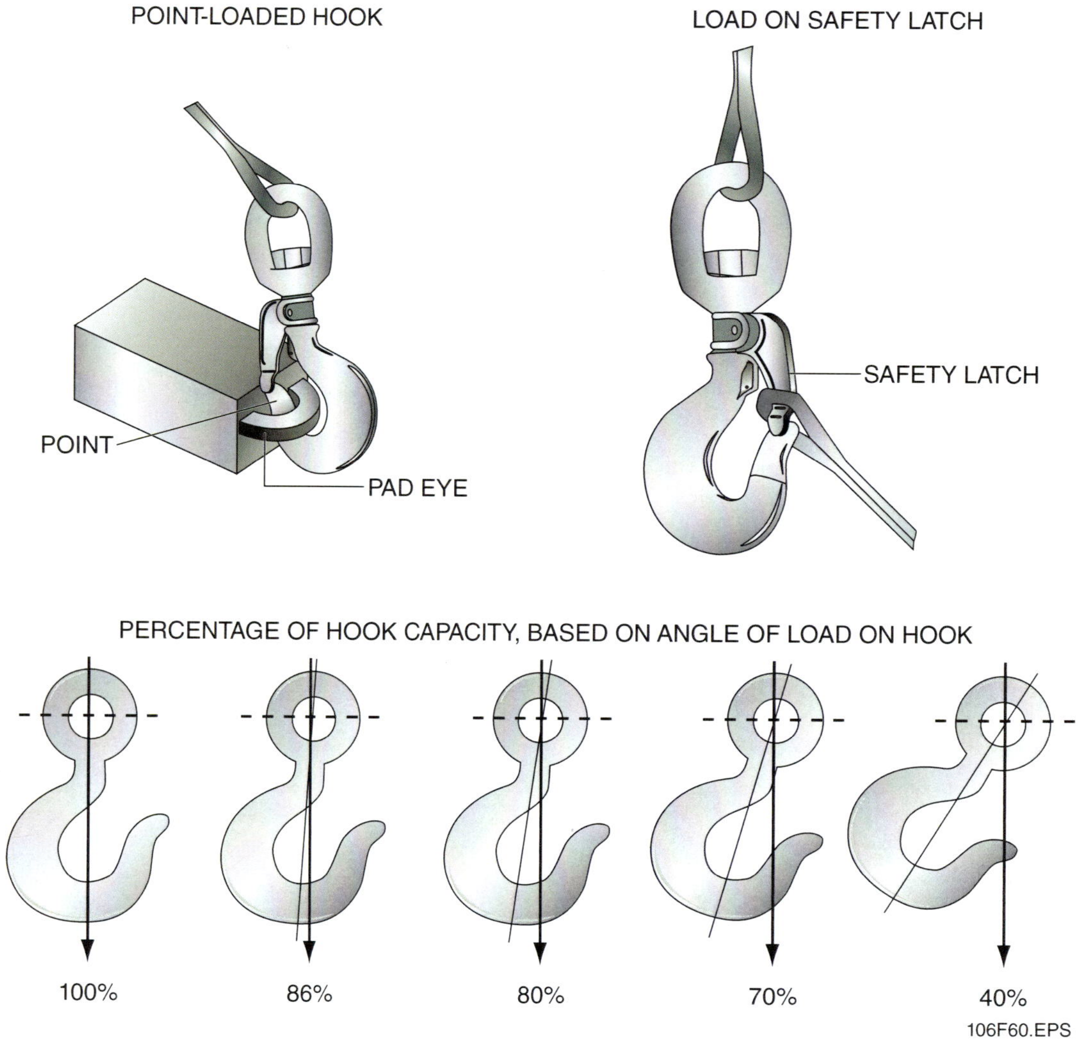

Figure 60 ◆ Point-loaded capacity reductions.

ON-SITE

Pre-Lift Safety Check

Use these guidelines to ensure your safety and the safety of others on site:

- Use the proper rigging equipment.
- Know how to inspect rigging equipment to ensure that it functions properly.
- Fully understand the principles involved in a lift so that you follow proper rigging procedures.
- Communicate with others effectively before, during, and after a lift.

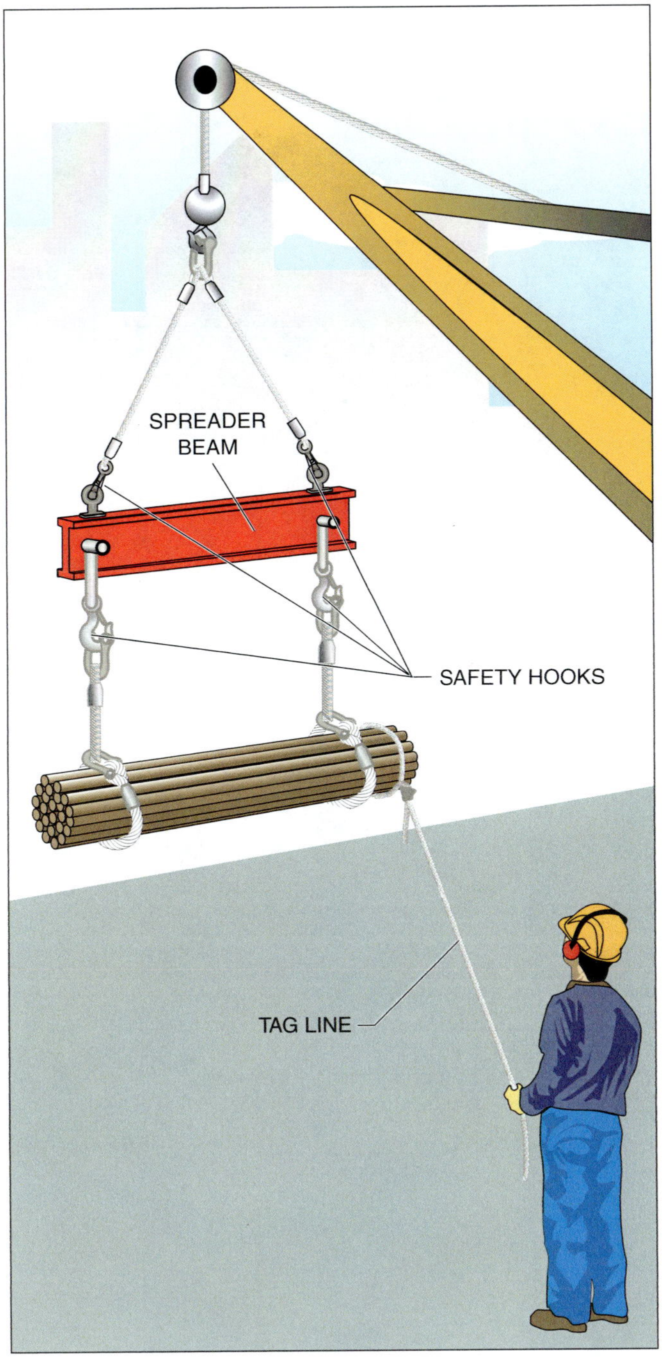

106F61.EPS

Figure 61 ◆ Safe and efficient load control.

or hand signals—to direct a load's movements. Using hand signals to communicate load navigation directions is an integral element in the safe transport and control of all loads. Established hand signals are used for communicating load navigation directions from the signal person to the crane operator. The hand signals used in rigging operations have been developed and standardized by the American National Standards Institute, or ANSI. **ANSI hand signals** have been developed and standardized for use with mobile cranes (*Figure 62*) and for use with overhead cranes (*Figure 63*). Standard hand signals, if used correctly and clearly understood by both the person signaling and the crane operator, provide clear and unmistakable instructions for the various combinations of movements required for the crane to transport the load safely.

7.4.2 Load Path, Load Control, and Tag Lines

Physical control of the load beyond the ability of the crane may be required. Tag lines are used to limit the unwanted or inadvertent movement of the load as it reacts to the motion of the crane or to allow the controlled rotation of the load for positioning. Tag lines are attached after the rigger verifies that the load is balanced.

Figure 62 ◆ Mobile crane hand signals. (1 of 2)

Figure 62 ◆ Mobile crane hand signals. (2 of 2)

Figure 63 ◆ Overhead crane hand signals.

CAUTION

Keep the swing path or load path clear of personnel and obstructions. Crane or hoist operators should watch the load when it is in motion. Never work under a suspended load. For more information about swing path safety, please refer to OSHA's *CFR 1926*.

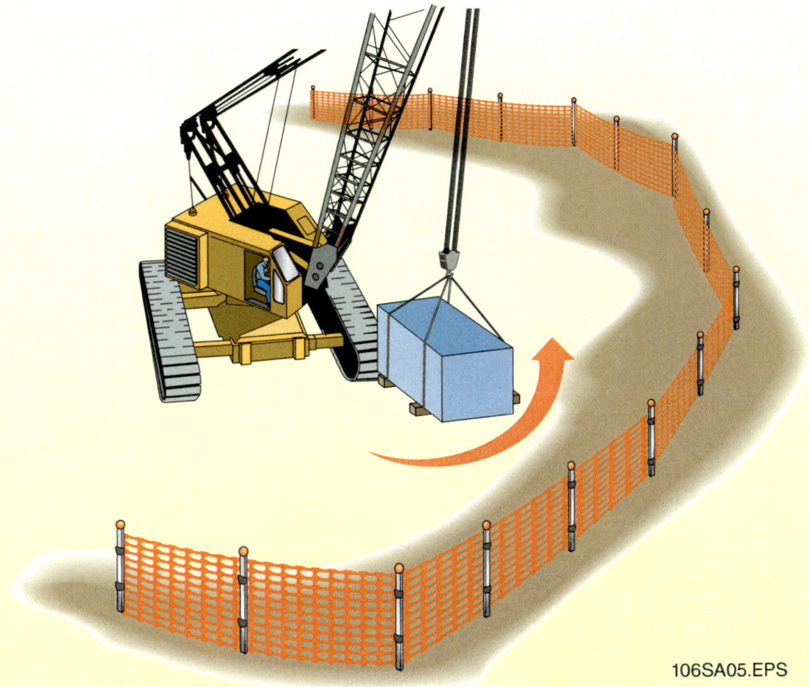

7.4.3 Load-Handling Safety

The safe and effective control of the load involves the rigger's strict observance of load-handling safety requirements:

- Keep the front and rear swing paths of the crane clear throughout the lift (*Figure 64*). Most people watch the load when it is in motion, and this prevents them from seeing the back end of the crane coming around. Using safety flags and roping off the area will help prevent accident and injury.
- Keep the landing zone clear of personnel other than the tag line tenders.
- Be sure that the necessary blocking and **cribbing** for the load are in place and set before you position the load for landing. Lowering the load just above the landing zone and then placing the cribbing and blocking is dangerous because the riggers must pass under the load. The layout of the cribbing can be completed in the landing zone before you set the load. Blocking of the load may have to be done after the load is set. In this case, do not take the load stress off the sling until the blocking is set and secured.
- Do not attempt to position the load onto the cribbing by manhandling it.
- Account for the speed and direction of any wind. To see which way the wind is blowing, look at the wind sock.

CAUTION

Landing the load and disconnecting the rigging are the final steps in any lift. Many rigging accidents and injuries occur at this stage because many riggers mistakenly consider the lift completed just before landing. They do not give their complete attention to the safe landing requirement.

CAUTION

Serious back injuries have resulted from personnel attempting to manually force a load into position or stop even a slight swing of the load. Injuries resulting in crushed hands and other limbs can easily occur should the load shift or swing during the final stages of landing. If the load must be landed with precise control, tag poles, additional tag lines, or hoisting devices should be used.

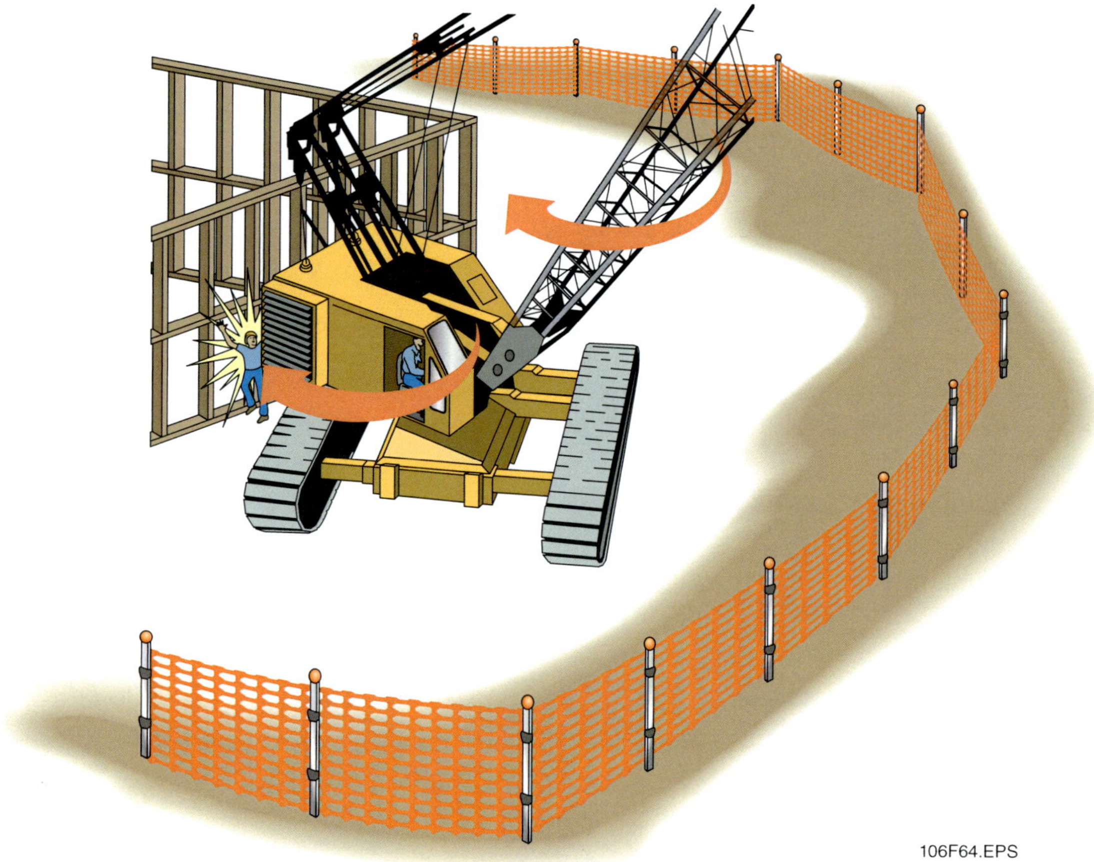

Figure 64 ◆ Rear swing path.

Review Questions

Section 7.0.0

1. A safe guideline for using a sling is to _____.
   a. shorten a sling by wrapping it around the hoist hook
   b. select the correct length of sling for the lift
   c. adjust the sling legs by knotting the length
   d. wire the links of a chain together to shorten the leg

2. Softeners for chain slings can _____.
   a. protect the load
   b. make the landing gentler
   c. make the load easier to handle
   d. decrease sling stress

3. A point-loaded hook's capacity _____ the closer the load moves toward the point of the hook.
   a. decreases drastically
   b. increases drastically
   c. stays the same
   d. decreases slightly

4. Additional physical control of a load may be achieved by using _____.
   a. manual labor
   b. sling angles
   c. tag lines
   d. wire rope

5. A navigational signal of one or both arms extended, palms down, and retracted in rapid motion indicates _____.
   a. hoist up
   b. move slowly
   c. retract boom
   d. emergency stop

Summary

Although rigging operations are complex procedures that can present many dangers, a lift executed by fully trained and qualified rigging professionals can be a rewarding operation to watch or participate in. This module has presented many of the basic guidelines you must follow to ensure your safety and the safety of the people you work with during a rigging operation.

The basic approach to rigging involves thorough planning before each lift and executing proper procedures during every lift. The information covered here offers you the groundwork for a safe, productive, and rewarding construction career.

Notes

Key Terms Quiz

Fill in the blank with the correct key term that you learned from your study of this module.

1. A simple rope-and-pulley system called a(n) __________ is used to lift loads.
2. __________, a type of rigging hardware, are available in three basic designs: unshouldered, shouldered, and swivel.
3. Use a single ring called a(n) __________ to attach multiple slings to a hoist hook.
4. The __________ is the distance between the master link of the sling to either the end fitting of the sling or the lowest point on the basket.
5. Also called blocking, __________ is material used to allow removal of slings after the load is landed.
6. The signal person uses __________ to communicate load navigation directions to the crane operator.
7. Using a choker hitch and forcing the choke down on loose bundles serves only to drastically increase the __________ on the choke leg.
8. An endless-loop synthetic web sling, also called a(n) __________, is made of a single-ply or multiple-ply sling formed into a loop.
9. The way the sling is arranged to hold the load is called the rigging configuration, or __________.
10. A(n) __________ hitch uses two or more slings to connect a load to a single hoist hook.
11. To form a choker hitch, wrap the sling around the load and pass one eye of the sling through a(n) __________ to form a loop around the load.
12. A(n) __________ uses a pulley system to give you a mechanical advantage for lifting a load.
13. A(n) __________ is used to move loads such as steel plates or concrete panels without the use of slings.
14. The __________ is the total amount of what is being lifted.
15. It is important to understand this concept: When _______ decreases, ________ increases.
16. The tension applied on the rigging by the weight of the suspended load is called the __________.
17. The __________ is the main connection fitting for chain slings.
18. To form an endless-loop web sling, the ends are __________ together.
19. __________ equals the lengthwise distance it takes for one strand of wire to make one complete turn around the core.
20. Standard eye-and-eye slings have eyes on the same __________, whereas twisted eye-and-eye slings have eyes at right angles to each other.
21. The __________ is the link between the load and the lifting device.
22. The maximum load weight that a sling is designed to carry is called __________.
23. Examples of __________ for synthetic slings include a missing identification tag, a puncture, and crushing.
24. Use a(n) __________ to attach a sling to a load.
25. Often found on a crane, a grooved pulley-wheel for changing the direction of a rope's pull is called a(n) __________.
26. Be sure to carefully attach hooks to the load to prevent binding of the hook in a(n) __________, which is a welded structural lifting attachment.
27. When slings are being pulled sideways by the load's weight, this __________ adds more stress to the sling.
28. The parts of the sling that reach from the attachment device around the object being lifted are called the __________.

29. A(n) __________ is a group of wires wound around a center core.

30. Riggers use a(n) __________ to limit the unwanted movement of the load when the crane begins moving.

31. If the __________ is showing, the sling is not safe for use.

32. A wire rope sling consists of high-strength steel wires formed into strands wrapped around a supporting __________.

33. An eyebolt is a piece of rigging hardware with a(n) __________, which means it has a series of spiral grooves cut into it.

34. To prevent the load from rotating freely, you must use some method of __________.

35. __________ slings are made of high-strength steel wires formed into strands wrapped around a core.

Key Terms

ANSI hand signals
Block and tackle
Bridle
Bull ring
Core
Cribbing
Eyebolt
Grommet sling
Hitch
Hoist
Lifting clamp
Load
Load control
Load stress
Master link
One rope lay
Pad eye
Plane
Rated capacity
Rejection criteria
Rigging hook
Shackle
Sheave
Side pull
Sling
Sling angle
Sling legs
Sling reach
Sling stress
Spliced
Strand
Stress
Tag line
Tattle-tail
Threaded shank
Warning-yarn
Wire rope

Profile in Success

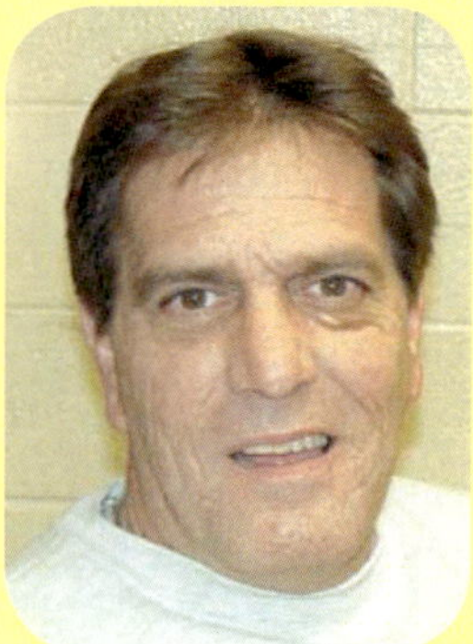

Richard W. Carr

Lead Technician
Constellation Energy Group, Power Source Generation Division
Baltimore, Maryland

Dick was born and raised in Baltimore. He majored in business administration at a local college preparatory school. After graduating, he took a job with Baltimore Gas and Electric (BG&E, now Constellation Energy Group) and has been with the company for 28 years. He has worked primarily as a machinery mechanic and plant technician, and recently became lead technician.

How did you become interested in the construction industry?

I started out as a business administration major, and I went on to become a technician in a power plant—you'll have to figure out that career path yourself!

I've always enjoyed fixing things at home, working with my hands, figuring out how radios and things like that worked. Growing up, though, I didn't have a career focus, a clear idea of what I wanted to do. Essentially, what drew me to BG&E was that it was a quality job with a quality company, in a field that I thought I could excel in. I liked the idea of a job for life, and the benefits were great. That sounded good to me, and the type of work was right up my alley.

When I started, I didn't really know what a power plant was or how electricity was made. I was only 21, and I didn't really know what I was getting myself into! But I'm still here. I've done well by it, and I think that they've done well by me.

I'm on the maintenance and mechanical side of the industry, doing fixes and repairs, as opposed to building things. It's much more integral with machinery and moving parts. The work involves taking care of things that need attention, as opposed to the other side of the construction industry where all that work is done ahead of time. I didn't have anything to do with the construction of the place where I work, but I keep the place working!

What are some of the things you do in your job?

I supervise roughly 20 men in a mechanical maintenance shop within a power plant. Our duties involve performing a schedule of maintenance work on the power-generating machinery. We have a shop where we dismantle smaller pieces for rebuild or inspection, but by and large most of the equipment—pumps, fans, motors, dampers, drives, and accessory equipment—is stationary. That means we have to go to it and work on it where it sits. This is a very hands-on type of job.

We have a very good preventive maintenance program that keeps us out of fighting fires all the time. Our job is to get to the problem before it gets to you, the customer. I'm sure you know the old saying, "If it ain't broke, don't fix it." Well, nowadays our motto is, "If it ain't broke, don't let it get broke!" Safety is our code, our number one concern.

What do you like most about your job?

I would say right off the top it is the sense of accomplishment, the pride of a job well done. My job is to make a machine work the way it was designed to. When I accomplish that, it's a good feeling. After all, it's kind of a strange industry we work in; we make a product you can't see. You can't wrap it up or put it in a box. I've been in this field for going on 30 years now, and I have always found it to be a very interesting business.

Another positive aspect of my job is the teamwork. We have scheduled outages, when a turbine or a boiler is taken offline for eight to twelve weeks for routine maintenance. We also have forced outages—for example, when a boiler develops a leak so large that it can't keep up operating pressure and it trips the generator. When an outage happens, whether it's scheduled or not, it takes a large group of people working toward that common goal to bring it back up online. I really enjoy working with a group of well-trained, interesting, and colorful people. When push comes to shove, everybody's there to do his or her part.

I really enjoy working with my hands. And you can use the knowledge and skills you pick up on the job in your household, on your car, or wherever you need to fix things. I get a lot of compliments around the house when I fix things or build a cabinet, for example. These skills can be applied to lots of things.

You mentioned the distinction between maintenance work and new construction. Can you elaborate a little about that?

Repair and maintenance is a relatively invisible activity. Most people don't stop to think about where electricity comes from. The only time they notice us is when our machines aren't working. But the lack of awareness goes beyond just the general public. We're finding that there's a lack of interest among students in this line of work as well. I can understand that, to some extent. You see people climbing utility poles or working down in a manhole in the street. When power lines go down in a storm, you see people out there putting them back up. But then when everything is fixed, they disappear and you forget about the generation end of the business. Power plants are tucked away out of sight.

Even if you end up in the construction end of the business, knowledge of the maintenance aspect can be really helpful. You should try to design and build things in such a way that they can be repaired. I don't know how many times we've been trying to repair a piece of machinery and someone has said in frustration, "Who on earth designed this thing?" Try to remember that people will be coming after you, and they're going to need to get access to the parts behind the wall.

I think that the maintenance part of our industry doesn't get as much exposure with younger people to get their interest. But I believe that if they knew that this kind of work was here, they'd probably be drawn to it. For a young person who likes working on cars or who is mechanically inclined, this type of work is right up their alley. And it's not like you only learn how to do one task or fix one type of machine. Years ago we had specialists, but now with cross-training we have people who can do many different tasks.

BG&E works with the school systems to promote the company's work-study program. Right now, we have two students who go to a tech school part time and get work experience here the rest of the time. The young people who have come through here as part of that program think it's really neat! To see students taking to it and enjoying it was good for me and for a lot of the other people here too. The average age of people in this field is going up, and hiring is down, so we're doing more with less. The need for more people is very real.

What do you think it takes to be a success in your trade?

I can relate my answer to the company where I've worked. Success is determined in large part by the basic job requirements because this company is very much focused on the basics: be prompt, be on time, have good work ethics, be willing to learn. I think those basics hold true in any field. You need to have a willingness to experience a new and interesting field of work.

What would you say to someone entering the trades today?

In my case, the best answer to that question is not so much what I would say, I think, as much as what people have said to me about this field. They tell me that they get a real sense of being part of something important when they belong to a group of talented, task-oriented people. For me, and for a lot of the people I work with, we don't do this job on a "need to do" basis, but on a "want to do" basis. The work is so multifaceted. You will face a new challenge every day, or develop different fixes for recurring problems. It challenges your skill.

The utility industry in general is a very good way to make a living. It's a challenging, thought-provoking job. It's very interesting, it's exciting, and it can make you tired, too. It can be really exhilarating to know that you're creating a product for hundreds of thousands of people who are depending on it. There's very seldom a dull day!

Profile in Success

This 25-Year-Old Has Mature Responsibilities on Power Plant Project

By Leah Hitchings

From building concrete canoes to building a $350-million power plant, Jason Krueger has performed all kinds of construction and engineering work, and all by the age of 25. Hired by Boldt Construction, the largest construction company in Wisconsin, while still attending University of Wisconsin, Madison, Krueger values his few years in the field more than his 17 years of education.

"Experience is what everyone is looking for," Krueger says.

Krueger's construction education began in high school, when he took weekend classes at UW Madison and participated in an Explorers program on construction. In college, he earned a degree in civil engineering while emphasizing a construction engineering and management program that is currently being made into a major. While busy building concrete canoes for the American Society of Civil Engineers' competitions, and participating in UW's Construction Club, Krueger decided to do a co-op program at Boldt Construction. For eight months, Krueger worked full-time for Boldt, where at first he remained in the office, "looking at drawings and getting familiar with their projects," he says.

After two months in the office, Krueger was placed on site at Appleton Medical Center in Appleton, Wis., where Boldt was building an ambulatory surgery addition.

"I felt pretty lost at first," Krueger says, "but I had a lot of support around me. They didn't treat me like some little kid."

Work on the medical center included a main drive-up canopy and an all-glass entrance, lobby, and waiting room. One of the first problems Krueger faced was laying out a radius for the large glass entrance. Because the architect had not detailed the radius well, Krueger used the skills he had learned in classes to lay it out himself.

Krueger's project manager for the medical center project and for all of his other projects at Boldt, Patrick Loughrin, remembers that even Krueger's first days on site showed that he was a good fit for Boldt.

"Jason was very enthusiastic and aggressive from the beginning," Loughrin says. "He was willing to learn and was always looking for things to accomplish. He didn't wait for things to be handed to him."

Learning was a big part of Krueger's first months at Boldt because most of Krueger's knowledge was from books, not field experience. However, Krueger quickly moved to being a project manager on another project, building a dining hall for a Boy Scout camp. Loughrin says that Krueger's first job as project manager went well, although he made some mistakes.

"Jason was a little too optimistic on some stuff," Loughrin says. "Some things needed closer attention to detail."

A project that followed, remodeling a shelter for abused women, was where Krueger "grew up a lot," Loughrin says.

Krueger believes that much of his early success was a result of his attitude with his supervisors. "A lot has to do with how you ask questions, and I always tried to respect them and be polite," he says.

When Boldt offered Krueger a full-time position, he eagerly accepted, even though he still had a year of college to finish. Boldt is headquartered in Appleton, Krueger's hometown, and he also liked their large size, "because there's a lot of opportunity to grow in the company." Krueger also liked that everyone in the company knows each other, and even top executives make a point of talking to everyone.

Now Krueger is focused on his largest project to date, a $350-million gas-fired turbine and steam generation facility in Beloit, Wis. Krueger is the civil field engineer for the plant, which is being built for Calpine Corp. Although he works 60 hours a week or more now, "It's well worth it," he says. "I'd never think of changing what I do."

Krueger still has the time to visit high schools and recruit for the construction industry, as well as return to UW for career fairs. Eventually Krueger would like to return to UW for mechanical and electrical training, but he does not plan on earning a master's degree.

"A master's degree won't help you get a job, while job experience is what's more important," he says. His business minor has been invaluable, however, because he deals with contracts and labor relations every day.

The long hours and challenging work have been rewarding for Krueger, because, he says, "I love what I'm doing."

Trade Terms Introduced in This Module

ANSI hand signals: Communication signals established by the American National Standards Institute (ANSI) and used for load navigation for mobile and overhead cranes.

Block and tackle: A simple rope-and-pulley system used to lift loads.

Bridle: A configuration using two or more slings to connect a load to a single hoist hook.

Bull ring: A single ring used to attach multiple slings to a hoist hook.

Core: Center support member of a wire rope around which the strands are laid.

Cribbing: Material used to either support a load or allow removal of slings after the load is landed. Also called blocking.

Eyebolt: An item of rigging hardware used to attach a sling to a load.

Grommet sling: A sling fabricated in an endless loop.

Hitch: The rigging configuration by which a sling connects the load to the hoist hook. The three basic types of hitches are vertical, choker, and basket.

Hoist: A device that applies a mechanical force for lifting or lowering a load.

Lifting clamp: A device used to move loads such as steel plates or concrete panels without the use of slings.

Load: The total amount of what is being lifted, including all slings, hitches, and hardware.

Load control: The safe and efficient practice of load manipulation, using proper communication and handling techniques.

Load stress: The strain or tension applied on the rigging by the weight of the suspended load.

Master link: The main connection fitting for chain slings.

One rope lay: The lengthwise distance it takes for one strand of a wire rope to make one complete turn around the core.

Pad eye: A welded structural lifting attachment.

Plane: A surface in which a straight line joining two points lies wholly within that surface.

Rated capacity: The maximum load weight a sling or piece of hardware or equipment can hold or lift. Also called lifting capacity, working capacity, working load limit (WLL), and safe working load (SWL).

Rejection criteria: Standards, rules, or tests on which a decision can be based to remove an object or device from service because it is no longer safe.

Rigging hook: An item of rigging hardware used to attach a sling to a load.

Shackle: Coupling device used in an appropriate lifting apparatus to connect the rope to eye fittings, hooks, or other connectors.

Sheave: A grooved pulley-wheel for changing the direction of a rope's pull; often found on a crane.

Side pull: The portion of a pull acting horizontally when the slings are not vertical.

Sling: Wire rope, alloy steel chain, metal mesh fabric, synthetic rope, synthetic webbing, or jacketed synthetic continuous loop fibers made into forms, with or without end fittings, used to handle loads.

Sling angle: The angle of an attached sling when pulled in relation to the load.

Sling legs: The parts of the sling that reach from the attachment device around the object being lifted.

Sling reach: A measure taken from the master link of the sling, where it bears weight, to either the end fitting of the sling or the lowest point on the basket.

Sling stress: The total amount of force exerted on a sling. This includes forces added as a result of sling angle.

Spliced: Having been joined together.

Strand: A group of wires wound, or laid, around a center wire, or core. Strands are laid around a supporting core to form a rope.

Stress: Intensity of force exerted by one part of an object on another; the action of forces on an object or system that leads to changes in its shape, strain on it, or separation of its parts.

Tag line: Rope that runs from the load to the ground. Riggers hold on to tag lines to keep a load from swinging or spinning during the lift.

Tattle-tail: Cord attached to the strands of an endless loop sling. It protrudes from the jacket. A tattle-tail is used to determine if an endless sling has been stretched or overloaded.

Threaded shank: A connecting end of a fastener, such as a bolt, with a series of spiral grooves cut into it. The grooves are designed to mate with grooves cut into another object in order to join them together.

Warning-yarn: A component of the sling that shows the rigger whether the sling has suffered too much damage to be used.

Wire rope: A rope made from steel wires that are formed into strands and then laid around a supporting core to form a complete rope; sometimes called cable.

Appendix

Rated Capacities

You should never exceed a sling's rated capacity. The rated capacity varies depending upon the type of sling, the size of the sling, and the type of hitch. Rigging operators must know the capacity of the sling they are using. Sling manufacturers usually have charts or tables that contain this information. A sample chart of rated capacities for web slings is shown in *Table A-1*.

Table A-1 Rated Capacities of Some Web Slings

FLAT EYE AND TWISTED EYE WEB SLINGS

TYPE III - FLAT EYES EACH END

TYPE IV - TWISTED EYE EACH END

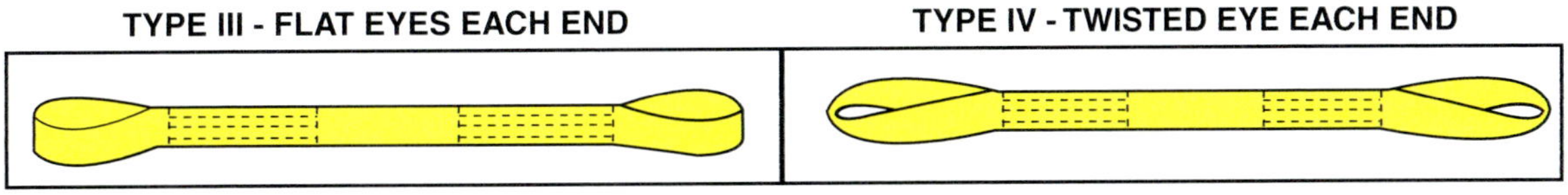

			RATED CAPACITIES IN POUNDS		
WIDTH	EYE WIDTH	EYE LENGTH	CHOKER	VERTICAL	BASKET
1"	1"	9"	1,280	1,600	3,200
1"	1"	9"	2,480	3,100	6,200
1"	1"	12"	4,960	6,200	12,400
2"	2"	9"	2,480	3,100	6,200
2"	2"	9"	4,960	6,200	12,400
2"	2"	12"	9,920	12,400	24,800
3"	1½"	9"	3,720	4,650	9,300
3"	1½"	12"	7,440	9,300	18,600
3"	1½"	18"	14,880	18,680	37,200
4"	1½"	12"	4,960	6,200	12,400
4"	1½"	12"	8,800	11,000	22,000
4"	2"	18"	17,600	22,000	44,000
6"	2"	14"	7,440	9,300	18,600
6"	2"	14"	13,200	16,500	33,000
6"	3"	20"	26,400	33,000	66,000
8"	3"	18"	9,920	12,400	24,800
8"	3"	18"	17,600	22,000	44,000
8"	4"	24"	35,200	44,000	88,000
10"	4"	22"	12,400	15,500	31,000
10"	4"	22"	22,000	26,500	55,000
10"	5"	28"	44,000	55,000	110,000
12"	5"	26"	14,880	18,600	37,200
12"	5"	26"	26,400	33,000	66,000
12"	6"	32"	52,800	66,000	132,000

106A01.EPS

Appendix

SLING ANGLES

The amount of tension on the sling is directly affected by the angle of the sling. *Figure A-1* shows the effect of sling angles on sling loading.

Figure A-1 ◆ Sling angles.

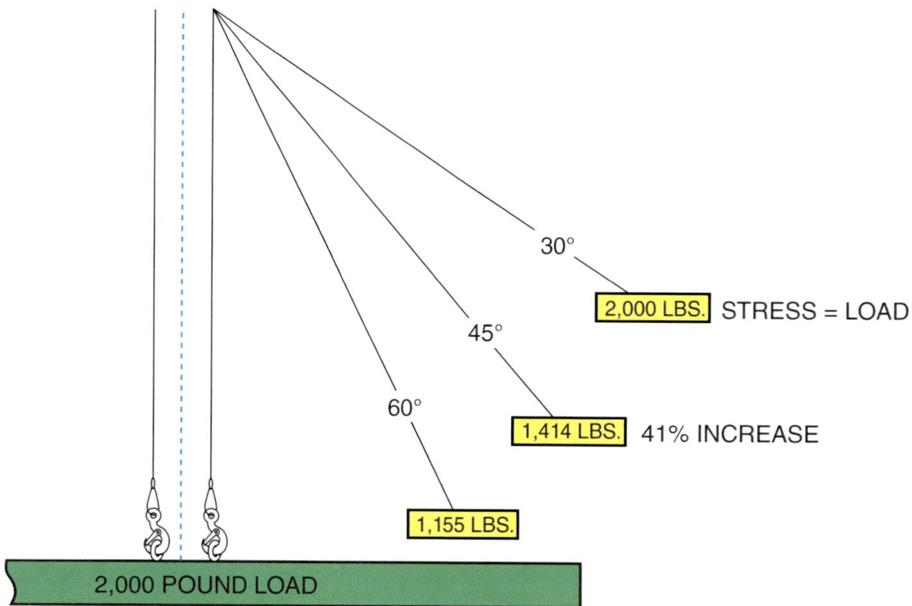

106A02.EPS

To actually determine the load on a sling, a factor table is used (See *Table A-2*).

Table A-2 Factor Table

SLING ANGLE	LOAD ANGLE FACTOR
30	2.000
35	1.742
40	1.555
45	1.414
50	1.305
55	1.221
60	1.155
65	1.104
70	1.064
75	1.035
80	1.015
85	1.004
90	1.000

106A03.EPS

Appendix

In the example shown in *Figure A-1*, two slings are being used to lift 2,000 pounds. When the slings are at a 45-degree angle, there is 1,414 pounds of tension on each sling. This can be determined mathematically by using the following equation:

$$\text{Sling Tension} = \frac{\text{Load (lbs)} \times \text{Load Angle Factor}}{\text{Number of Legs}}$$

Therefore:

$$\text{Sling Tension} = \frac{2{,}000 \times 1.414}{2}$$

$$\text{Sling Tension} = \frac{2{,}828}{2}$$

$$\text{Sling Tension} = 1{,}414$$

Use the weight of the load (2,000 pounds) and multiply it by the corresponding angle factor in *Table A-2*. Then, divide it by the number of slings (2). The result is 1,414 pounds of tension on each sling.

Additional Resources

This module is intended to present thorough resources for task training. The following reference works are suggested for further study. These are optional materials for continued education rather than for task training.

Bob's Rigging and Crane Handbook, Latest Edition. Bob DeBenedictis. Leawood, KS: Pellow Engineering Services, Inc.

High Performance Slings and Fittings for the New Millennium, 1999 Edition. Dennis St. Germain. Aston, PA: I & I Sling, Inc.

Mobile Crane Manual, 1999. Donald E. Dickie, D. H. Campbell. Toronto, Ontario, Canada: Construction Safety Association of Ontario.

Rigging Manual, 1997. Toronto, Ontario, Canada: Construction Safety Association of Ontario.

NCCER CURRICULA — USER UPDATE

NCCER makes every effort to keep its textbooks up-to-date and free of technical errors. We appreciate your help in this process. If you find an error, a typographical mistake, or an inaccuracy in NCCER's curricula, please fill out this form (or a photocopy), or complete the online form at **www.nccer.org/olf**. Be sure to include the exact module ID number, page number, a detailed description, and your recommended correction. Your input will be brought to the attention of the Authoring Team. Thank you for your assistance.

Instructors – If you have an idea for improving this textbook, or have found that additional materials were necessary to teach this module effectively, please let us know so that we may present your suggestions to the Authoring Team.

NCCER Product Development and Revision
13614 Progress Blvd., Alachua, FL 32615
Email: curriculum@nccer.org
Online: www.nccer.org/olf

❑ Trainee Guide ❑ Lesson Plans ❑ Exam ❑ PowerPoints Other ______________________

Craft / Level: Copyright Date:

Module ID Number / Title:

Section Number(s):

Description:

Recommended Correction:

Your Name:

Address:

Email: Phone:

Glossary of Trade Terms

ANSI hand signals: Communication signals established by the American National Standards Institute (ANSI) and used for load navigation for mobile and overhead cranes.

Apparatus: An assembly of machines used together to do a particular job.

Arc: The flow of electrical current through a gas such as air from one pole to another pole.

Arc welding: The joining of metal parts by fusion, in which the necessary heat is produced by means of an electric arc.

Block and tackle: A simple rope-and-pulley system used to lift loads.

Bridle: A configuration using two or more slings to connect a load to a single hoist hook.

Bull ring: A single ring used to attach multiple slings to a hoist hook.

Combustible: Capable of easily igniting and rapidly burning; used to describe a fuel with a flash point at or above 100°F.

Competent person: A person who can identify working conditions or surroundings that are unsanitary, hazardous, or dangerous to employees and who has authorization to correct or eliminate these conditions promptly.

Concealed receptacle: The electrical outlet that is placed inside the structural elements of a building, such as inside the walls. The face of the receptacle is flush with the finished wall surface and covered with a plate.

Confined space: A work area large enough for a person to work, but arranged in such a way that an employee must physically enter the space to perform work. A confined space has a limited or restricted means of entry and exit. It is not designed for continuous work. Tanks, vessels, silos, pits, vaults, and hoppers are examples of confined spaces. See also *permit-required confined space.*

Core: Center support member of a wire rope around which the strands are laid.

Cribbing: Material used to either support a load or allow removal of slings after the load is landed. Also called blocking.

Cross-bracing: Braces (metal or wood) placed diagonally from the bottom of one rail to the top of another rail that add support to a structure.

Dross: Waste material resulting from cutting using a thermal process.

Electrical distribution panel: Part of the electrical distribution system that brings electricity from the street source (power poles and transformers) through the service lines to the electrical meter mounted on the outside of the building and to the panel inside the building. The panel houses the circuits that distribute electricity throughout the structure.

Excavation: Any man-made cut, cavity, trench, or depression in an earth surface, formed by removing earth. It can be made for anything from basements to highways. See also *trench.*

Extension ladder: A ladder made of two straight ladders that are connected so that the overall length can be adjusted.

Eyebolt: An item of rigging hardware used to attach a sling to a load.

Flammable: Capable of easily igniting and rapidly burning; used to describe a fuel with a flash point below 100°F.

Flash: A sudden bright light associated with starting up a welding torch.

Flashback: A welding flame that flares up and chars the hose at or near the torch connection. It is caused by improperly mixed fuel.

Flash burn: The damage that can be done to eyes after even brief exposure to ultraviolet light from arc welding. A flash burn requires medical attention.

Flash goggles: Eye protective equipment worn during welding operations.

Flash point: The temperature at which fuel gives off enough gases (vapors) to burn.

Grommet sling: A sling fabricated in an endless loop.

Ground: The conducting connection between electrical equipment or an electrical circuit and the earth.

Guarded: Enclosed, fenced, covered, or otherwise protected by barriers, rails, covers, or platforms to prevent dangerous contact.

Hand line: A line attached to a tool or object so a worker can pull it up after climbing a ladder or scaffold.

Glossary of Trade Terms

Hazard Communication Standard (HazCom): The Occupational Safety and Health Administration standard that requires contractors to educate employees about hazardous chemicals on the job site and how to work with them safely.

Hitch: The rigging configuration by which a sling connects the load to the hoist hook. The three basic types of hitches are vertical, choker, and basket.

Hoist: A device that applies a mechanical force for lifting or lowering a load.

Lanyard: A short section of rope or strap, one end of which is attached to a worker's safety harness and the other to a strong anchor point above the work area.

Lifting clamp: A device used to move loads such as steel plates or concrete panels without the use of slings.

Load: The total amount of what is being lifted, including all slings, hitches, and hardware.

Load control: The safe and efficient practice of load manipulation, using proper communication and handling techniques.

Load stress: The strain or tension applied on the rigging by the weight of the suspended load.

Lockout/tagout: A formal procedure for taking equipment out of service and ensuring that it cannot be operated until a qualified person has removed the lockout or tagout device (such as a lock or warning tag).

Management system: The organization of a company's management, including reporting procedures, supervisory responsibility, and administration.

Master link: The main connection fitting for chain slings.

Material safety data sheet (MSDS): A document that must accompany any hazardous substance. The MSDS identifies the substance and gives the exposure limits, the physical and chemical characteristics, the kind of hazard it presents, precautions for safe handling and use, and specific control measures.

Maximum intended load: The total weight of all people, equipment, tools, materials, and loads that a ladder can hold at one time.

Mid-rail: Mid-level, horizontal board required on all open sides of scaffolding and platforms that are more than 14 inches from the face of the structure and more than 10 feet above the ground. It is placed halfway between the toeboard and the top rail.

Occupational Safety and Health Administration (OSHA): An agency of the U.S. Department of Labor. Also refers to the Occupational Safety and Health Act of 1970, a law that applies to more than more than 111 million workers and 7 million job sites in the country.

One rope lay: The lengthwise distance it takes for one strand of a wire rope to make one complete turn around the core.

Pad eye: A welded structural lifting attachment.

Permit-required confined space: A confined space that has been evaluated and found to have actual or potential hazards, such as a toxic atmosphere or other serious safety or health hazard. Workers need written authorization to enter a permit-required confined space. See also *confined space.*

Personal protective equipment (PPE): Equipment or clothing designed to prevent or reduce injuries.

Plane: A surface in which a straight line joining two points lies wholly within that surface.

Planked: Having pieces of material 2 or more inches thick and 6 or more inches wide used as flooring, decking, or scaffolding.

Proximity work: Work done near a hazard but not actually in contact with it.

Qualified person: A person who, by possession of a recognized degree, certificate, or professional standing, or by extensive knowledge, training, and experience, has demonstrated the ability to solve or prevent problems relating to a certain subject, work, or project.

Rated capacity: The maximum load weight a sling or piece of hardware or equipment can hold or lift. Also called lifting capacity, working capacity, working load limit (WLL), and safe working load (SWL).

Rejection criteria: Standards, rules, or tests on which a decision can be based to remove an object or device from service because it is no longer safe.

Glossary of Trade Terms

Respirator: A device that provides clean, filtered air for breathing, no matter what is in the surrounding air.

Rigging hook: An item of rigging hardware used to attach a sling to a load.

Right angle: An angle that measures 90 degrees. The two lines that form a right angle are perpendicular to each other. This is the angle used most in the trades.

Right triangle: A triangle that includes one 90-degree angle.

Scaffold: An elevated platform for workers and materials.

Shackle: Coupling device used in an appropriate lifting apparatus to connect the rope to eye fittings, hooks, or other connectors.

Sheave: A grooved pulley-wheel for changing the direction of a rope's pull; often found on a crane.

Shoring: Using pieces of timber, usually in a diagonal position, to hold a wall in place temporarily.

Side pull: The portion of a pull acting horizontally when the slings are not vertical.

Signaler: A person who is responsible for directing a vehicle when the driver's vision is blocked in any way.

Slag: Waste material from welding operations.

Sling: Wire rope, alloy steel chain, metal mesh fabric, synthetic rope, synthetic webbing, or jacketed synthetic continuous loop fibers made into forms, with or without end fittings, used to handle loads.

Sling angle: The angle of an attached sling when pulled in relation to the load.

Sling legs: The parts of the sling that reach from the attachment device around the object being lifted.

Sling reach: A measure taken from the master link of the sling, where it bears weight, to either the end fitting of the sling or the lowest point on the basket.

Sling stress: The total amount of force exerted on a sling. This includes forces added as a result of sling angle.

Spliced: Having been joined together.

Stepladder: A self-supporting ladder consisting of two elements hinged at the top.

Straight ladder: A nonadjustable ladder.

Strand: A group of wires wound, or laid, around a center wire, or core. Strands are laid around a supporting core to form a rope.

Stress: Intensity of force exerted by one part of an object on another; the action of forces on an object or system that leads to changes in its shape, strain on it, or separation of its parts.

Switch enclosure: A box that houses electrical switches used to regulate and distribute electricity in a building.

Tag line: Rope that runs from the load to the ground. Riggers hold on to tag lines to keep a load from swinging or spinning during the lift.

Tattle-tail: Cord attached to the strands of an endless loop sling. It protrudes from the jacket. A tattle-tail is used to determine if an endless sling has been stretched or overloaded.

Threaded shank: A connecting end of a fastener, such as a bolt, with a series of spiral grooves cut into it. The grooves are designed to mate with grooves cut into another object in order to join them together.

Toeboard: A vertical barrier at floor level attached along exposed edges of a platform, runway, or ramp to prevent materials and people from falling.

Top rail: A top-level, horizontal board required on all open sides of scaffolding and platforms that are more than 14 inches from the face of the structure and more than 10 feet above the ground.

Trench: A narrow excavation made below the surface of the ground that is generally deeper than it is wide, with a maximum width of 15 feet. See also *excavation.*

Trencher: An excavating machine used to dig trenches, especially for pipeline and cables.

Warning-yarn: A component of the sling that shows the rigger whether the sling has suffered too much damage to be used.

Glossary of Trade Terms

Welding shield: (1) A protective screen set up around a welding operation designed to safeguard workers not directly involved in that operation. (2) A shield that provides eye and face protection for welders by either connecting to helmet-like headgear or attaching directly to a hard hat; also called a welding helmet.

Wind sock: A cloth cone open at both ends mounted in a high place to show which direction the wind is blowing.

Wire rope: A rope made from steel wires that are formed into strands and then laid around a supporting core to form a complete rope; sometimes called cable.

Industrial Coating and Lining Application Specialist Level One

Core Index

Index

Industrial Coating and Lining Application Specialist

69101-09
Introduction to the Trade

69102-09
Surface Preparation

69103-09
Industrial Coatings

69104-09
Coating Application

69105-09
Health and Safety, Debris Management, Containment, and Ventilation

69101-09

Introduction to the Trade

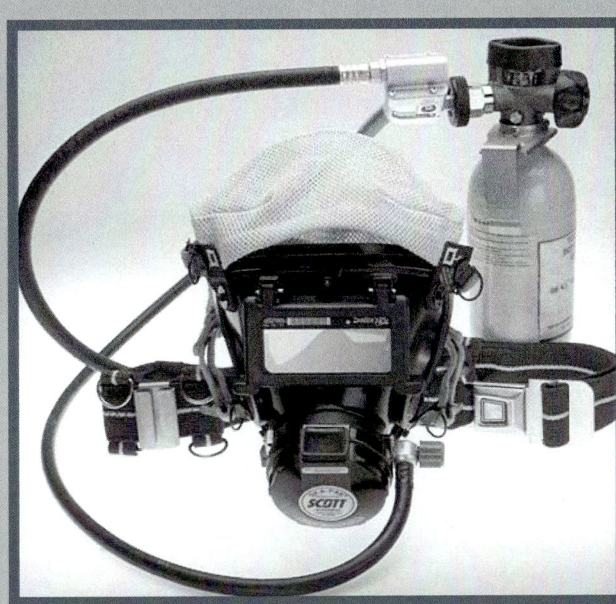

69101-09

Introduction to the Trade

Overview

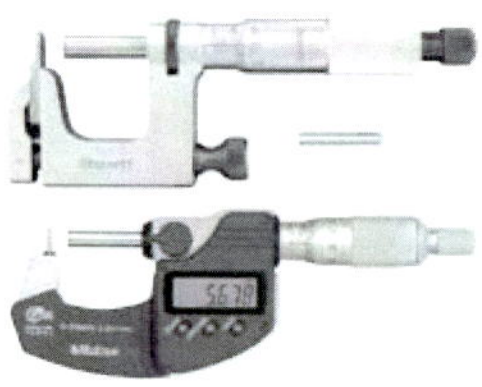

This training module is the first in a series of five modules that teach coating application skills. In this first module, *Introduction to the Trade*, the trainee will discover the composition and purpose of different industrial coatings. Additionally, this module defines and discusses the basic components of corrosion and the use of coatings as corrosion preventatives. Also, this module discusses the reasons for selecting a particular type of coating and introduces the various coating and surface preparation tools. Finally, this module covers the responsibilities of a person working in the industrial coatings profession, as well as keys to opportunities and success within the coatings industry.

Table of Contents

Figures

Tables

Objectives

When you have completed this module, you will be able to do the following:

1. Define the composition and purpose of different industrial coatings.
2. Discuss some of the causes of premature failure of coatings.
3. Define and list the components of process control.
4. Define quality control and quality assurance.
5. State the purpose of preparing test sections for demonstration processes.
6. Explain some of the preparation and application methods and how some coatings are used.
7. Identify the responsibilities of a person working in the industrial coatings profession.
8. State the personal characteristics of a professional.
9. Explain the importance of safety in working with industrial coatings.

Trade Terms

Cathodic protection
Corrosion
Corrosion inhibitors
Electrostatic spray
Fireproofing coatings
Galvanic corrosion
Material safety data sheet (MSDS)
NACE International
NACE-SSPC standards
NCCER
Plural component spray
Powder coatings
Process control
Protective coatings
Qualified person
Surface profile
Thermal spray

Required Trainee Materials

1. Pencil and paper
2. Appropriate personal protective equipment

Prerequisites

Before you begin this module, it is recommended that you successfully complete *Core Curriculum*, Modules 00101-04 and 00106-04.

This course map shows all of the modules in the first level of the Industrial Coating and Lining Application Specialist curriculum. The suggested training order begins at the bottom and proceeds up. Skill levels increase as you advance on the course map. The local Training Program Sponsor may adjust the training order.

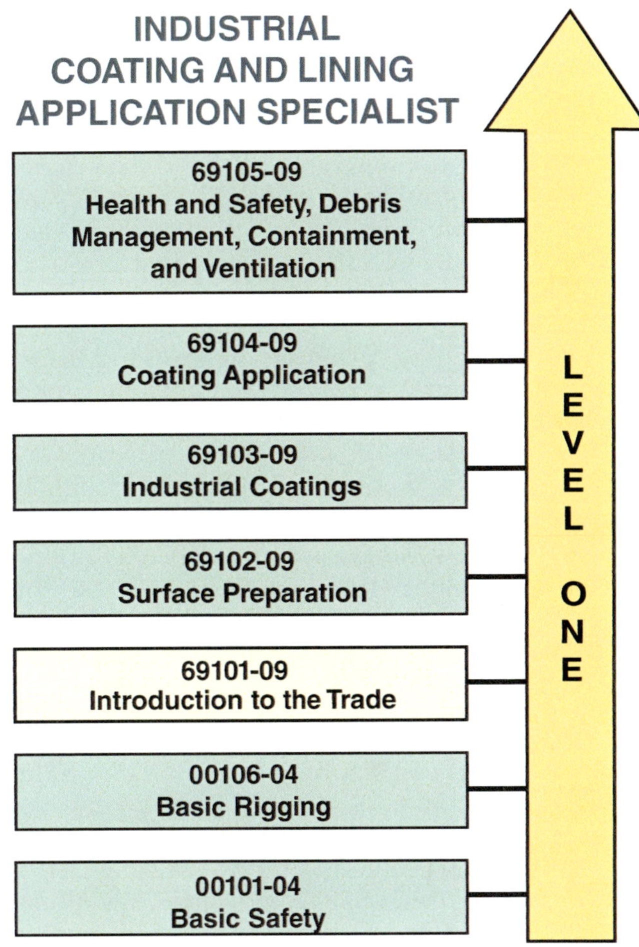

101CMAP.EPS

1.0.0 ◆ INTRODUCTION

This training program is brought to you by **NCCER** and **NACE International**, and is designed and developed by subject matter experts in the industrial coatings industry. The primary goal of NCCER is to standardize craft training throughout the country so that both employers and employees will benefit from the training. NACE International's mission is to protect people, assets, and the environment from the effect of **corrosion**, and NACE strives to do this through training and publication of corrosion control technical information. Together, NCCER and NACE bring you a program that fully supports the NACE No. 13/SSPC-ACS-1, *Industrial Coating and Lining Application Specialist Qualification and Certification.*

This written standard defines levels of training for all industrial painting personnel. It includes criteria for the knowledge, skills, and experience required of an application specialist for preparing surfaces and applying **protective coatings** to steel and concrete surfaces of complex industrial structures.

This is the only competency-based paint applicator training program that is accredited by NCCER and developed to serve a global workforce of painter personnel employed to prevent corrosion of industrial structures. A competency-based program requires that the trainee demonstrate the ability to safely perform specific job-related tasks in order to receive credit. This approach is unlike other apprentice training programs that only focus on completion of a certain number of hours in the classroom and on the job.

The goal of the NCCER-NACE Industrial Coatings Applicator Training Program is to set a globally recognized standard for training of paint applicator personnel so that all benefit from the training, no matter their location or the industrial structures being painted. A trainee in an NCCER-accredited program can become part of a National Registry for craft workers, and can receive a certificate for each level of training completed (see *Appendix A*). If the trainee applies for a job with any participating contractor, a transcript of training will be available. The continuity of this program is portable. That is, trainees who begin employment with another participating contractor may continue their NCCER-NACE training where they left off, with an uninterrupted transcript from the National Registry.

This NCCER-NACE program is developed in accordance with Department of Labor Office of Apprenticeship requirements for related classroom training (*29 CFR 29*: et seq.), and as such, can also be registered as formal apprenticeship training. In addition, many technical schools and colleges are using the program.

There are three skill levels in this program. This first level, *Industrial Coating and Lining Application Specialist Level One* (hereafter referred to as Coating Applicator) requires the trainee to demonstrate a basic knowledge of safety, **process control**, surface preparation of material to be coated, and coating application. Additionally, the Coating Applicator course requires that the trainee demonstrate the correct procedure for all of the above, with the understanding that the competencies are geared toward an entry-level worker. This program includes these specialty areas:

- Coating concrete, which requires protection against mechanical weathering from friction, impact and the water freezing cycle. Concrete should be sealed to protect it from acidic material and water.
- High pressure water cleaning and ultra high-pressure waterjetting are cleaning procedures that use water under pressure to clean and remove foreign surface material from any surface material that can withstand the water jets.
- **Electrostatic spray**, which is the process that uses charged coating particles and a grounded workpiece. The particles stick to the workpiece by way of an electric bond.
- **Plural component spray**, which is a mixture of multiple components delivered to the nozzle and applied to the workpiece in precise ratios. Two part epoxies and polyurethane coatings are examples of coatings that are applied with plural component spray.
- **Powder coatings**, which may be applied using fluidized bed or dipping systems, cloud chambers and electrostatic spraying.
- Specialty coatings, which include coatings designed for a particular circumstance or environment.
- **Thermal spray**, which are coatings applied as molten granules sprayed against a workpiece. The coating particles flatten themselves by impact and stick to the surface. One application of thermal spray is in coatings applied to some ceramics.
- **Fireproofing coatings** are a category of coatings that are unique. Some are called intumescents; they turn to foam when exposed to temperature above a certain level. Other fireproofing coatings are really fire resistant. Fire resistant coatings penetrate beneath the surface of a porous material and prevent heat transfer and ignition. Finally, a third type of fireproofing coating contains fire resistant fibers.

2.0.0 ◆ PREVIEW OF INDUSTRIAL COATINGS

Industrial coatings are used everywhere. Every industrialized area has a need for this profession. Every day, new types of coatings and processes are developed and skilled workers are required to apply these coatings.

There are four primary types of paint and coatings products used to preserve, protect, and beautify:

- Interior paints, primers, sealers and varnishes are used in commercial buildings and private homes.
- Coatings that are factory applied to manufactured goods as part of the production process.
- A third type is industrial coatings, such as marine paints, high-performance maintenance coatings, and automotive refinish paints; and
- Finally, allied paint products, including putties, paint and varnish removers, paint thinners, pigment dispersions, and paint brush cleaners.

Industrial coatings and services are needed for a wide variety of reasons. Some coatings provide protection from weather exposure and rust. Still others disperse heat and reduce heat transfer, reduce friction, and retain or shed chemical petroleum products. Some types of coatings can be thinned to nearly any viscosity, and can be used as dips or sprays. Some powder coatings provide good scratch resistance and protect against damage from chemicals. Some types of coatings create a hard and long lasting surface. Finally, some coatings are flexible, durable, and resistant to UV radiation.

3.0.0 ◆ DEFINITION OF A COATING

A coating is a clear or pigmented film-forming material that protects the surface to which it is applied from the effects of the environment. Many chemicals are used in the formulation of a coating. Sometimes specialty chemical components are added to a coating to allow it to meet specified conditions.

Conventional coatings have two major components: pigment and vehicle. A pigment is composed of fine solid particles added during the manufacture of a coating. Pigments are substantially insoluble in the liquid portion of the coating and are used to impart color, corrosion control, or decorative properties. A vehicle is a combination of resin and solvent and composes the liquid portion of a formulated coating.

3.1.0 Reasons for Coating

There are a number of reasons for using coatings. They include the enhancement of visual appearance, mechanical protection, and protection from rust, outdoor elements, biological attacks, and chemical attacks.

3.1.1 Visual Appearance

Coatings present attractive colors and surfaces for aesthetic value. Coatings may also be color-coded to indentify contents of structures or systems, or used to provide warnings of hazardous areas.

3.1.2 Mechanical Protection

Coatings reduce wind erosion, friction and wear. Coatings are used to provide electrical and heat insulation, as well as sound insulation or deadening. Coatings also reduce water penetration on a surface. Coatings can help to smooth a surface of a material, making the flow of a product such as a liquid through a piping system, more efficient.

3.1.3 Corrosion Protection

NACE defines corrosion as "the deterioration of a material, usually a metal that results for a reaction with its environment." Typically, the most important reason for coating an industrial assembly is to protect the primary metal structure from the effects of corrosion. Later in this training module corrosion is discussed further. For now, corrosion is usually an electrochemical reaction that occurs when the surface of a metal interacts with its environment, such as rainwater, seawater, and different chemicals. Coatings provide a barrier between the metal and the environment and slow the corrosion process.

3.1.4 Biological Protection

Coatings are used on ships' hulls to protect them from barnacles, teredos, zebra mussels, and other marine organisms that can damage the hull or reduce the hydrodynamic efficiency (see *Figure 1*) of the hull moving through the water.

Coatings on pipeline systems can also serve the purpose of protecting the metallic piping from microbiological influenced corrosion (MIC). This form of corrosion can occur in the underground pipelines as they are attacked by soil microorganisms. Finally, coatings on porous materials, such as wood or plaster, protect the materials from contamination by biological or chemical agents.

101F01.EPS

Figure 1 ◆ Ship's hull prepared for coating.

3.1.5 Chemical Protection

Coatings are a critical component in the protection of both the outside and the inside of many structures. Structures that hold chemicals or other materials, including above and below ground water and fuel tanks, ballast tanks on ships, and storage tanks at various processing facilities, they all require corrosion protection on the inside (a lining) and the outside (a coating). Coatings provide this corrosion protection, and also insure that rust from and unprotected surface does not contaminate the stored product.

3.1.6 Environmental Protection

Coatings serve as a barrier, reducing the environment exposure of the metal assembly to which they are applied. Many different environments present special environment exposure hazards and they are generally recognized as chemical and marine; chemical and high humidity; marine and high humidity; chemical and low humidity, and rural and low humidity. For example:

- Chemical and marine environments are found in offshore oil and gas production platforms.
- Chemical and high humidity environments are found in near coast chemical and petroleum refineries.
- Marine and high humidity environments are found in splash zones in offshore platforms.
- Chemical and low humidity environments are found in inland production including chemical plants and power generation facilities.
- And finally, rural and low humidity environments are found in inland areas where the highway and bridge infrastructure compose the largest commercial structures.

4.0.0 ◆ CORROSION OVERVIEW

As previously mentioned, NACE International defines corrosion as the deterioration of material, usually a metal, because of a reaction with its environment. This definition is very general, and it recognizes that materials other than steel, such as concrete, wood, and plastics, deteriorate or corrode.

Corrosion and its product, rust, are the result of natural processes that return refined materials of a higher energy state to their original lower energy state. Just about any material, if exposed to the elements long enough, will corrode. For instance, iron (steel), exposed to air and water, will rust in a few hours, sometimes minutes, depending on the situation.

Everyone in industry today (the plant owner, refinery operator, mechanical engineer, chemical engineer, and so on) is affected by corrosion and, as a result, looks for ways to prevent material from returning to its natural, unusable state. The desire to protect all types of materials –wood, concrete, steel, cast iron, stainless alloys, aluminum, plastics, ceramics, and so on – promotes the need for a basic understanding of corrosion.

4.1.0 Basic Chemistry and Corrosion

All corrosion of iron at normal ambient conditions is an electrochemical process. Simply put, this means that the process involves the transfer of ions and electrons across a surface. Transfer of electrons implies the generation of a current (corrosion current). Both electrons (through a metallic conductor) and ions (through an electrolyte) carry the corrosion current.

Corrosion is established as direct current (DC) circuits. DC circuits are defined by the relationship called Ohms Law, $E = IR$, where E is the driving voltage of the circuit, I is the current magnitude, and R is the resistance of the circuit. The greater the current flow in the corrosion circuit, the greater the metal loss.

4.2.0 The Corrosion Cell and Its Components

The corrosion cell is made up of four parts: the electrolyte, the anode, the return path (also known as the metallic pathway), and the cathode. The electrolyte is a liquid or corrosive medium that conducts electricity. The majority of electrolytes are based on water. The return path is a metallic pathway that carries negative charges, in the form of electrons, between the anode and the cathode. The anode is that part of the metal that corrodes (i.e., dissolves in the electrolyte). The anode is generally steel and is generally coated to help protect it from corrosion. The cathode is the less-active area on the steel where corrosion does not take place.

All four of the above components must be present for corrosion to occur. Removing one or more of them will prevent corrosion from occurring. It is not always possible or practical to remove these components, but the attempt at removal is referred to as corrosion control.

The corrosion cell is illustrated in *Figure 2*.

4.3.0 Forms of Corrosion

Corrosion may take place generally or locally. If a piece of uncoated steel is immersed in a suitable electrolyte it is likely to create generalized corrosion over a large surface area. If a piece of steel is also in contact with a different metal then the point of contact between the steel and the other metal will likely generate the more severe localized corrosion. Corrosion may take many forms, including the following:

- Generalized corrosion, in which large surfaces or entire objects corrode. An example of generalized corrosion is a piece of sheet steel that is exposed to the atmosphere. The sheet steel will rust over its entire surface. The most useful prevention of generalized corrosion is a good surface coating.
- Localized corrosion typically includes pitting and crevice corrosion. Pitting is a type of localized corrosion in which deep localized pits form in the surface of the structure. These pits weaken the integrity of the structure. Crevice corrosion occurs at places where there is a small, enclosed space where an electrolyte is likely to stay in constant contact with the surface of the structure.
- **Galvanic corrosion** occurs when two unlike metals are touching one another in the presence of an electrolyte (usually water). Recall the previous description of the corrosion cell. Galvanic corrosion is a classic example of a corrosion cell. When galvanic corrosion occurs, one of the two metals deteriorates. Specifically, it is the more negative metal that corrodes. The metal that will corrode first can be determined by its place on a list of metals (*Figure 3*), running from the least likely to corrode to the most likely. The metals least likely to corrode are also described as the more noble metals.

  If a less noble metal is placed in contact with a more noble metal in the presence of an electrolyte, the less noble metal will corrode and the more noble metal will not. Thus, if a copper penny is lying on a piece of cast iron in salt water, the cast iron will probably corrode in the area close the copper. Copper is more noble than cast iron.

  The preferred way to prevent galvanic corrosion is to place a non-conductive material between the two unlike metals.
- Environmental cracking is a form of corrosion that happens relatively quickly. It is a combination of tension and corrosion and appears as a set of cracks at right angles to the direction of maximum tension. Environmental cracking can cause serious failures.
- Corrosion in concrete and steel rebar (*Figure 4*) that is used to reinforce concrete is a topic of special interest because concrete is the most widely used building material. Concrete forms must of the world's highway infrastructure. Concrete is porous so it allows water and chemicals to penetrate its surface. Water may freeze and split concrete. Chemicals may break apart

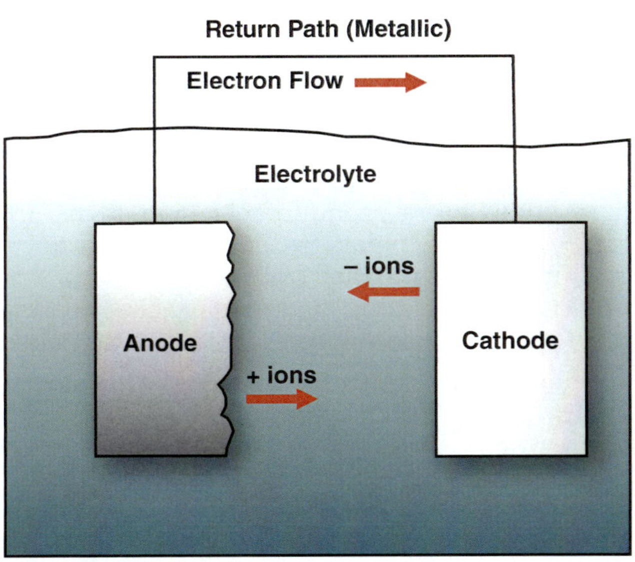

Figure 2 ◆ Corrosion cell.

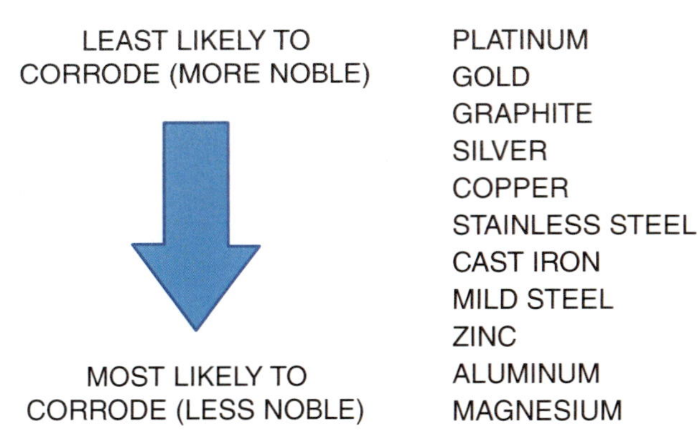

Figure 3 ◆ List of metals.

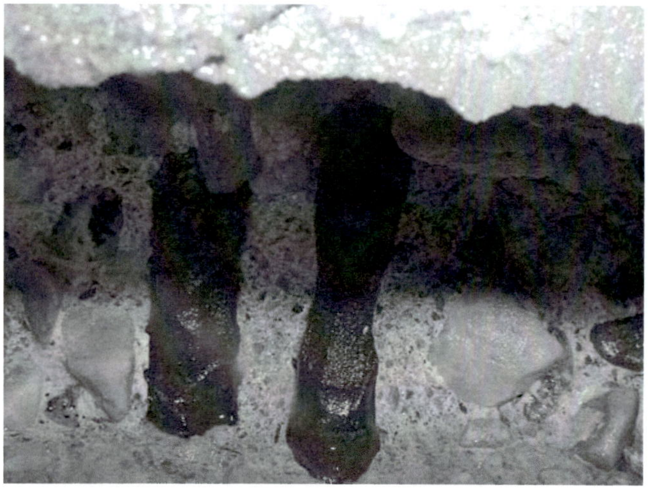

101F04.EPS

Figure 4 ◆ Concrete reinforcement bars in the bottom of a cracked column.

the chemical structure of the concrete. Steel rebar may rust from exposure to chemicals and water and the rust product may expand the size of the reinforcing bar and break or spall the concrete around it. To reduce the porosity of concrete several coating systems have been developed for the surface of concrete. Additionally, coating systems for rebar are now widely used.

A thorough discussion of the many forms of corrosion is beyond the scope of this class. The preceding text is simply a good reference for the professional coating applicator. A level one coating and lining applicator must possess basic knowledge of industrial coatings and linings and should be able to identify rust. It is also important to know where to take special precautions to prevent the failure of the coating or lining that is being applied.

4.4.0 Techniques for Corrosion Control

No single method or technique can control all corrosion problems. Each situation must be examined individually. However, several options can be used to control corrosion, including:

- Protective coatings
- Material selection
- Design
- Cathodic protection
- Corrosion inhibitors
- Alteration of the environment

4.4.1 Protective Coatings

Protective coatings provide corrosion protection by one or a combination of three methods. They can act as a barrier, an inhibitor, and/or a sacrificial anode.

Sacrificial coatings generally contain a zinc compound. These coatings function as anodes and corrode before the steel surface that they protect.

Barrier coatings keep electrolytes (generally water) from reaching the surface of the material they protect. They effectively removed the electrolyte. If you recall, the electrolyte is one of the four required components that compose a corrosion cell. By removing the electrolyte from the corrosion cell, corrosion will be slowed. It is impossible to completely eliminate corrosion by use of a barrier coating. Barrier coatings are semi-permeable even if they are applied perfectly. Perfect application of coating is always the goal of any applicator, but it is a goal that is not possible to achieve.

Inhibitive coatings are usually primers, such as zinc phosphate. These particular primers react with moisture and with steel to form a thin passivating layer.

4.4.2 Material Selection

If designers select slow-corroding materials (i.e., more corrosion resistant) and do not allow dissimilar metals to contact one another, corrosion may be retarded. If it is not possible to use more corrosion-resistant materials, then they must be electrically insulated.

4.4.3 Design

Design can help to slow the corrosion process. The key lies in an understanding of the effects of shapes and coatings. Coatings tend to pull away from sharp edges as they dry, thus; reducing coating thickness at sharp angles. This fact makes sharp edges more prone to corrosion. Sharp inside angles tend to gather dust and dust attracts moisture. Moisture that continually comes into contact with a coating will damage that coating over time. Corrosion will result from a damaged coating.

Welded joints are easier to coat than bolted joints. Welded joints, properly smoothed through grinding, are much smoother than bolted joints. Therefore, welded joints are easier to coat that bolted joints.

4.4.4 Cathodic Protection

Cathodic protection (CP) is another corrosion control technology used to prevent corrosion of underground pipelines and storage tanks, rebar in road surfaces and submerged pilings, ships' hulls, offshore platforms, storage tank bottoms, and other structures.

Recall from the discussion on galvanic corrosion that there are four essential elements of a corrosion cell:

- Anode
- Cathode
- Metallic pathway or conductor
- Electrolyte or moisture

Recall also that there is an electrical difference between the anode and the cathode, and that corrosion occurs at the anode. CP, then, involves manipulating the electrical current in order to make a particular surface or structure cathodic, thereby protecting it from corrosion (hence the term cathodic protection). CP is achieved through the use of sacrificial anodes and/or electrical current controllers.

4.4.5 Corrosion Inhibitors

Corrosion inhibitors are chemical substances that can be added to an environment. The chemical nature of these substances is such that the corrosion reaction is slowed.

4.4.6 Alteration of the Environment

Finally, alteration of the environment can reduce the corrosion rate. An example is the use of dehumidifiers to prepare tanks for abrasive blasting. At relative humidities below 40 percent, the rate of corrosion approaches zero.

4.5.0 Potential Causes of Coatings Failure

Coating failure occurs when the coating no longer provides the service or protection for which it was intended.

Premature coating failure may be defined as deterioration of the coating system, or corrosion of the substrate, more rapidly than would be expected in a particular service condition.

Some of the most common causes of coatings failures are discussed in the paragraphs below.

4.5.1 Poor Surface Profile

A **surface profile** is difficult, if not impossible to see with the naked eye. Surface profile is defined as the difference between the peaks and valleys on a surface. While the peaks and valleys are not visible, they are important to achieve good sticking power for a coating.

When the surface profile of the substrate is too smooth, the coating may be unable to bind to the surface. At the same time, if the surface is too rough, the coating has to be very thick to cover all of the exposed material. This includes by extension the presence of protrusions, including rivets and bolts as well as bumps in the surface itself. Specific application techniques are used to deal with protrusions, such as first coating the intersection of the protrusion and the surface (for example, the place where the rivet or bolt meets the face of the plate) and then coating the surface and the rest of the protrusion. Tools used for determining the condition of the surface profile include the surface comparator, replica tape, and the anvil micrometer (*Figure 5*).

4.5.2 Surface Contamination

The surface to be coated must be clean before the coating can be applied. Rust, grease and oil, mill scale, or other contaminants can produce coating failure if they are not removed prior to coating. A number of techniques are available for eliminating visible contaminants from a surface, including waterjetting, pressure washing, chemical cleaning, and solvent cleaning.

4.5.3 Condensation

The presence of condensation is one of the criteria in scheduling coatings. The presence of dew at early morning can prevent coatings from sticking to some materials (*Figure 6*). The other related problem with water and ferrous metals is rusting, especially flash rusting after pressure washing or waterjetting. Rusting can begin within minutes of the completion of a cleaning.

5.0.0 ♦ TOOLS OF THE TRADE

All trades have a range of tools with which the craft workers do their jobs. These include hand tools and power tools, but also include documentation tools and production tools. When workers apply coatings, various methods of application are used. Each will be discussed in detail in following chapters, but an introduction occurs here.

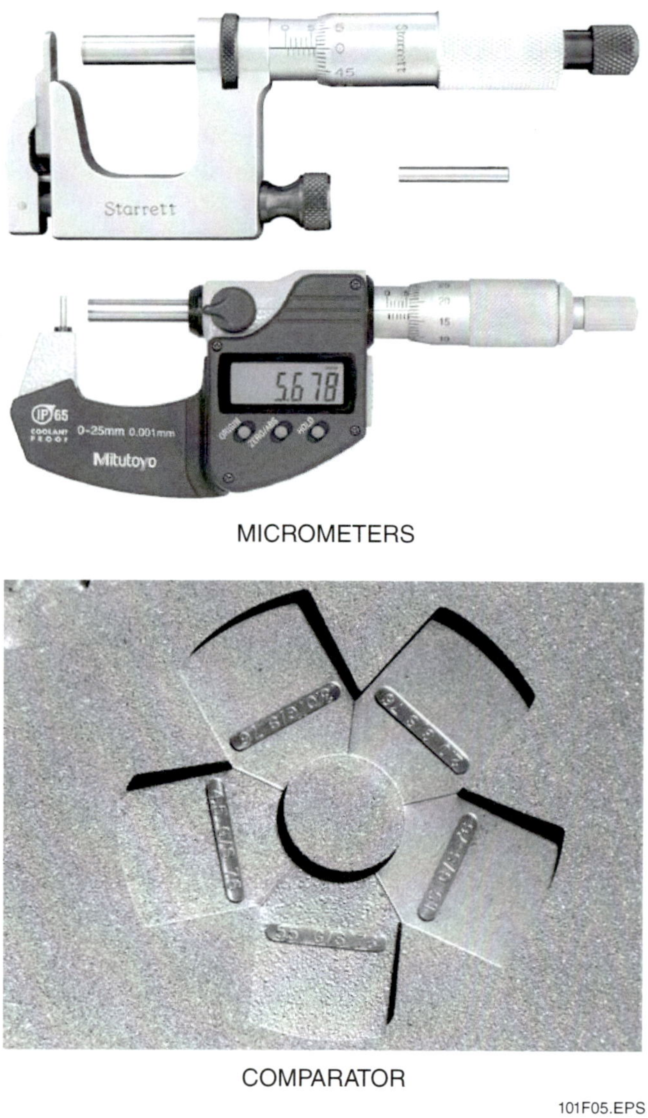

Figure 5 ◆ Equipment for testing surface profile.

Every tool has a unique set of personal protective equipment (PPE) that must be worn in order to operate the tool in a safe manner. Gloves, eye, hearing, and foot protection are usually a minimum on a job site. Remember that working safely is never an option.

5.1.0 Hand Tools for Surface Preparation

5.1.1 Hand Abrasives

The industrial coatings technician uses many kinds of abrasives. Some hand tools used for abrasive work are wet and dry abrasive pads. The particular type depends on the specific application. To roughen coats of paint to improve adhesion, steel wool or emery cloth are common forms of hand abrasives.

5.1.2 Scrapers and Putty Knives

One of the most universal hand tools is the putty knife (*Figure 7*). It is used as a scraper or application tool in nearly every trade. There are many variations of this tool, including the five-in-one, or various types of scrapers. Scrapers and putty knives are used in various ways to remove or apply material in different contexts.

5.1.3 Wire Brushes

The wire brush (*Figure 8*) is used either to make cuts on a surface to prepare for a coating application, or to remove surface corrosion or soil before applying a coating.

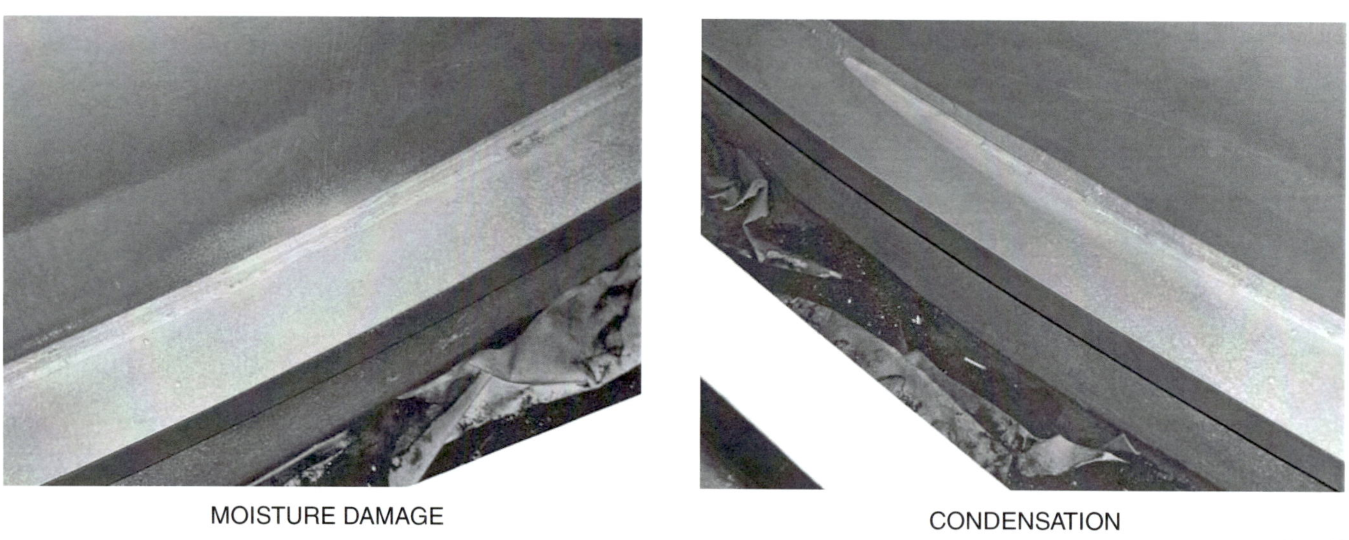

Figure 6 ◆ Condensation and the result.

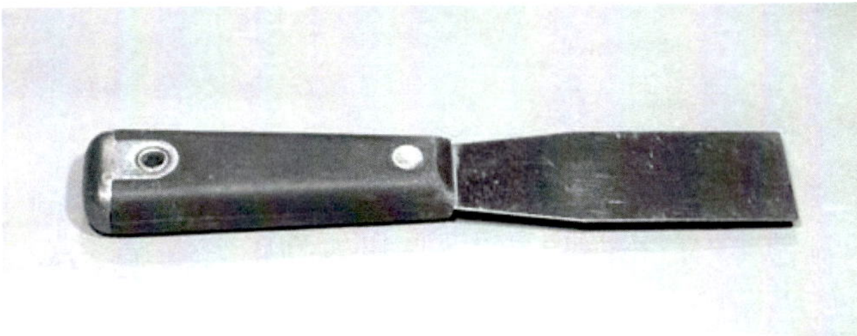

101F07.EPS

Figure 7 ◆ Putty knife.

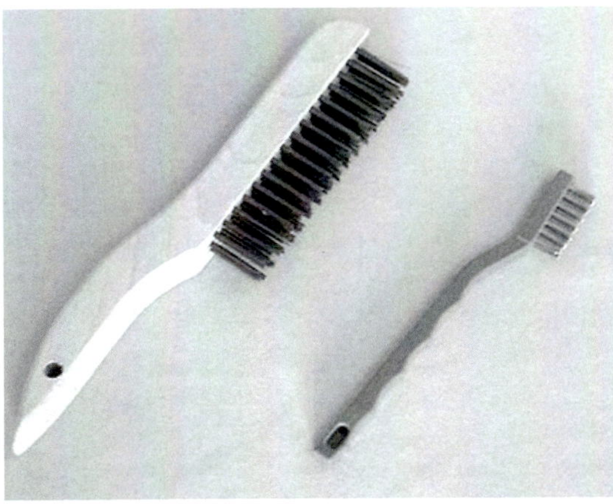

101F08.EPS

Figure 8 ◆ Wire brushes.

5.1.4 Chipping Hammer

Chipping hammers (*Figure 9*), such as the ones used to remove welding slag, are also used for chipping loose, relatively brittle coatings from a surface to be coated.

5.2.0 Power Tools for Surface Preparation

Power tools are often used in the coatings industry, and special ones are developed for almost any operation. The question of whether to use hand or power tools depends on the scale of the operation. When the power tool is the better choice, some precautions must be taken.

All power tools require safe practices. Any tool that can remove paint can remove skin just as easily. Wire wheels and grinders may catch in angles or on irregularities and spin out of control. The wire wheels can throw pieces of spring steel wire with such force as to penetrate clothing and skin, and grinders and sanders can cut through flesh and bone. The sparks from grinders are capable of igniting flammable materials.

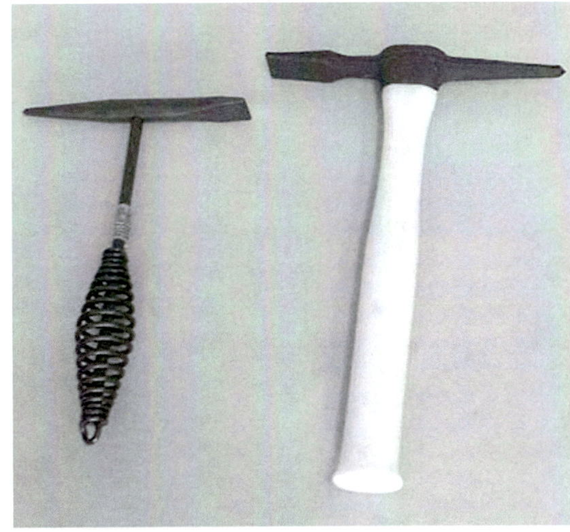

101F09.EPS

Figure 9 ◆ Chipping hammers.

5.2.1 Grinders and Sanders

Power sanders and side grinders (*Figure 10*) are generally used to remove coatings or to clean up and smooth surfaces. They are fairly fast-acting. Always remember that they can cause injury at the slightest contact while running. Make sure the tool is stopped before setting it down.

5.2.2 Power Wire Brushes

There are several versions of the wire wheel or power wire brush (*Figure 11*). They are either run on something like a drill or on a side grinder machine. These are powerful machines, capable of removing coatings or paint.

5.2.3 Needle Guns and Air Chisels

Needle guns (*Figure 12*, top) are usually found in surface preparation work. They can efficiently remove coatings or weld slag in small areas. They are most commonly pneumatic tools, similar to the air chisel (*Figure 12*, bottom).

5.2.4 Vacuum Shrouds

Vacuum shrouds are normally attached to grinders or other surface preparation tools, allowing dust and chips to be contained safely. This is critical in cases where toxic substances, such as lead-based paints, are being removed.

101F10.EPS

Figure 10 ◆ Side grinder.

101F11.EPS

Figure 11 ◆ Power wire brushes.

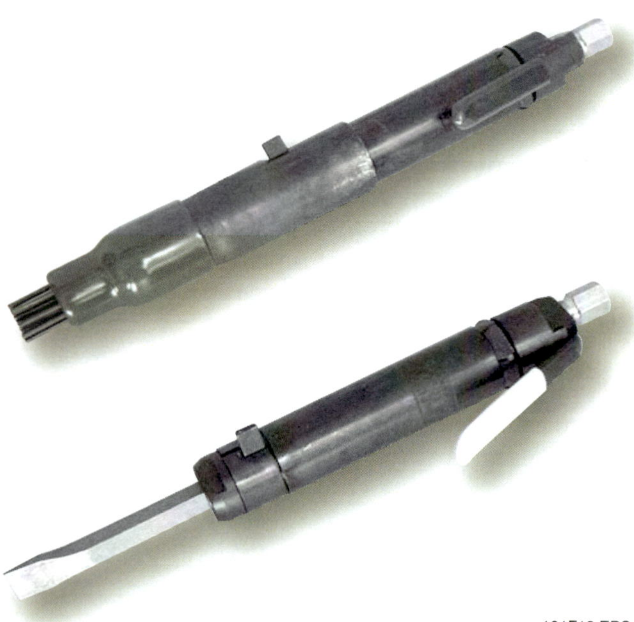

101F12.EPS

Figure 12 ◆ Needle guns.

6.0.0 ◆ PROCESS CONTROL

A key component of a successful process of tasks, sometimes called process control, is a written procedure to maintain consistency from job to job. All jobs should have a written work plan that utilizes the process control procedure. Because no two jobs are exactly alike, the work plan may vary, but procedures in the work plan should be based on the process control plan.

Some items that should be included in the process control plan include both a startup check list and a daily check list. The following is a list of essential plans:

1. *Maintenance Plan* – Prior to starting each job, there should be a check list for every piece of equipment to ensure all equipment is working properly. This should include mechanical equipment (compressors, air coolers, blast pots, etc., as well as non-mechanical equipment such as lighting, air hoses, blast nozzles, etc.).
2. *Scaffolding Plan* – Whether the contractor is doing its own scaffolding or using a subcontractor, scaffolding should be checked prior to use to make sure it is safe and complies with OSHA or other local requirements.
3. *Containment Plan* – The containment plan should be approved by a Professional Engineer (PE), Certified Industrial Hygienist, or other **qualified person** to make sure it complies with the specifications.
4. *Abrasive Blasting Plan* – Procedures for checking grit, recycle plan, air contamination check, type and size of grit, blast nozzle condition, hose condition, etc.
5. *Mixing and Application Plan* – Check list to make sure proper thinner is used, proper amount, pot life, induction time, etc.
6. *Quality Control Plan* – Inspection plan to be followed by the contractor. This should be similar but separate from the inspection plan.

Sometime during a project, a demonstration of one or more of the project processes will be performed. For example, a contractor will be asked to prepare an area of the steel to the specified surface preparation/blast requirement so that all involved can agree that the cleanliness prepared is what is being specified. Besides demonstrating the ability of the contractor to meet the specifications, this area can be preserved and be referred back to if there are any disagreements during the project.

A page from a sample work plan is included in *Appendix B*.

6.1.0 Quality Programs

As the work is being done in accordance with the work plan, another part of process control is implemented: quality control (QC) and quality assurance (QA). Coatings must be applied in a manner that meets contractor and owner requirements as well as applicable standards. Professional high quality work is verified first by the applicator through a process of quality control. Qualified inspectors then follow the quality control inspections with a quality assurance inspection.

All parts of the project are under the QC umbrella and are the responsibility of the applicator first and ultimately the contractor. Work that is done must meet or exceed the minimum standards set out by the contract and/or project specifications. This work must be observed, tested, and documented (reported) to verify that requirements are met. This responsibility includes workforce training and experience, equipment and materials, and the quality of the final product, including every aspect of the surface preparation and painting process.

Some common standards used for industrial coatings are from ASTM International, Code of Federal Regulations (CFR), Federal Standards (FEC-STD), NACE International, The Society for Protective Coatings (SSPC), and the International Organization for Standardization (ISO).

Quality assurance is an audit process to verify that the work performed by the contractor meets established standards, and that the QC report accurately reflects the work. The QA process is the responsibility of the owner or the owner's representative. QA can be done by spot testing, by observing the QC process, or by reviewing QC documentation.

The exact sequence of events will vary from job to job. However, there are certain places (hold points) in every job where inspection is required when that part of the project is completed. For example, after surface preparation the work should be inspected and, if necessary, rework performed. Before the contractor can continue to the next step, the QA/QC inspectors must accept the surfaces. *Table 1* is an example of a set of hold point inspections for a specific job.

7.0.0 ♦ COATING PROFESSIONAL ROLES

There are various levels of expertise and responsibilities within the industrial coating profession; the following is a brief description of some titles on the job site. It cannot be stressed forcefully enough that, no matter the job title, personal safety is always of paramount importance. Many job sites may specify a role of Safety Engineer or Safety Specialist, but safety is always a personal responsibility.

7.1.0 Safety

All people on a job site must understand and follow plant and contractor safety rules. Your job site and your contractor may have different safety rules.

7.2.0 Helper

A helper must know when and how to use the required PPE (such as respirators, eye protection, hearing protection, etc.) Additionally, a helper must understand plant and contractor safety rules. Also, a helper shall:

- Safely use hand or power tools to perform required surface preparation.
- Be able to apply coatings by brush (such as stripe coating).

7.3.0 Abrasive Blaster

An abrasive blaster shall have experience in the maintenance coating/painting field. And in addition, shall:

- Be able to set up equipment, using proper tools.
- Be able to differentiate between blast finishes White Metal, Near-White Metal, Commercial, and Brush-Off Blast Cleaned Surface Finish.
- Achieve proper, efficient flow of abrasive, use properly sized nozzle and hose, and check for nozzle wear.
- Recognize and correct blasting deficiencies.
- Know and use proper signaling techniques.

Table 1 Division of QC and QA Hold Point Inspections

Hold Point Inspection	Quality Control	Quality Assurance
Ambient conditions (pre-preparation)	TD	TD
Pre-surface preparation conditions (welds, rust, edges, etc.)	VO	VO
Pre-surface preparation equipment conditions (grease, dust, etc.) and/or SP-1	V	V
Surface preparation equipment operation and abrasive size/type	VD/TP	RD/VO/TP
Abrasive cleanliness	TP	RP/TP
Compressed air cleanliness	TD	RD/TP
Surface cleanliness	TO	TO
Surface profile	TO	TO
Soluble salt contamination	TO	RO/TP
Soluble salt remediation	VO	VO
Ambient conditions (pre-mixing and application)	TD	TD
Mixing and materials (batch nos., pot life, thinner, etc.)	VO	RO/TP
Application equipment operation (pressures, agitation, type, etc.)	VD/TP	RD/VD/TP
Compressed air cleanliness (spray application)	TD	RD/TP
Wet film thickness	TO	RO/TP
Stripe coat, caulking, intercoat cleanliness, etc.	VO	RO/VO
Dry film thickness	TO	TO
Visual appearance (runs, drips, sags, etc.)	VO	VO
Repairs	VO	VO

Legend:

Frequency of Verification	Type of Verification
D = Daily	V = Visual
O = Occurrence-based	T = Testing
P = Periodic	R = Review of QC report or testing

101T01.EPS

7.4.0 Painter/Coating Applicator

A coating applicator shall have experience in the maintenance coating/painting field, and in addition shall:

- Possess the ability to perform the duties of the Helper.
- Be able to properly mix coatings.
- Know how to select hand application tools and demonstrate proper use techniques.
- Be able to use a wet film thickness gauge.
- Know how to set up spray equipment.
- Know spray equipment requirements and demonstrate proper spraying techniques.
- Be able to safely assemble, operate, and use rigging necessary to perform the work.
- Possess the ability to perform the duties of the abrasive blaster.

7.5.0 Linings Applicator

The linings applicator shall have a minimum of experience in the maintenance coating/painting field and lining experience. In addition, the Linings Applicator shall:

- Possess the ability to perform the duties of the painter.
- Be able to apply specialty coatings requiring high skill level and applicator ability (e.g., phenolic, high-heat coatings, etc.).
- Be knowledgeable in and able to set up, run, and maintain dehumidification, heating, and ventilation equipment.

7.6.0 Foreman/Lead Man

The foreman/lead man should be fully responsible for successful completion of all field activities, and be responsible to ensure that planning and documentation for safety, production, and quality are performed as required. The foreman is responsible for the contractor's effectiveness for the duration of the job. The foreman must assure that all work is done properly and in accordance with specifications.

The responsibilities of the foreman include, but are not necessarily limited to, the following:

- Performing pre-job safety checks and conduct crew safety meetings on a regularly scheduled basis.
- Performing work in accordance with specifications.
- Communicating with, and directing the efforts of, work personnel.
- Maintaining quality assurance (QA) and quality control (QC) (including inspection and documentation) of the work throughout all phases of the project.
- Keeping up-to-date, accurate time records, daily QC/work reports, and ensure that manhours, materials, and quantities are properly reported in accordance with the planning and control section of the costing specification.
- Using coating inspection equipment such as wet- and dry-film thickness gauges, profile gauge, psychrometer, humidity and dew point charts, coating and surface thermometers, etc.

7.7.0 Superintendent

The superintendent shall assume full responsibility for crew effectiveness over the entire job from start to finish, assuring that all work is done properly and in accordance with the specifications. The superintendent should possess the capabilities and responsibilities for planning, scheduling, estimating, and project management of the project. The superintendent shall be responsible for ensuring all safety and environmental requirements are met. The superintendent will be responsible for verifying that quantities, hours, and cost information are reported correctly.

8.0.0 ◆ KEYS TO PROFESSIONAL SUCCESS

A successful application specialist produces finished products of high quality, in a timely manner and on budget. This means that the job is completed by the deadline, and that it does not cost more than what was originally expected.

Successful applicators must be flexible enough to adapt their methods of working to the particular needs of each job. They must also improve their knowledge of the different tools and materials needed to accomplish certain jobs. Over time, successful applicators will improve their ability to do their jobs on time, on budget, and to the satisfaction of their customers. Satisfied customers mean repeat business.

Keep current with the latest methods, materials, and equipment in the coatings profession. This knowledge will help you and the company you work for stay competitive. You should never stop learning. When you complete your apprenticeship, you will simply be moving into a bigger classroom: the world.

Safety on the job site is a personal responsibility. Always demonstrate a high level of professionalism. Coating applicators are part of a proud heritage, and are vital to modern society. Coating applicators should create customers who think well not only of individual work, but of the profession in general.

Industrial coating professionals stick to a code of professional ethics. Ethics are principles and values that guide conduct.

Successful coatings applicators exhibit all of the following qualities:

- Positive attitude
- Willingness to learn
- Attentiveness to rules and regulations
- Promptness and reliability

THINK ABOUT IT

Late for Work

Showing up on time is a basic requirement for just about every job. Your employer is counting on you to be there at a set time, ready to work. While legitimate emergencies may arise that can cause you to be late for or even miss work, starting a bad habit of consistent tardiness is not something you want to do. What are the possible consequences that you could face as a result of tardiness and absenteeism?

8.1.0 Positive Attitude

Motivation, attention to detail, and high energy describe a person with a positive attitude. These people understand the importance of safety and are flexible to change. A positive attitude will help the professional coating applicator fulfill obligations and take pride in the job.

A positive work environment is a product of positive attitude and leads to greater productivity. Employers pay attention to job place attitude.

8.2.0 Willingness to Learn

A good coatings applicator never stops learning. New machines, tools, materials, and methods are constantly being introduced. Even experienced applicators need to be willing to learn about these advancements. Good coatings applicators keep their methods and tools up-to-date. Workers with current or advanced skills help their company compete successfully with other companies. The better a company competes, the more profit it makes. Profitable companies provide jobs, benefits, and salaries to the professional coatings applicator workforce.

8.3.0 Attention to Rules and Regulations

Employers want workers who respect the rules and regulations that apply to their company, their work, and their profession. Rules and regulations exist to keep people safe and to keep a project on schedule. Companies establish rules pertaining to work times, administrative issues, and professional conduct. Local, state, and federal authorities develop regulations that apply to a wide range of issues, including safety and health, materials, tools and equipment, and legal issues.

People can work well together only if they all understand the work that needs to be done, when it needs to be done, and who will perform the work. Rules and regulations are necessary in any work situation. The rules might vary from employer to employer, so it is important to comprehend all pertinent regulations for each and every job.

8.4.0 Promptness and Reliability

Professionals are always prompt and reliable when they report to work. A prompt employee shows up to work on time; a reliable worker shows up every day. Employees who are chronically late or absent have poor work habits. Professionals should always strive to exhibit the highest level of professionalism.

8.5.0 Career Opportunities

The construction industry employs more people and contributes more to the nation's economy than any other industry. Society will always need new homes, roads, airports, hospitals, schools, factories, and office buildings, and all of these structures will need coatings. This means there will always be a source of well-paying jobs and career opportunities for construction trade professionals, including application specialists. Both large and small contracting firms employ application specialists. Application specialists who master the principles of small-business management and develop and hone their entrepreneurial skills can be self-employed.

Application specialists can choose from a number of career paths (*Figure 13*) depending on their skills and interests. Where each will go will depend on the pride and professionalism each person shows. How well applicators succeed will depend on each person's willingness to develop skills and obtain training. With time and experience, application specialists progress from trainees to one or more different specialties. Following are some of the many positions and careers in the industrial coatings profession.

INSIDE TRACK

Tips for a Positive Attitude

Here is a short checklist of things that you should keep in mind to help you develop and maintain a positive attitude:

- Remember that your attitude follows you wherever you go.
- Helpful suggestions and compliments are much more effective than negative ones.
- Look for the positive characteristics about your teammates and supervisors.
- Focus on the good things about your job and the place where you work.

INSIDE TRACK

Ethical Principles for Members of the Construction Trades

Honesty: Be honest and truthful in all dealings. Conduct business according to the highest professional standards. Faithfully fulfill all contracts and commitments. Do not deliberately mislead or deceive others.

Integrity: Demonstrate personal integrity and the courage of your convictions by doing what is right even when there is great pressure to do otherwise. Do not sacrifice your principles for expediency, be hypocritical, or act in an unscrupulous manner.

Loyalty: Be worthy of trust. Demonstrate fidelity and loyalty to companies, employers, fellow craftspeople, and trade institutions and organizations.

Fairness: Be fair and just in all dealings. Do not take undue advantage of another's mistakes or difficulties. Fair people display a commitment to justice, equal treatment of individuals, tolerance for and acceptance of diversity, and open-mindedness.

Respect for others: Be courteous and treat all people with equal respect and dignity regardless of sex, race, or national origin.

Law abiding: Abide by laws, rules, and regulations relating to all personal and business activities.

Commitment to excellence: Pursue excellence in performing your duties, be well-informed and prepared, and constantly endeavor to increase your proficiency by gaining new skills and knowledge.

Leadership: By your own conduct, seek to be a positive role model for others.

Foremen are leaders who direct a single team of application specialists. Large construction projects with more than one team require supervisors who oversee the work of all teams. They are responsible for assigning, directing, and inspecting the work of each team.

The safety manager is another key on-site person. Safety managers are responsible for the safety and health of the workers on a project. They are responsible for developing a safety plan and procedures, training workers in safety procedures, and ensuring that the project is in compliance with applicable health, environmental, and safety regulations.

Project managers, often called project administrators, control the scope and direction of an industrial coatings business. Managers ensure that the right workers are being assigned to each task. Managers are also responsible for handling project accounting, which involves managing payroll, taxes, and employee benefits; tracking project finances; and providing administrative support for the contractor. Large contracting firms may have several project managers.

Estimators work for contractors and building supply companies. They assess the materials and labor required to complete a project. Contractors submit bids for jobs based on these estimates. Estimators must have a thorough understanding of construction methods as well as knowledge of the availability and prices of materials and supplies.

Estimators are experienced application specialists with superior mathematical abilities and the patience to prepare detailed, accurate estimates. Today's estimators also need solid computer skills. Estimators have a great deal of responsibility, because errors in estimates can cost the contractor significant amounts of money. Depending on the size and type of the business, the estimating job may also be done by the contractor or the project manager.

Construction contractors and owners establish and run contracting businesses. Typically, they hire apprentices, application specialists, and crew leaders. Depending on the size of their business, contractors may work with the crew or may manage the business full-time. Small contractors may have only one or two employees who manage the business, prepare estimates, obtain supplies, and perform the actual contracted work. Subcontractors and contractors/owners may specialize in particular types of coatings.

Inspectors may work for a contractor or for a municipal jurisdiction such as a local or state government. Inspectors ensure that all work follows applicable codes and quality requirements. Inspectors review and approve work at various stages of construction, depending on the installation.

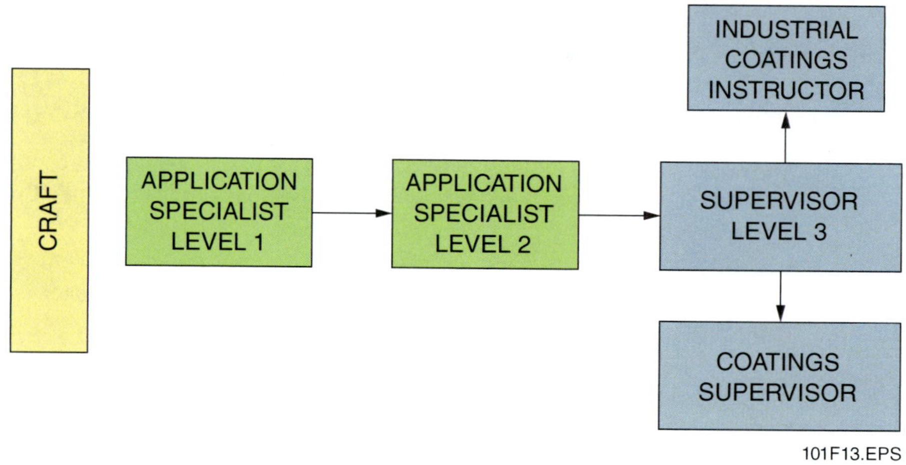

Figure 13 ◆ Opportunities in the coatings industry.

Instructors teach trainees to become the next generation of application specialists, managers, and inspectors. Many dedicated and experienced application specialists find that becoming an instructor is a natural evolution in their career.

No matter what career path you choose, you will be embarking on a lifelong learning process. Effective application specialists along any career path need to keep up-to-date with new tools, materials, and methods. If you choose to become a manager or if you intend to start your own construction business someday, you will need to learn the appropriate skills and continue to hone your skills as an application specialist.

Becoming a trainee is your first step down an exciting and challenging career path. Remember, every successful manager and business owner started as a trainee. Successful application specialists have at least one thing in common: a willingness to keep on learning. That process of learning begins with training.

9.0.0 ◆ FORMAL CONSTRUCTION TRAINING

The NCCER/NACE Industrial Coating and Lining Application Specialist curriculum provides trainees with industry-driven training and education. It adopts a purely competency-based teaching philosophy. This means that trainees must demonstrate to the instructor that they possess the understanding and the skills necessary to complete the tasks that are covered in each module before they can advance to the next stage of the curriculum.

When the instructor is satisfied that a trainee has successfully demonstrated the required knowledge and skills for a particular module, that information is sent to NCCER/NACE and kept in the NCCER Registry. The Registry can then confirm training and skills for workers as they move from state to state, company to company, or even within a company.

Whether you enroll in the NCCER/NACE program or another Office of Apprenticeship-approved program, ensure that you work for an employer or sponsor who supports a nationally standardized training program that includes credentials to confirm your skill development.

9.1.0 NACE-SSPC Standards

NACE International and The Society for Protective Coatings (SSPC) have developed a joint standard that outlines the requirements for qualifications and certifications for an industrial coating and lining application specialist, referred to as Application Specialist.

Application Specialist qualification and certification covers all aspects of surface preparation and coating application for steel and concrete surfaces of complex industrial structures. Training is received via classroom and on-the-job experience. To be qualified and certified, the application specialist must learn a certain body of knowledge and be able to pass applicable tests.

The body of knowledge required includes:

- Environmental, safety, and health concerns and requirements
- Process control (specifications, codes and standards, quality assurance, and quality control)
- Materials (consumables, surface preparation materials, coating materials, material properties, characteristics and utilization, and **material safety data sheets [MSDS]**)

- Surface preparation (purpose of surface preparation, chemical cleaning, hand and power tools, waterjetting, abrasive blasting, surface preparation standards, environmental controls, substrate repairs, and surface contaminants)
- Application (storage, mixing, environmental conditions and controls, conventional spray, airless spray, thermal spray, trowel, brush, roller, tapes, heat-shrinkable coatings, powder coatings, plural-component spray, curing, recoat and overcoat, defects, and repairs)

Specialty areas of the body of knowledge include:

- Installation of polymer coatings and surfacing on concrete floors and secondary containment
- High-pressure water cleaning and ultra-high-pressure waterjetting;
- Electrostatic spray
- Plural-component spray
- Powder coatings
- Specialty pipeline coatings
- Thermal spray coatings
- Fireproofing

9.2.0 Qualification Levels for Application Specialist

The NACE-SSPC standard outlines three levels of qualification for application specialist, with additional specialty qualifications for Level Two:

- *Level One* – Basic knowledge of industrial coatings and linings
- *Level Two* – Detailed knowledge and skills of industrial coatings and linings. Additional specialty qualifications are Level Two knowledge plus detailed knowledge and skills of specialty areas as outlined above
- *Level Three* – Level Two plus basic supervisory knowledge and skills, basic training knowledge and skills, and basic communication knowledge and skills

Level One is the entry level, with application specialists working with and supervised by Level Two and Level Three application specialists. Level Two is for experienced application specialists working independently. Level Three is for application specialists responsible for planning, oversight, evaluation, and supervision of industrial coating and lining of complex industrial structures.

9.3.0 Qualification Requirements for Application Specialist

Level One – Successful completion of a skills assessment program measuring reading text, document use, and numerical literacy; successful completion of examinations and skills test on the body of knowledge; demonstration of the ability to read, write, and communicate in the language of the workplace.

Level Two – Successful completion of all Level One requirements. Qualify to take the Level Two exam by (1) or (2) or (3):

1. Successful completion of two years of a three-year apprentice program
2. Five years of related work experience that includes a minimum of 5,000 hours of verifiable work experience and a minimum of 3,000 hours of coatings-related technical training
3. Eight years (minimum) of related work experience that includes a minimum of 8,000 hours of verifiable work experience and successful completion of the Level Two examination

Level Three – Qualify to take the Level Three exam by (1) or (2):

1. Five years minimum of related work experience after achieving qualification to Level Two
2. Five years minimum of verifiable work experience in inspection or supervision of coating-related work for complex industrial structures after completion of the requirements for Level Two examinations and successful completion of the Level Three examination

Review Questions

1. Reasons for applying coatings include visual appearance, protection from chemical attacks, protection from biological attacks, and _____.
   a. temporal protection
   b. mechanical protection
   c. lightning protection
   d. clinical protection

2. Coatings on porous materials other than concrete do not protect the materials from _____.
   a. wind erosion
   b. galvanometers
   c. biological contamination
   d. thermal contamination

3. Corrosion takes place in metal because the refined metal is in a(n) _____.
   a. high energy state
   b. low energy state
   c. depressive state
   d. impure state

4. The four required components of the corrosion cell are the anode, the cathode, the metallic path, and the _____.
   a. cathodic path
   b. metal plate
   c. inhibitive path
   d. electrolyte

5. The types of coatings discussed in this module include inhibitive, sacrificial, and _____.
   a. barrier
   b. blockade
   c. comprehensive
   d. adhesive

6. When rebar corrosion takes place, it _____.
   a. shrinks the rebar
   b. cracks the concrete
   c. has no effect on the concrete
   d. strengthens the concrete

7. Where there are sharp changes in the surface, coatings _____.
   a. work better
   b. should be applied last
   c. should not be applied
   d. tend to thin, compared to flat surfaces

8. Corrosion inhibitors properly added to an environment can _____.
   a. increase corrosion
   b. reinforce corrosion
   c. dissolve corrosion
   d. retard corrosion

9. Quality control is considered to be the job of _____.
   a. the applicator first, and ultimately the contractor
   b. the owner first, and ultimately the applicator
   c. the owner's inspector, and ultimately the applicator
   d. the safety engineer, and ultimately the contractor

10. Process control is getting the crew to _____.
    a. change approaches for each job
    b. stop on time
    c. do things the same way
    d. use good time management

Summary

The industrial coating applicator specialist must master a number of skills, including the correct use of tools and an understanding of different types of coatings, as well as the correct use of safety equipment. The applicator must also develop responsible work habits, including punctuality and safety consciousness. The career possibilities are very promising for the applicator who pursues excellence.

Notes

Trade Terms Introduced in This Module

Cathodic protection: The use of an electric current or a sacrificial anode to protect a metallic object from corrosion.

Corrosion: The deterioration of a material, usually a metal, that results from a reaction with its environment.

Corrosion inhibitors: Substances that can be added to the environment to retard or prevent corrosion.

Electrostatic spray: Coatings that use electrostatic charges of particles to be sprayed to coat an object.

Fireproofing coatings: Coatings that, due to their composition, are fireproof or fire-retardant.

Galvanic corrosion: The electrochemical deterioration of a metal when two dissimilar metals are placed in contact with the presence of a liquid that can carry a current.

Material safety data sheet (MSDS): A document that must accompany any hazardous material. The MSDS identifies the substance and details the exposure limits, the physical and chemical characteristics, the kind of hazard it presents, precautions for safe handling, storage, and use, and specific control and disposal issues.

NACE International: An organization whose mission is to protect people, assets, and the environment from the effect of corrosion.

NACE-SSPC standards: Written industry codes that guide blasters, applicators, supervisors, and owners in the use of minimum best practices for successful protective coating systems. The written codes are voted on and approved by hundreds of industry experts.

NCCER: An organization whose primary goal is to standardize construction craft training throughout the country so both employers and employees will benefit from the training, no matter their location.

Plural component spray: Coatings using several components mixed at the nozzle.

Powder coatings: Coatings applied as a powder and then treated to produce a film of coating.

Process control: A means of employing established work procedures, methods, and tools for improving quality and reducing variation.

Protective coatings: A coating defined by its protective, rather than aesthetic, properties. Protective coatings can include sacrificial, barrier, or inhibitive coatings.

Qualified person: A person who is trained to perform a critical or dangerous task. For example, a safety technician may be a qualified person to certify the safety of scaffolding.

Surface profile: Graphic representation of a surface condition.

Thermal spray: Coatings applied as a heated spray.

Appendix A

Samples of NCCER/NACE Training Credentials

NATIONAL CENTER FOR CONSTRUCTION EDUCATION AND RESEARCH

BUILDING TOMORROW'S WORKFORCE

August 27, 2009

Sample Student
National Center for Construction Education and Research
3600 NW 43rd St Bldg G
Gainesville, FL 32614-1104

Dear Sample,

On behalf of the National Center for Construction Education and Research, I congratulate you for successfully completing NCCER's Contren® Learning Series program. I also congratulate you for choosing construction as a career.

You are now a valuable member of one of our nation's largest industries. The skills you have acquired will not only enhance your career opportunities, but will help build America.

Enclosed are your credentials from the National Registry. These industry-recognized credentials give you flexibility in planning your career and ensure your achievements follow you wherever you go.

To access your training accomplishments through the Automated National Registry, follow these instructions:

1. Go to www.nccer-anr.org.
2. Click the "Individuals" button.
3. Enter the NCCER card number, located on front of your wallet card or transcript, and your PIN.
   Note: The default PIN is the last four digits of your SSN. You may change your PIN after you login.
4. First-time users will be directed to answer a few security questions upon initial login.
5. Contact the registry department with any questions.

NCCER applauds your dedication and wishes you the best in your future endeavors.

Sincerely,

Donald E. Whyte

Donald E. Whyte
President, NCCER

Enc.

3600 NW 43rd St, Bldg G ◦ Gainesville, FL 32606 ◦ P 352.334.0911 ◦ F 352.334.0932 ◦ www.nccer.org

Affiliated with the University of Florida

nccer

National Center for Construction Education and Research

This is to certify that

Sample Student

has achieved NACE/NCCER certification for successful completion of

Industrial Coating & Lining Specialist I

CertifiedPlus!

in the National Center's National Craft Assessment and Certification Program on this First day of September, 2007

Tony Keane
Executive Director, NACE International

Donald E. Whyte
President, NCCER

THE STANDARD FOR DEVELOPING CRAFT PROFESSIONALS

13614 Progress Boulevard, Alachua, Florida 32615 • p. 888.622.3720 f. 386.518.6255 • www.nccer.org

Official Transcript

January 17, 2012

NCCER Card #: 1720726
Trainee Name: John Q Smith
Sponsor: Austin Industrial Incorporated
Address: 2801 E 13th St
La Porte, TX 77571

Current Employer/School:
Solomon Plumbing Company

Module	Description	Instructor	Training Location	Date Completed
00101-04	Basic Safety	Kevin Jenkins	Solomon Plumbing Company	2/20/2008
00102-04	Introduction to Construction Math	Dave Buck	Building Trades Institute, LLC	8/8/2008
00103-04	Introduction to Hand Tools	Kevin Jenkins	Solomon Plumbing Company	1/1/2008
00104-04	Introduction to Power Tools	Dave Buck	Building Trades Institute, LLC	8/8/2008
00105-04	Introduction to Blueprints	Kevin Jenkins	Solomon Plumbing Company	3/20/2008
00106-04	Basic Rigging	Dave Buck	Building Trades Institute, LLC	8/8/2008
00108-04	Basic Employability Skills	Rod Blackburn	Utility Contractors, Inc.	3/15/2009
02101-05	Introduction to the Plumbing Profession	Kevin Jenkins	Solomon Plumbing Company	3/22/2008
26101-02	Electrical Safety	Don Whyte	National Center for Construction Education &	7/29/2002
26102-02	Hand Bending	Don Whyte	National Center for Construction Education &	7/29/2002
26103-02	Fasteners and Anchors	Don Whyte	National Center for Construction Education &	7/29/2002
26104-02	Electrical Theory One	Don Whyte	National Center for Construction Education &	7/29/2002
26105-02	Electrical Theory Two	Don Whyte	National Center for Construction Education &	7/29/2002
26106-02	Electrical Test Equipment	Don Whyte	National Center for Construction Education &	7/29/2002
26107-02	Introduction to the National Electrical Code	Don Whyte	National Center for Construction Education &	7/29/2002
26108-02	Raceways, Boxes, and Fittings	Don Whyte	National Center for Construction Education &	7/29/2002
26109-02	Conductors	Don Whyte	National Center for Construction Education &	7/29/2002

Page 1

Donald E. Whyte
President, NCCER

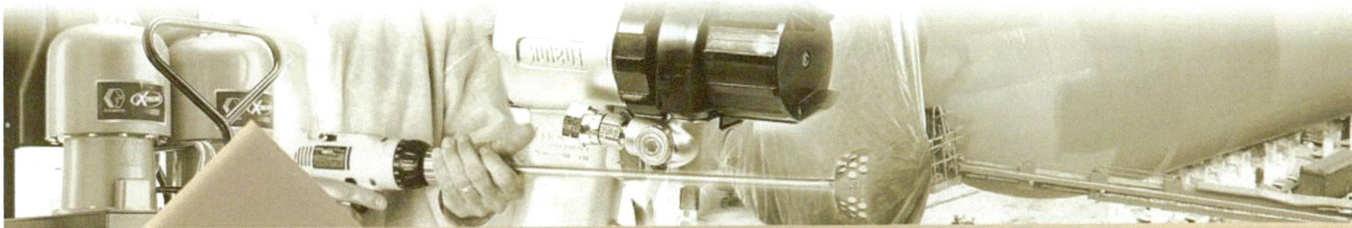

Appendix B

Sample Work Plan

Specifications

Vessel 411P Back Row Flake Storage Bin

Item #	Specification Description		Full/ Spot	Surface Preparation	Full/ Spot	Primer	DFT Mills	Full/ Spot	finish	DFT Mills	Notes
1	Internal shell,	Step 1	Full	Water wash	n/a	n/a	n/a		n/a		800 to 2000 psi
	cone and ceiling	Step 2	Full	SP-10	Full	Plasite 7122HAR	6-8	Full	Plasite 7122HAR	6-8	Both coats white

Comments/Observations: Minimum inspection requirements

1. Visually inspect blasted surface for cleanliness (SSPC-10), record surface profile and give approval to paint.
2. Record atmospheric conditions, surface temperature, coating batch numbers, observe and ok mixing of coatings before painting.
3. Record wet film and dry film thickness readings.
4. Observe and record low voltage holiday detection.
5. Report all observations and data recording to ________________ and ________________ in writing.

* Respirators-blast hood and compressed air when blasting. Full face and bottled air when painting. 1/2 mask OV mixing.

** Dimensions of vessel are: height 67', diameter 14'; cone is 12" in diameter.

Scope of Work

1. Mobilize equipment (blast pot, air drier, blasting abrasive, blast hoses, air hoses, spider, scaffolding, breathing air bottles).
2. Rig equipment on the vessel (build support scafford, hang spider, set up breathing air bottles and attach hoses, stretch out hoses).
3. Secure 2-4 gas LEL meters and 1-life line yoyo.
4. Request confirmed space entry permit.
5. Enter vessel by way of spider and build scaffold.
6. Enter vessel by way of scaffold and pressure wash interior with 800 to 2000 psi demin water. Face shield and wet suit.
7. Enter vessel by way of scaffold. Put on air supplied blast hood.
8. Abrasive blast shell to an SSPC-SP10 quality. Measure profile. Profile should be 3 to 4 mils. Test Text tape can be used for measuring profile.
9. Visually inspect and measure cleanliness of blasted surface. A soluble salt test kit can be used to measure and record cleanliness of surface.
10. Determine if surface is clean enough to paint or if additional washing and abrasive blasting is required.
11. De-rig blast equipment and set up spray painting equipment (2 gallon conventional spray pot, hose and spray gun).
12. Obtain 7122har (three-component coating) from storage.
13. Mix material per manufacturer's requirements (mix liquid and pigment first, then add cure). Use 1/2 mask respirator when mixing.
14. Allow mixed material to sit for 30 minutes before applying.
15. Use standard spray paint guns (Devilbiss or Binks)
16. Pour coating into spray pot, close pot, pull trigger and adjust spray pattern (8-12 inch fan width).
17. Put air supplied full face respirator.
18. Enter vessel and brush strip welds and irregularities with 7122har, thinned 30%.
19. Pull in spray gun and apply a mist coat on the surface. Allow 1 to 2 minutes for solvent to flash then apply a full wet coat to a wet film thickness of 10 to 12 mils. Spray gun should be held not more than 10 inches from the work. Exit vessel.
20. Allow coating to air dry overnight.
21. Measure dry film thickness. It should be 6 to 8 mils.
22. Mix and apply second coat of 7122har (same as the first coat). Let dry overnight and measure and record final DFT (12 to 15 mils).
23. Visually inspect and perform holiday detection on 100% of the finished surface.
24. Hand sand and apply additional coating to all holidays.
25. Remove and clean all paint equipment.
26. Remove scaffold.
27. Enter vessel using spider. Hand sand and apply more coating on all damaged areas.
28. Exit vessel.
29. Clean up, de-rig and move out of area.

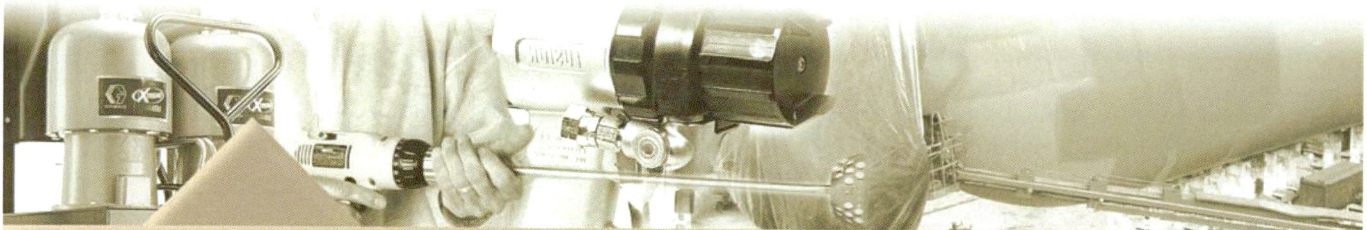

Profiles in Success

Gil Rogers

Manager, Field Services
CSI Coating Systems (2005) Inc.
Nisku, Alberta, Canada

How did you choose a career in the industrial coatings field?
Like many folks in the coatings field, I followed in the footsteps of my father, who was a painter.

What types of training have you been through?
Apprenticeship in the Painting and Decorating Trade, achieving Journeyman Status in 1979.
NACE Coating Inspector Certification, 1996
SSPC Protective Coating Specialist, 2000
NACE Protective Coating Specialist, 2002

What kinds of work have you done in your career?
Journeyman Painter; Coatings Supervisor; Project Manager; Industrial Coating Sales; Coating Inspector; Coating Consultant; Field Manager

Tell us about your present job and what you like about it.
I am currently working as the Field Manager for CSI Coating Systems (2005) Inc. My responsibilities include initiating client contact, maintaining those relationships, reviewing quotes, attending job-related meetings, hiring new employees, setting up employee training programs, managing conflict, initiating capital acquisitions, and supporting equipment maintenance programs.

The most enjoyable part of my current job is the challenge of meeting deadlines and the variety of tasks I am required to perform. I enjoy interacting with my staff and clients and appreciate the opportunity to meet many new people. I also enjoy the challenge of finding ways to apply emerging technology to our work.

What factors have contributed most to your success?
I have always been interested in learning and have taken advantage of every learning opportunity afforded to me. This has allowed me to make significant progress in this industry. Whenever I don't know something, I stick with the problem until I find the answer. Often it leads to more questions about new products and techniques. I think of every challenge as an opportunity to learn more about this business.

What advice would you give to those new to the field?
Learn, learn, and learn. Take advantage of those that know more than you do. Ask questions. The only dumb question is the one you don't ask. Find a mentor and never let him or her out of your sight!

You will be successful if you always strive to keep up with the new developments in the industry. Learn something new every day. If you aren't learning something new every day, you are doing something wrong.

Tell us about some of the accomplishments you have had in your career.
I have worked on many interesting projects including a major bridge recoating project which later became the subject of a documentary on the Discovery channel. I have worked underground and 300 feet in the air. I've worked on ships and tanks, painted mansions, jails, and farm buildings.

Since 1999 I have been an instructor in the NACE Coating Inspector Program. This has allowed me to instruct over 1,000 students in 10 different countries. My travels have taken me outside of North America to South America, Southeast Asia, Africa, Europe, and the Middle East.

I am currently involved in the development of the joint NACE-NCCER Applicator Training Program. This program is near and dear to my heart, and I believe it has the potential to change our industry in a significant way. I welcome yet another learning opportunity.

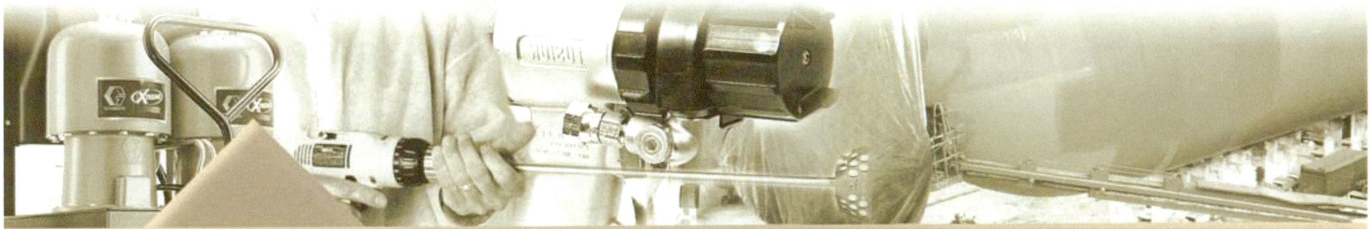

Additional Resources and References

Additional Resources

The following reference works are suggested for further study. These are optional materials for continued education rather than for task training.

NACE International, www.nace.org

The Protective Coating User's Handbook, Dr. Louis D. Vincent, 2004. Houston, Texas: NACE Press.

Figure Credits

NACE International, 101F02, Appendix B

Rogers & Associates Consulting Services, 101F01, 101F04, 101F06

L.S. Starrett Company, 101F05 (top)

Mitutoyo, 101F05 (middle)

Mike O'Brien, 101F05 (bottom)

Topaz Publications, Inc, 101F07-101F12

KTA-Tator, Inc., 101T01

NCCER CURRICULA — USER UPDATE

NCCER makes every effort to keep its textbooks up-to-date and free of technical errors. We appreciate your help in this process. If you find an error, a typographical mistake, or an inaccuracy in NCCER's curricula, please fill out this form (or a photocopy), or complete the online form at **www.nccer.org/olf**. Be sure to include the exact module ID number, page number, a detailed description, and your recommended correction. Your input will be brought to the attention of the Authoring Team. Thank you for your assistance.

Instructors – If you have an idea for improving this textbook, or have found that additional materials were necessary to teach this module effectively, please let us know so that we may present your suggestions to the Authoring Team.

NCCER Product Development and Revision

13614 Progress Blvd., Alachua, FL 32615

Email: curriculum@nccer.org

Online: www.nccer.org/olf

❑ Trainee Guide ❑ Lesson Plans ❑ Exam ❑ PowerPoints Other ____________________

Craft / Level: ____________________ Copyright Date: ____________________

Module ID Number / Title: ____________________

Section Number(s): ____________________

Description: ____________________

____

____

____

____

Recommended Correction: ____________________

____

____

____

____

Your Name: ____________________

Address: ____________________

____

Email: ____________________ Phone: ____________________

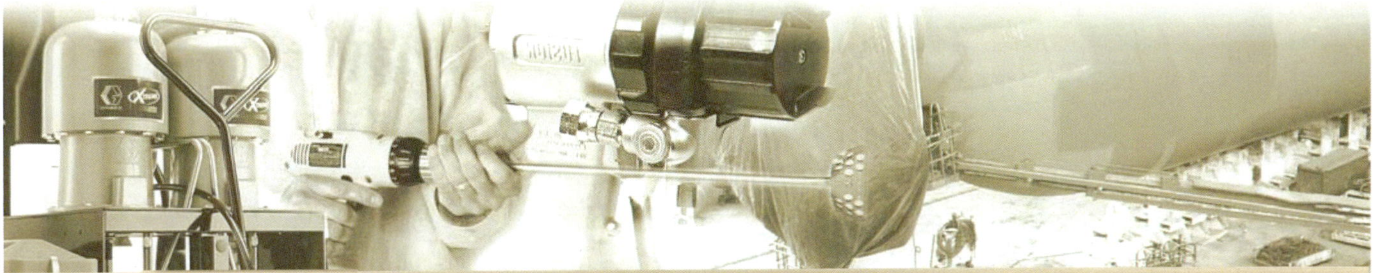

Industrial Coating and Lining Application Specialist Level One

69102-09

Surface Preparation

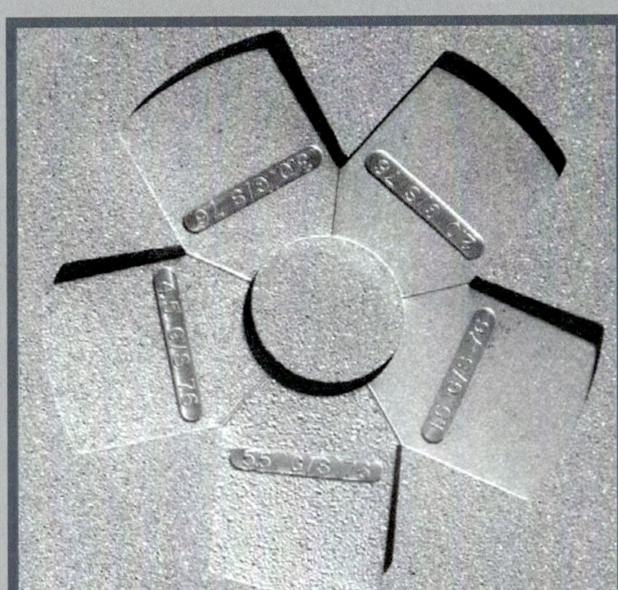

69102-09

Surface Preparation

Overview

Preparing a surface for coating can be the most time consuming, costly, and vital part of the industrial coating process. The goal of surface preparation is to provide the coating with a surface to which it can adhere to provide a long-lasting finish. Even the smallest bit of residue from greases, salt, chemicals, grit, or fingerprints can cause the coating to fail. Surface preparation is the first step in preserving the life of costly equipment, buildings, or fixtures. Coatings may appear attractive, be eye-catching, or even make something unattractive less obvious, but their most important job is to preserve the surface to which they are applied. An industrial coatings craftworker will work with many different types of surfaces, such as steel, metal, concrete, and synthetics. In this module, you will learn about cleaning these surfaces using hand and power tools, as well as by dry abrasive blasting and waterjetting.

Table of Contents

Figures

Tables

Objectives

When you have completed this module, you will be able to do the following:

1. Define surface preparation and list reasons for its importance.
2. Define the goals of surface preparation.
3. Recognize the surface conditions of steel substrates, mill scale, and steel fabrication surface defects.
4. State surface preparation methods and reasons for coating galvanized steel, weathering steel and other metals.
5. List properties of concrete, types of concrete surfaces, reasons for assessing and repairing concrete surfaces prior to surface preparation, and methods of surface preparation of concrete surfaces.
6. State how, when, and why to use solvents, hand tools, and power tools for cleaning.
7. State how, when, and why dry abrasive blasting is done.
8. State how, when, and why water cleaning is done.
9. Explain reasons for surface preparation and consequences of poor preparation methods.
10. State when overcoating is used, and how surfaces are prepared for overcoating.

Trade Terms

Aggregate	pH
Alloy	Porous
Efflorescence	Saponify
Emulsify	Stripe coat
Flash rust	Traps
Laitance	Water-miscible
Mill scale	Weathering

Required Trainee Materials

1. Pencil and paper
2. Appropriate personal protective equipment

Prerequisites

Before you begin this module, it is recommended that you successfully complete *Core Curriculum*, Modules 00101-04 and 00106-04; and *Industrial Coating and Lining Application Specialist Level One*, Module 69101-09.

This course map shows all of the modules in the first level of the Industrial Coating and Lining Application Specialist curriculum. The suggested training order begins at the bottom and proceeds up. Skill levels increase as you advance on the course map. The local Training Program Sponsor may adjust the training order.

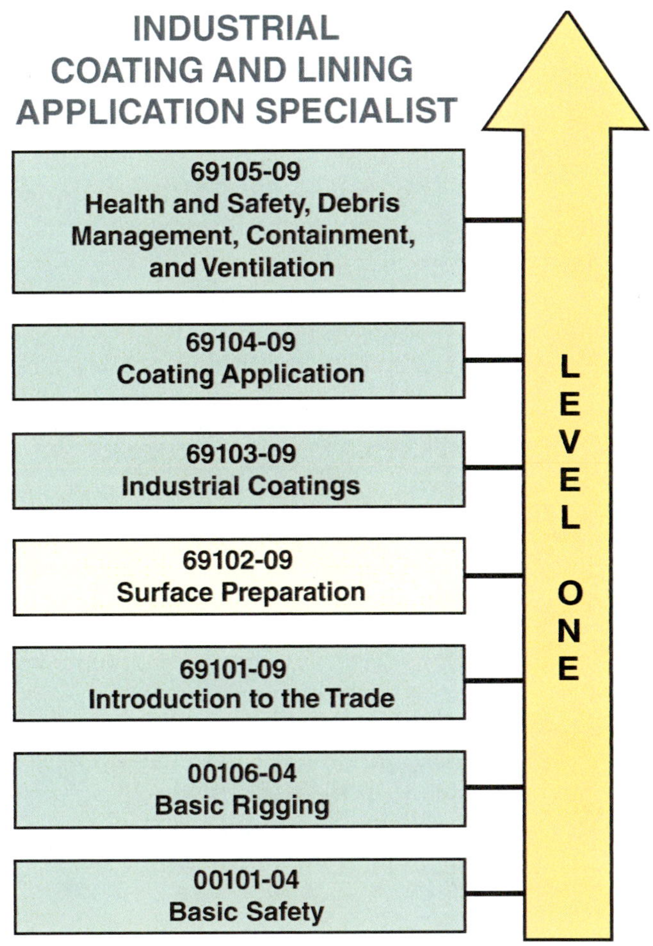

102CMAP.EPS

1.0.0 ◆ INTRODUCTION

You must complete surface preparation before applying the coating. Surface preparation is one of the main causes of paint failure. Coatings adhere best to a rough surface. In this module, you will learn about preparing surfaces using hand tools, power tools, dry abrasive blasting, and waterjetting. The general rule of surface preparation is: the better the degree of surface preparation, the longer the service life of the coatings being applied.

Some of the factors that can reduce the service life of a coating are:

- Residues of oil, grease, and soil, which can prevent adhesion or mechanical bonding of the paint to the surface.
- Residues of various chemical salts, which can induce corrosion, causing a coating to fail.
- Rust on the surface, which interferes with the coating bonding to the surface.
- Loose or broken **mill scale**, which can cause early coating failure, and adherent mill scale, which can cause later failure.
- Rust scale, or pack rust, which cannot be coated over and cannot maintain adhesion to steel.
- Surface profile/anchor pattern, which may be so deep that peaks cannot be adequately protected with paint, or which is not deep enough, causing coating failure because of loss of adhesion.
- Sharp ridges, burrs, edges, or cuts from mechanical cleaning equipment, which prevent adequate thickness of coatings over the irregularities.
- Surface condensation, if coated over, may result in blistering and delamination failure.
- Old coatings, which may have poor adhesion, may be incompatible with new coatings, or may be too deteriorated for recoating.
- Surface preparation of concrete is a special topic to be addressed later in this module.

2.0.0 ◆ SURFACE PREPARATION STANDARDS

It was necessary for the industry to set standards in order to reduce confusion between owners, specifiers, and applicators. The biggest standards organizations dealing with coatings in the United States are ASTM, Society for Protective Coatings (SSPC), and NACE International. The rest of the world commonly uses the standards of the International Organization for Standardization (ISO).

Many industry standards will be cited throughout this module. *Table 1* shows common surface preparation standards.

3.0.0 ◆ SURFACE PREPARATION VISUAL STANDARDS

NACE and SSPC have published several visual guides that show the different conditions of steel surfaces before and after cleaning. These guides will help you to figure out whether the surface has been prepared properly.

It is important to remember that the visual guides only describe the pictorial standard. The primary source of information is still the written standards given below:

- *SSPC-VIS 1, Guide and Reference Photographs for Steel Surfaces Prepared by Dry Abrasive Blast Cleaning*
- *SPC-VIS 2, Standard Method of Evaluating Degree of Rusting on Painted Steel Surfaces*
- *SSPC-VIS 3, Guide and Reference Photographs for Steel Surfaces Prepared by Hand and Power Tool Cleaning*
- *NACE VIS 7/SSPC-VIS 4, Guide and Reference Photographs for Steel Surfaces Prepared by Waterjetting*
- *NACE VIS 9/SSPC-VIS 5, Guide and Reference Photographs for Steel Surfaces Prepared by Wet Abrasive Blast Cleaning*

The International Organization for Standardization (ISO) standard 8501-1 contains photographic representations of four grades of blast cleaning:

- Sa 3, Blast Cleaning to Visually Clean Steel
- Sa 2.5, Very Thorough Blast Cleaning
- Sa 2, Thorough Blast Cleaning
- Sa 1, Light Blast Cleaning
- St 3, Very Thorough Hand or Power Tool Cleaning
- St 2, Thorough Hand or Power Tool Cleaning

Table 2 provides a comparison of abrasive blast cleaning standards.

Table 1 Common Surface Preparation Standards

SURFACE	PREPARATION
NACE No. 1/SSPC-SP 5	White Metal Blast Cleaning
NACE No. 2/SSPC-SP 10	Near-White Metal Blast Cleaning
NACE No. 3/SSPC-SP 6	Commercial Blast Cleaning
NACE No. 4/SSPC-SP 7	Brush-Off Blast Cleaning
NACE No. 5/SSPC-SP 12	Surface Preparation and Cleaning of Metals by Waterjetting Prior to Recoating
NACE No. 6/SSPC-SP 13	Surface Preparation of Concrete
NACE No. 8/SSPC-SP 14	Industrial Blast Cleaning
SSPC-SP 1	Solvent Cleaning
SSPC-SP 2	Hand Tool Cleaning
SSPC-SP 3	Power Tool Cleaning
SSPC-SP 11	Power Tool Cleaning to Bare Metal
SPC-SP15	Commercial-Grade Power Tool Cleaning

Table 2 Comparison of Abrasive Blast Cleaning Standards

STEEL STRUCTURES	SSPC	NACE	SWEDISH STANDARD SIS 05 5900-1967/ISO 850-11988
White Metal	SSPC-SP 5	NACE # 1	Sa 3
Near White Metal	SSPC-SP 10	NACE # 2	Sa2.5
Commercial Blast	SSPC-SP 6	NACE # 3	Not Applicable
Brush Off Blast	SSPC-SP 7	NACE # 4	Sa 1
Industrial Blast	SSPC-SP 14		Sa 2

4.0.0 ◆ DETERMINING CONDITION OF SUBSTRATES

The first step in surface preparation is to determine the condition of the substrate that is to be painted. The VIS (visual) standards allow you to identify the rust grade of the substrate before cleaning and compare with photographic examples shown in NACE/SSPC VIS standards or ISO standards. The publications that these visual standards appear within are *SSPC-VIS 1 Guide and Reference Photographs for Steel Surfaces Prepared by Dry Abrasive Blast Cleaning*, and *ISO 8501-1:1988 Preparation of steel substrates before application of paints and related products – Visual assessment of surface cleanliness – Part 1: Rust grades and preparation grades of uncoated steel substrates and of steel substrates after overall removal of previous coatings.*

5.0.0 ◆ COMMON SUBSTRATES

5.1.0 Steel

Steel is the most often used metal. It is low-cost and available in a variety of shapes and sizes. Very high temperatures are used to form steel. As the hot steel cools, it reacts with oxygen in the air. The reaction generates mill scale on the surface of the newly formed steel. Mill scale is a black or dark blue oxide layer that has a changing bond to the steel surface. It is often mistaken for rust.

5.2.0 Galvanized Steel

Galvanized steel is steel that has been dipped into a bath of molten zinc (*Figure 1*). The zinc bonds to the steel surface, forming a corrosion-preventive zinc-rich coating. Before you begin coating, the surface must be correctly cleaned and prepared. If

102F01.EPS

Figure 1 ◆ Close-up view of galvanized steel substrate.

done right, the coating will bond to the zinc. This can be done by **weathering** the surface for several months (six or more). Weathering roughens the zinc surface so coatings can adhere. Because the surface of new galvanized steel is very smooth, it can be prepared for coating by washing it with an acid etching solution. Then, it is rinsed with clean water. The surface can also be prepared with the application of a brush blast, which roughens the surface.

5.3.0 Concrete

Concrete is the most widely used building material in the world. The method for preparing concrete is similar to the method used on steel, but the steps are very different. Concrete is full of pores and is often reinforced with a steel bar. This steel bar will rust when exposed to liquid. Coating concrete is meant to keep liquid from reaching the steel bar. This will slow corrosion.

5.4.0 Weathering Steel

Weathering steel has a special trait: the initial surface corrosion becomes a protective barrier. This barrier prevents more corrosion. Weathering steel should be coated when it is used in places full of moisture, such as drainage areas. It is best to coat weathering steel used in the following environments:

- Coastal areas
- Areas with a high amount of rainfall, high humidity, or constant condensing fog
- Areas where salts are routinely used to de-ice roads
- Areas where the structure can be exposed to concentrated chemical vapors
- Drainage areas

When weathering steel is to be painted, the recommended surface preparation is solvent cleaning (detailed in SSPC-SP 1), followed by Near-White Metal Blast Cleaning (detailed in NACE No. 2/SSPC-SP 10).

5.5.0 Aluminum

Aluminum and stainless steel usually do not need to be coated to protect them from corrosion. The surfaces of these metals oxidize when exposed to air, forming a protective barrier. However, these metals corrode when exposed to chemicals. In these cases, aluminum can be prepared for coating by degreasing and thoroughly rinsing with water in accordance with SSPC-SP 1, Solvent Cleaning. When a higher surface profile is desired, the surface can be prepared by wet or dry abrasive blasting with fine-particle sand or plastic abrasives.

5.6.0 Stainless Steel

There are many types of stainless steel, and some can corrode. Stainless steel made in China usually has higher iron content and may corrode. When the surface is unblemished, stainless steel is usually prepared for coating by degreasing and a good water rinse. When there is rust on the surface, the stainless steel may need to be scrubbed by hand or blasted with wet (or dry) abrasives. When stainless steel is prepared by abrasive blasting, non-ferrous abrasives should be used to avoid corrosion caused by embedment. These pieces can lead to corrosion after the surface is coated. If stainless steel is to be power cleaned by hand, the brushes should be stainless steel.

6.0.0 ◆ STEEL SUBSTRATE COMMON FABRICATION AND DESIGN DEFECTS

Many structures are not designed with coating in mind. Fabrication defects and design flaws can make coating difficult (*Figure 2*). Applicators can aid the surface preparation operation by noticing and solving design defects.

102F02.EPS

Figure 2 ◆ Steel substrate.

6.1.0 Design and Fabrication Defects

Some common design and fabrication defects that affect the coating process include:

- Hard-to-reach or inaccessible areas
- Rivets, bolts, or other connectors
- Welds
- Gaps (particularly skip welds or surfaces close together)
- Overlapping surfaces (e.g., roof plates in water tanks)
- Angle iron badly oriented or in complex arrangements
- Threaded areas
- Dissimilar metals
- Sharp edges, particularly on corners or rough cut plate
- Construction aids

6.1.1 Hard-to-Reach Areas

Hard-to-reach areas are difficult to prepare for surface preparation and coating. These areas need special attention.

6.1.2 Riveted and Bolted Construction

Riveted and bolted construction can leave gaps and areas that are difficult to clean and coat. It is difficult to protect bolts in this condition. Riveted and bolted construction should be replaced by welding if possible.

6.1.3 Welds

Welds generally create rough, discontinuous areas on a plane surface and may have many sharp edges. Often, welds are not cleaned correctly, leaving weld spatter, slag, and acid flux leftover. If there is a large amount of weld spatter in one area, use a power grinder to remove it. If these imperfections are not cleaned away, corrosion may occur faster.

Especially in tanks and vessels, weld areas should be cleaned, ground smooth (or flushed depending upon the operating conditions), and **stripe coated**. It is not always recommended to grind welds.

Skip Welds

Skip welds occur when welding is broken. This means spot welding instead of continuous welding. If welding is not continuous, seal the overlap joints with a compound. The compound should work with the coating as written in:

- NACE SP0178
- Fabrication Details
- Surface Finish Requirements
- Proper Design Considerations for Tanks and Vessels to Be Lined for Immersion Service.

Epoxy and urethane-based caulks have been used successfully.

Overlapping plates and roof plates are usually skip-welded, creating an area you can't reach for coating. Moisture can move between the plates and corrosion can occur.

Weld Spatter

Weld spatter (*Figure 3*) is when small pieces of molten metal splash onto the surface near the weld. Weld spatter must be removed by hand or power tools. Abrasive blasting does not effectively remove weld spatter.

Rough Welds

Rough welds should be ground to remove sharp edges and spikes. Use hand or power tools. Abrasive blasting will not smooth rough welds.

CAUTION

Sometimes it is not desirable to grind welds smooth, because grinding can cause the weld to lose strength. The project specification may prohibit the use of a grinder, especially on the walls of pressure vessels or pressure pipes. Check with your supervisor before you use a grinder on any job.

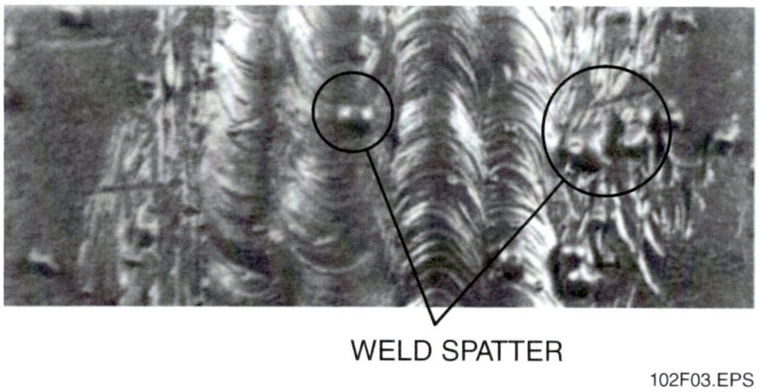

Figure 3 ◆ Weld spatter.

6.1.4 Laminations

Laminations and similar defects must be corrected to expose areas beneath them.

6.1.5 Sharp Edges and Angles

Use hand or power tools to deal with sharp edges and angles. Coatings can bridge over the base of a sharp bend or angle rather than smoothly sticking to the surface. This creates a void beneath the coating that can trap moisture and lead to corrosion. Because of this, sharp angles and edges must be ground to a smooth radius as explained in the coating specification or procurement documents. Make sure the area is smooth before applying coating. Use a brush to apply a stripe coat on the angle or bend. This will make sure the area gets enough coverage.

7.0.0 ◆ PRE-CLEANING

Surface cleanliness means figuring out how much of the visible contaminants (such as original mill scale, rust, and paint) have been removed from the surface. Also, figure out how much invisible contamination is present. These are usually in the form of salts. The applicator must be careful to not introduce contaminants to bare steel.

7.1.0 Visible Contaminants

- *Mill Scale* – It can be hard to tell the difference between mill scale and steel. One way to figure out if mill scale is present is to apply an acidic solution of copper sulfate to a blast-cleaned surface. On clean steel, the area will appear a bright copper color. On mill scale, the area will appear black.
- *Oil* (including the oil from your bare fingertips) – There are tests that can determine whether oil is present on the surface. Use a solvent to test for the presence of surface oil. On a clean surface, solvent will flow continuously. If oil is present, it will break into droplets. Ensure that the solvent used will work with the coating to be applied later. Ultraviolet light will light up traces of some hydrocarbon oils, grease, and fingerprints on the surface. Most oils will shine, but some will not. A clean rag wiped across the surface should pick up any oil or dirt.

CAUTION

Looking directly into unshielded ultraviolet lamps can cause severe eye damage, including blindness. Consult the safety engineer or other knowledgeable person about any ultraviolet lamp you may consider using.

- *Spent abrasives* – They might create contaminating dust, and you must make sure none is present. One way to test for leftover dust is by putting clear tape on the surface and removing it to check how much dust sticks to the tape.
- Mud or dirt
- Sweat

7.2.0 Non-Visible Surface Contamination-Soluble Salts

Corroded steel is harder to clean than new steel. The surface of corroded steel is rougher than new steel. Steel that has been exposed to corrosion in the presence of certain contaminants (*e.g.*, sulfates, chlorides) might be difficult to clean well. Even though the surface may appear to be clean and free of corrosion, it might contain enough nonvisible contamination to make a surface unfit for coating.

In extreme cases, after blasting, heavily contaminated areas will absorb moisture from the air. They will change to a dark color and corrode very fast. This is called **flash rusting**. The effect can sometimes be seen minutes after the blast cleaning process (especially when there is a lot of humidity). This is a sign that the surface is contaminated with soluble salts (such as chloride salts).

If soluble salts are left on the surface of steel when coated, it leads to a process called osmosis. Osmosis can cause delamination of coatings, blistering, and under-coating corrosion. Below is a brief explanation of osmosis, semi-permeable membranes, and why it is important to remove salts.

- *Osmotic pressure* (*osmotic cell*) – An osmotic cell forms when you have a semi-permeable membrane with liquids on both sides. If the liquids have different concentrations of soluble salts, through the process of osmosis, an osmotic cell will form.
- *Semi-permeable membrane* – A semi-permeable membrane is one that allows some molecules to pass through and in a certain phase (vapor or liquid), but does not allow other molecules to pass through. In coating applications, paint can act as a semi-permeable membrane. How permeable the paint is depends on various properties of the paint including type, hardness, porosity, and thickness.

It is essential that all soluble salts be completely removed from steel before coating. All coatings, no matter how perfect in every other way, are permeable (allowing liquids or gasses to pass through) to some degree. This coating will create a semi-permeable membrane with a higher salt concentration on the substrate side of the coating. This higher salt concentration will naturally draw moisture to it, just like the salt in a salt shaker draws moisture from the air. Moisture, often in a vapor (gaseous) state, may pass through the coating and create corrosion under the coating. When this happens, the water vapor becomes the electrolyte (a liquid that conducts electricity); this contributes to the formation of a corrosion cell beneath the coating. The coating will fail as a result of this corrosion. To stop this from happening, it is important to remove the soluble salts from the surface of the workpiece prior to coating.

7.3.0 Testing for Soluble Salts

A number of spot tests have been made to test for soluble salt substances. The tests use chemistry and electrical conductivity measurements to determine the presence and concentration of many soluble salts (such as chloride, potassium, and ferrous salts).

7.4.0 Consequences of Coating over Dust, Oil or Other Contaminants

Painting over oil can have consequences. Oil is a lubricant, which stops substances, including coatings, from sticking. If there is oil between the coating and the steel, the coating will not stick. Also, if water-based coatings are being applied, fisheyes may form because oil and water do not mix. This will cause voids in the coating.

8.0.0 ◆ SUBSTRATE SOLVENT CLEANING

8.1.0 Solvent Cleaning, SSPC-SP 1

Solvent cleaning is a method for removing all visible oil, grease, soil, drawing and cutting compounds, and other soluble contaminants from steel surfaces. Some coatings can be removed with solvents.

A variety of solvent cleaning materials may be used, including:

- Organic solvents
- Alkaline cleaners
- Acidic cleaners
- Detergents
- Emulsion cleaners
- Water/steam

See *Table 3* for personal protective equipment generally required for solvent cleaning.

Table 3 Personal Protective Equipment for Solvent Cleaning

PPE	Conditions
Eye protection	All times
Face protection	Flying debris, handling caustic liquids
Protective work gloves	Handling liquids, removing debris
Respirator	Dusty areas, closed area, excessive vapors
Protective clothing	Handling liquids, removing debris

8.1.1 Organic Solvents

Organic solvents include kerosene, turpentine, naphtha, mineral spirits, toluene, and xylene. They clean metal by dissolving and diluting the oil and grease contamination on the surface. Organic solvents cannot remove inorganic materials (such as chlorides, sulfates, weld flux and mill scale). Today, cost considerations and regulations concerning the use of organic solvents have become more restrictive. They are primarily used only in highly controlled situations. As with all cleaning materials, be sure to consult the product material safety data sheet (MSDS) for safety requirements. MSDSs are supplied to employers by chemical manufacturers.

8.1.2 Alkaline Cleaners

Alkaline cleaners include trisodium phosphate (TSP) and sodium hydroxide. They **saponify** (convert to soap) certain oils, greases, and surface-active components. They also wash away other contaminants. The soaps formed by these cleaners can be washed away with water. As with all cleaning materials, be sure to consult the product MSDS for safety requirements. Alkaline cleaners can burn skin.

8.1.3 Acidic Cleaners

Acidic cleaners are usually made of fairly strong acids (such as phosphoric acid) with small quantities of surfactants (substances that promote easy spreading), **water-miscible** solvents (solvents that mix easily with water), organic wetting and **emulsifying** agents.

Acidic cleaners remove soil by chemical attack and dissolving the reaction products. They are used to remove corrosion products or for other special purposes. To dispose of used acid cleaners, you need to collect them for processing. Acidic cleaners can pose health and safety hazards. Like alkaline cleaners, acid can burn skin. Be sure to consult the product MSDS for safety requirements.

8.1.4 Detergents

Detergents emulsify and dissolve oils and dirt so they can be washed away with water. After the surface is washed with a detergent, it must be rinsed with plenty of clean water to avoid leaving a detergent film on the surface.

The use of organic solvents has become restrictive and the alkaline/acidic cleaners can be dangerous. This has led to more use of detergents (especially biodegradable formulations) to remove oil, grease, and other contaminants from surfaces. Be sure to consult the product MSDS for safety requirements.

8.1.5 Emulsion Cleaners

Emulsion cleaners are sold as proprietary products and should be used only as instructed by the manufacturer. These cleaners contain oil-soluble soaps or emulsifying agents, buffering salts, dispersants, and inhibitors, along with kerosene or some type of mineral spirits. These materials are generally sprayed onto the surface to be cleaned. They function by wetting, emulsifying, dispersing, and dissolving the contaminants.

8.1.6 Water/Steam

Water/steam is a very powerful solvent and may be used to clean a surface very effectively. Water blast cleaning and waterjetting will be presented in subsequent sections of the module.

9.0.0 ◆ SUBSTRATE CLEANING USING HAND AND POWER TOOLS

9.1.0 Hand Tool Cleaning, SSPC-SP 2

Hand tool cleaning is a method of preparing steel surfaces with non-powered tools to remove loose rust, mill scale, paint, and other damaging matter. Hand tool cleaning is not intended to remove adherent mill scale, rust, and paint. Mill scale, rust, and paint are considered adherent if they cannot be removed by lifting with a dull putty knife.

Hand tool cleaning is one of the oldest methods of cleaning. It is generally used when power-operated equipment is not available, or when work is inaccessible to power tools. It is also used when the job is too small to warrant the use of power tools.

Hand cleaning tools include wire brushes, scrapers, chisels, knives, and chipping hammers.

CAUTION

Never use hand tools on surfaces that can generate sparks when flammable liquids are in use. The sparks will ignite the vapors.

9.2.0 Power Tool Cleaning, SSPC-SP 3

Power tool cleaning is a method of preparing steel surfaces with power-assisted hand tools. It removes all loose rust, mill scale, paint, and other damaging matter. It is not intended to remove adherent mill scale, rust, and paint. These are considered adherent if they cannot be removed by lifting with a dull putty knife.

Power tool cleaning is frequently used in maintenance operations. Besides removing loose mill scale, rust, and paint, this method is also used to remove weld flux, weld spatter, and laminations. It also smoothes rough welds and rounds out gouges before abrasive blast cleaning.

Power tool cleaning tools include rotary wire brushes, impact tools, and grinders and sanders.

9.3.0 Power Tool Cleaning to Bare Metal, SSPC-SP 11

In 1989, SSPC adopted a new standard: SP 11 (Power Tool Cleaning to Bare Metal). This standard calls for power tool cleaning to produce a bare metal surface and to retain a profile when a clean, roughened, bare metal surface is desired. SSPC-SP 11 should be used when abrasive blast cleaning is not possible or allowed.

When viewed without magnification, metallic surfaces prepared with SSPC-SP 11 should be free of all visible oil, grease, dirt, dust, mill scale, rust, paint, oxide, corrosion products, and other foreign matter. Slight residues of rust and paint might be left in the lower portion of pits, if the original surface is pitted.

9.4.0 Commercial Grade Power Tool Cleaning, SSPC-SP 15

This standard covers the requirements for power tool cleaning to provide a commercial-grade power-tool-cleaned steel surface, and to retain or produce a minimum 25 µm (1.0 mil) surface profile. When viewed without magnification, a commercial-grade power tool cleaned steel surface should be free of all visible oil, grease, dirt, rust, coating, oxides, mill scale, corrosion products, and other foreign matter, except as noted.

Random staining should be limited to no more than 33 percent of each unit area of surface as defined. The unit of surface area is a 3 ft^2 (0.3 m^2) area. Slight residues of rust and paint may also be left in the bottoms of pits if the original surface is pitted.

This standard differs from SSPC-SP 11, Power Tool Cleaning to Bare Metal, in that stains of rust, paint, or mill scale may remain on the surface.

9.5.0 Rotary Wire Brushes

Rotary wire brushes (*Figure 4*) are designed in two general types: straight (or in-line) and rotary. They are used to loosen mill scale, rush and paint. They can also be used to smooth rough welds and round out gouges before abrasive blast cleaning.

Overworking the surface with a wire brush may be harmful because excessive burnishing can develop a polished surface. This would provide a poor anchor for most coatings. Rotary power brushes can easily spread oil and grease over the surface. Therefore, solvent cleaning is an essential step before power wire brushing.

9.6.0 Impact Tools

Rotary impact tools are new tools on the market. They combine the easy use and speed of a rotary tool, but produce the profile made by impact tools. The tool looks like a rotary wire brush, but it has sharpened steel bristles that are made of hardened steel. The bristles are bent at an angle, so that the tips impact the surface, cleaning and generating a surface profile at the same time. There is an accelerator bar that flings the tips against the steel, generating a profile of 2.5 to 3.0 mil (0.06 to 0.08 µm), which is well above the requirements of an SSPC SP-11.

Scrapers and special chisels can be mounted into pneumatically or electrically-powered impact tools. Rotary impact tools work by using a cutting or chipping action. These types of tools are best used when heavy deposits of rust scale, mill scale, old paint, weld flux, slag, and other brittle products are removed from the metal.

Chipping/scaling hammers are the most widely used type of impact tool. A chisel is inserted into the power tool. The impact from an air- or electrically-operated piston is sent to the chisel, and then to the surface to be cleaned. Be careful, because this tool can easily cut into the surface, removing sound metal and leaving sharp burrs that can cause the coating to fail early.

Scaling or chipping hammers are usually a slower, more expensive method of cleaning the surface. But, in some cases (such as when considerable rust scale or heavy paint formation must be removed), they prove to be economical.

102F04.EPS

Figure 4 ◆ Power brush.

102F05.EPS

Figure 5 ◆ Grinders and sanders.

A needle gun, or needle scaler, has many hardened steel rods. The rods are vibrated against the surface to break apart the existing coating or rust. Needle guns are quite effective on welds, corners, and irregular surfaces. However, they tend to polish, rather than just to clean, the metal surface.

Rotary scalers can be used to remove rust and scale on large areas. Be careful to prevent the cutters of these scalers from cutting the metal to such an extent that metallic points extend far above the surface. This will cause early coating failure due to insufficient coverage.

9.7.0 Grinders and Sanders

Grinders and sanders are often used to prepare surfaces for coating (*Figure 5*). These may be the same tools as used for power wire brushing. Just substitute the wire brushes for sanding discs or grinding wheels. Grinding is good for removing weld spatter, smoothing weld seams, or rounding off sharp edges or corners. They are also frequently used to repair minor fabrication defects.

To achieve a deep surface profile using a power sander, use a coarse grit sanding disc. Sanding discs can produce extremely good anchor profiles (anchor profile refers to the surface pattern), with complete removal of rust and mill scale. However, such cleaning methods are very expensive for large areas.

Also available are pneumatically and electrically-operated sanders that have a flat pad or abrasive surface that comes in contact with the metal to be cleaned. Some of the motor-driven sanders have an orbital motion.

Power tool cleaning in accordance with this standard produces a greater degree of cleaning than SSPC-SP 3 or SSPC-SP 15. However, surfaces prepared according to SP 11 should not be compared to surfaces prepared by abrasive blast cleaning.

10.0.0 ◆ SUBSTRATE PREPARATION METHOD BY DRY ABRASIVE BLASTING

The most common method of surface preparation is by dry abrasive blasting. This method is typically used on metals, but can be used on other surfaces. Dry abrasive blasting applies a stream of small abrasive particles under pressure to a surface. This removes rust, mill scale, or other contaminants.

Use the proper size, hardness, and angularity of abrasive to achieve the required results. The larger the size, the greater the profile. Angular media works best to remove heavy layers of paint and corrosion. Rounded media works best on mill scale and light contamination.

The effectiveness of the blasting process also depends on the volume and pressure of air passing through the nozzle (or nozzles, if more than one set-up is being used) when compression of air takes place. Adequate pressure is critical to an effective blast process. If the abrasive is not being sent forward at sufficient pressure, much more time is required to accomplish the job.

10.1.0 Dry Abrasive Blasting Equipment

Although there are a variety of equipment brands with different features available for dry abrasive blasting, the basic required equipment components are:

- Compressor
- Moisture separator and air drying equipment
- Metering valves
- Blast pot
- Hoses and couplings
- Nozzle
- Safety equipment

10.2.0 Compressor

Energize the compressor, check for leaks, overheating, unusual noises or vibrations before beginning work. Continue to watch the air compressor throughout the shift (*Figure 6*). Check noticeable problems immediately.

10.3.0 Compressor Size

In the blast cleaning operation, the air compressor must be able to deliver sufficient air volume to maintain the required pressure at the nozzle. This volume is measured in cubic feet per minute (ft^3/min or cfm) or in cubic meters per minute (m^3/min). Although large industrial compressors supply compressed air for other processes, no other industry process requires such large quantities of compressed air as blast cleaning. It is vital to use compressors of adequate size. Undersized compressors might have the following problems:

- Wasted time waiting for an undersized compressor to build up required pressure
- Inability to add new tools to the system
- Strain on the compressor during peak loads
- Time loss caused by air-starved equipment operating inefficiently
- Increased possibility of breakdowns because of excessive motor use
- Excessive heat and condensation

10.4.0 Quality of Compressed Air

During abrasive blasting, the abrasive is loaded into the blast pot and forced under pressure from the blast machine through the blast hose. The operator is dressed in protective clothing with a hood designed for dry abrasive blasting. This hood is supplied with air from the compressor that feeds the blast pot, or a separate one. Therefore, because the compressed air is often used as the operator's breathing equipment while working, it is critical that the breathing air be fresh, pure, and free of contaminants.

The air supply used for breathing must be filtered and monitored for carbon monoxide. Within North America, breathing air must be as specified in *CSA G-7.1, Commodity Specification for Air [Grade D]*, or *ANSI Z88.2, American National Standard for Respiratory Protection*.

Oil vapor or water droplets carried by compressed air can contaminate the surface being cleaned. In addition, oil and moisture can contaminate the abrasives themselves, clogging the system. Therefore, compressors need to be equipped with devices to scrub the air of oil, moisture, vapors, and other contaminants, including the following:

- Oil and moisture **traps** (*Figure 7*)
- Charcoal filters to purify the air
- Carbon monoxide detectors to monitor the air to the operator's hood
- Driers to remove water from the compressed air
- Absorbent filters that absorb moisture in compressed air
- Cooling system to extract water from the compressed air by cooling it (cooler air holds less water than warmer air)

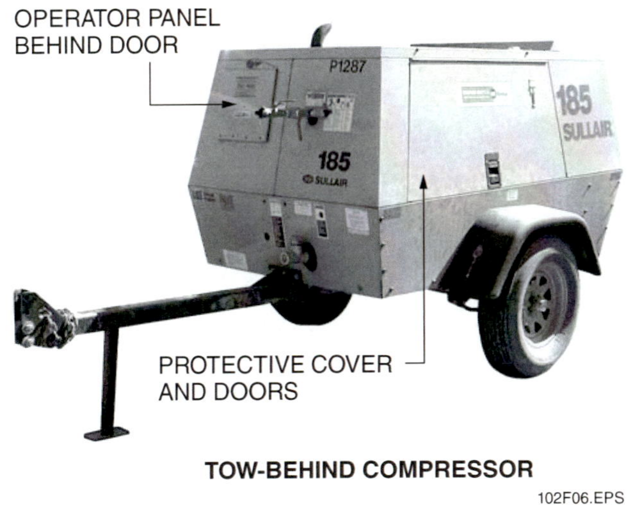

Figure 6 ♦ Air compressor.

10.4.1 Air Blotter Test

The procedure to determine whether air is suitable for blasting or painting is described in *ASTM D 4285, Standard Test Method for Indicating Oil or Water in Compressed Air*.

102F07.EPS

Figure 7 ◆ Traps.

The test procedure is as follows:

- Allow blast equipment to reach operating conditions.
- Discharge air to remove any condensation in the system.
- Use blotter paper or a clean white cloth.
- Turn on the air; if it is blast air, do not turn on the abrasives.
- Hold paper or cloth 24 in (60 cm) from the nozzle for 1 minute.
- Examine cloth for signs of oil or dirt; a UV light can be used for careful examination.

10.5.0 Blast Pot

The blast pot should be maintained and possibly emptied daily. It should be kept dry to avoid contamination of the abrasive.

10.6.0 Hoses and Couplings

All hoses and couplings should be of sufficient size and type to match the compressor output. Two kinds of hoses are used: air supply hoses and blast abrasive hoses.

The air supply hose carries air from the compressor to the abrasive blast unit. The recommended size should be three to four times the nozzle orifice. On lines longer than 100 ft (30 m), the inner diameter of the hose should be four times the blast nozzle orifice size. The large size helps eliminate loss of air pressure.

Hose components should be connected with externally fitted couplings, wired together with whip checks to prevent the hose ends from whipping around in the event of accidental separation.

Because of pressure drop in longer hoses, they should be kept as short as possible. Preferably, the blast hose length (the distance from pot to blast nozzle) should be no more than 50 ft (15 m) per section. When greater lengths are necessary, the hose should be kept in a straight line as much as possible; any bends should be wide and have a smooth radius.

10.7.0 Nozzles

The speed of dry abrasive blasting is directly related to the size of nozzle orifice (*Figure 8*). The larger the orifice, the more air used. So, the nozzle selected depends on the capacity of the compressor and hoses.

Maintenance of nozzle size is very important and can be a problem if normal cast iron nozzles are used, because these wear down rapidly. Nozzles made of wear-resisting **alloys** or materials such as tungsten carbide or ceramics may cost more initially, but will require less maintenance.

Longer nozzles result in more velocity and a more concentrated blast pattern, so they are preferable on tough cleaning jobs. However, use of a nozzle too large for the work may result in power being wasted by undue blasting. Larger nozzles also require a greater volume of air (hence a larger air compressor) to work properly.

For many years, all nozzles had a straight internal bore with a small throat opening. The straight nozzles have an outlet abrasive velocity of about 349 km/h (217 mph or 315 ft/s [96 m/s]). In addition, straight bore nozzles tend to spread the abrasive in a large blast pattern, with more concentration in the center and less at the edges. In 1954, the Venturi shape was adapted to abrasive blast nozzles. This shape has a large entrance throat, tapers gradually into a short straight section in the middle, and then flares at the outlet end. The Venturi shape allows for abrasive velocity up to 724 km/h (450 mph, 660 ft/s[200 m/s]) and an almost equal impact over the entire surface. The Venturi nozzle is the most effective shape for tough cleaning jobs.

Figure 8 ♦ Typical blast nozzles.

10.8.0 Relationship Between Pressure, Temperature, and Volume

The Universal Gas Law states that Pressure × Volume/Temperature = Constant. While it is not necessary to understand the source of the formula, it does have some important ramifications for blasting and painting. For the most part, temperature can be considered a constant. The important thing to remember about temperature is that increasing the air temperature increases the volume of air, and cooler temperatures decrease the volume of air. As a result, as the temperature increases during the day, a larger volume of air will be required to maintain the same pressure. Because warm air is less dense, more air is required to achieve the same pressure.

More important is the relationship between volume and pressure. It is important to maintain pressure at the blast nozzle. As the blast nozzle wears, the area increases according to the formula $A = \pi r^2$ where $\pi = 3.14$ and r is the radius of the blast nozzle.

Table 4 shows the relationship between nozzle size, cross sectional area, and required air pressure.

Going from a #4 nozzle to a #6 nozzle doubles the area of the nozzle and increases the volume cubic foot per minute (CFM) of air passing through the nozzle by two.

Going from a #4 to a #8 is 4 times the area and increases the CFM by four.

This has a direct result on blasting efficiency. As the area increases by the square of the radius, the volume of air passing through the nozzle increases, and as a direct result, the pressure decreases. Each 1 psi (7 kPa) drop in air pressure gives about a 1.5 percent decrease in efficiency.

Table 4 Relationship Between Nozzle Size, Cross Sectional Area, and Required Air Pressure

Nozzle Size	Diameter	Area	Required CFM at 100 psi
4	0.25	0.2	81
5	0.31	0.3	137
6	0.38	0.4	196
7	0.44	0.6	254
8	0.50	0.8	338

Thus, a 10 percent drop in psi (kPa) = 15 percent loss of efficiency.

A $\frac{1}{16}$ increase in nozzle orifice size (going from a #3 to a #4 nozzle) can increase the volume of air required to maintain the same pressure by 25 percent. To ensure enough CFM to the blast nozzle, the blast hose diameter should be 3 to 4 times the diameter of the nozzle being used.

Three things that can adversely affect the pressure at the nozzle are the length of the air hose, the number of bends in the air hose, and the elevation of the blast nozzle above the compressor.

To maintain efficient blast operations, the diameter of the blast nozzle and the nozzle pressure should be performed regularly. The nozzle aperture gauge and the nozzle pressure gauge (*Figure 9*) are used to make these determinations. Remove the end cap of the gauge and remove the wax marker. Use the marker to mark both rounded sides of the gauge.

1. From the back of the blast nozzle, insert the gauge fully into the nozzle and twist gently one full turn when it is fully inserted.
2. Remove the nozzle and read where the break is in the wax mark. This is the nozzle orifice size.

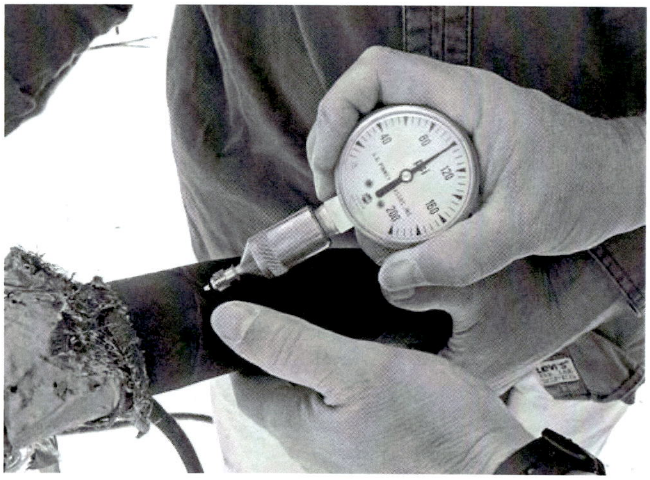

102F09.EPS

Figure 9 ◆ Nozzle pressure gauge.

3. Bend the hose about 6 in (15 cm) from the nozzle.
4. While the abrasive is flowing, insert the needle of the nozzle pressure gauge into the bent area at a 45-degree angle with the tip toward the nozzle. Make sure the needle fully penetrates the hose.
5. With the tip near the center of the air stream, read the pressure. When you remove the needle the hole will close.

11.0.0 ◆ SURFACE PROFILE

Abrasive blasting roughens the substrate (surface profile), creating a surface to which coatings can bond tightly. Surface roughness is the average height difference between the peaks and valleys. This is important because it increases the surface area of the substrate, which means more bonding surface for the coating. It also increases the amount of paint needed to cover a surface because the valleys must be filled in before the coating builds over the peaks. Abrasive blasting is one of the most important methods to ensure you have a proper surface for painting.

11.1.0 Surface Profile Measurement Methods

There are currently three commonly used ways to determine surface profile. Each one has advantages and disadvantages. The following three methods are detailed in *ASTM D 4417, Standard Test Method for Field Measurement of Surface Profile of Blast Cleaned Steel*:

1. Surface Profile Visual Comparator (D 4417 Method A) (*Figure 10*)
2. Surface Profile Gauge Profilometer (D 4417 Method B)
3. Replica Tape (D 4417 Method C)

11.2.0 Method A: Surface Profile Visual Comparator

There are several different visual comparators. These comparators are cast in metal to approximate surface profiles up to 4 mil (0.1 mm). The comparators are held up to the blast profile and usually viewed with a lighted magnifier (5X or 10X). The texture and roughness of the comparator are compared to the blasted steel. Because the profile changes based on the media used for blasting, comparators are sold as sand, shot, or grit. Comparators, when used by an experienced inspector, should yield an accuracy of approximately 0.5 mil (0.01 mm) within the range of the comparator. Because not all grit gives the same profile, this is probably the most problematic comparator to use.

11.3.0 Method B: Surface Profile Gauge

This gauge has a wide base so it can rest on the peaks (*Figure 11*). It also has a needle that measures the depth of a valley. Because it has a single needle, only one peak-to-valley measurement is made with each reading. To determine the substrate's average surface profile in a particular area, several measurements must be made, then averaged. (ASTM specifies 10 measurements per spot). The meter must be zeroed on a smooth surface, such as glass. With a dial gauge, readings must be recorded as they are made.

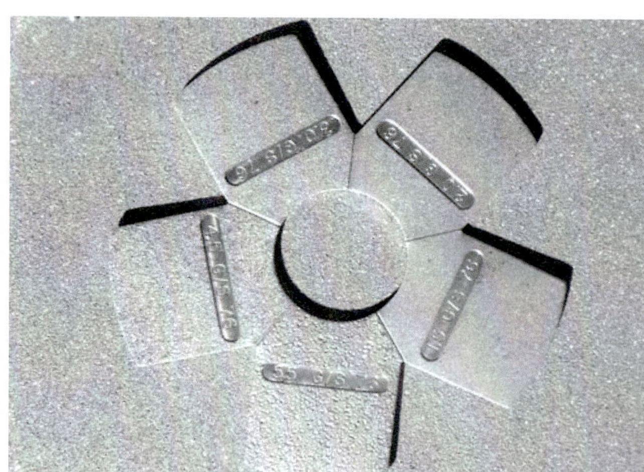

102F10.EPS

Figure 10 ◆ Comparator.

New gauges are digital and will record average, minimum, and maximum readings. They can be downloaded to a computer for a permanent record. Many top models also have built-in wireless (Bluetooth). This is the most cost-effective method for large numbers of readings and is relatively fast.

11.4.0 Method C: Replica Tape

Replica tape (*Figure 12*) is probably the most commonly used method to determine surface profile. The tape has a compressible foam layer with a plastic covering. A burnishing tool is used to rub the foam into the profile. The foam takes on the shape of the profile and is measured with a spring micrometer.

When using replica tape, make sure the area is clean and representative of the area being tested.

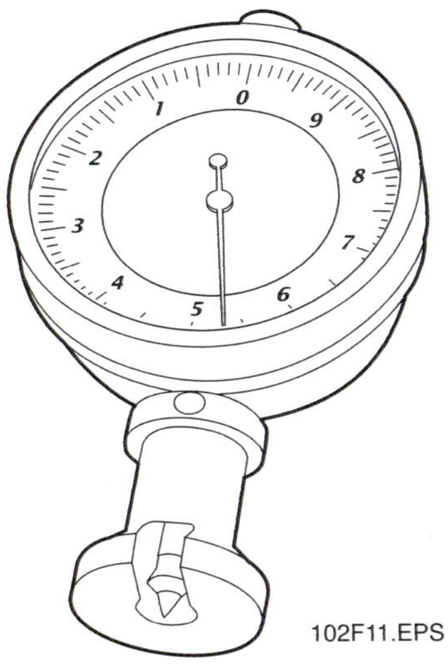

102F11.EPS

Figure 11 ◆ Profile gauge.

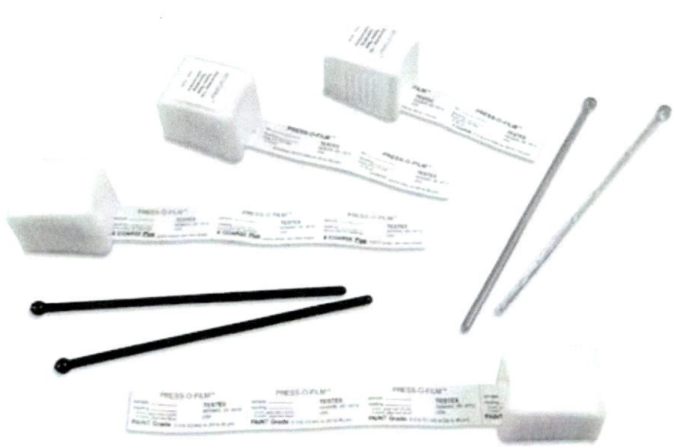

102F12.EPS

Figure 12 ◆ Replica tape.

If the tape is not rubbed with sufficient pressure, the reading will not be correct.

Replica tape is most commonly available under the trade name Testex Tape and comes in several grades:

- Elcometer© 122 B, Testex Tape, 50 Tape Roll Coarse, 0.8 to 2 mil (0.02 to 0.05 mm)
- Elcometer© 122 C, Testex Tape, 50 Tape Roll Extra Coarse, 1.5 to 4.5 mil (0.038 to 0.11 mm)
- Elcometer© 122 D, Testex Tape, 50 Tape Roll Paint Grade, 1.3 to 3.3 mil (0.033 to 0.084 mm)
- Elcometer© 122 F, Testex Tape X, Coarse Plus, 1.5 to 8 mil (0.038 to 0.2 mm)

12.0.0 ◆ DRY ABRASIVE SELECTION FACTORS

The type of abrasive used must be the type specified. Some or all of the following considerations should be taken into account when a blasting abrasive is selected:

- The kind of surface to be cleaned
- Size and shape of object to be cleaned
- Type of cleaning facility: outdoor, indoor with cabinet, or blast room
- Existing surface conditions
- Conditions desired after cleaning
- Types of coating to be applied
- Whether or not the abrasive is to be recycled
- Desired surface profile
- Size of abrasive
- Shape of abrasive
- Density of abrasive
- Hardness of abrasive
- Friability (brittleness) of abrasive

12.1.0 Classification and Characteristics of Dry Blast Abrasives

Abrasives are materials that do the actual work in abrasive blast cleaning. The two major classifications of abrasives are metallic and non-metallic.

Table 5 lists the two major classifications of abrasives and some types of abrasives found in each category. Choice of abrasive is generally determined by the specification and may be detailed within the coating manufacturer's application instructions or on the technical data sheets for a specific product. *Table 6* details the surface profiles the various abrasives yield.

12.2.0 Size of Dry Abrasive Sieve Test

Sieves are used to measure the grain size of abrasives (*Figure 13*). The larger the number is of the sieve, the smaller the particle. The number of the sieve indicates the number of holes per inch of screen. A No. 10 sieve has 10 holes per linear inch (0.4 cm), so particles smaller than a tenth of an inch (0.10 in [0.04 cm]) can pass through. A No. 200 sieve has 200 holes per inch (2.5 cm) so only very small particles can pass through.

Abrasives are often referred to by a size ratio, so 30/60 sand means that all of the particles need to pass through a No. 30 but not a No. 60 sieve.

A batch test is performed to check that a new batch (or lot) of abrasive meets the requirements in the specification.

12.3.0 Cleanliness of Abrasives

- *Lead content* – A representative sample of cleaned work-mix ferrous metallic abrasive shall be tested for lead. This needs to be done by a laboratory.
- *Water-soluble contaminants* – A representative sample of cleaned work-mix ferrous metallic abrasive shall be tested for conductivity. A sample of the abrasive is taken and mixed with water. The conductivity of the water is checked and a high number indicates the presence of soluble salts.
- *Oil content* – There shall be no presence of oil either on the surface of the water or as an emulsion. A UV light can be used to help detect oil.
- *pH* – The pH of the water should be checked and be near neutral. Abrasive on both the low or high side can indicate contamination.

Table 5 Classification and Types of Abrasives

		NON-METALLIC		
Metallic	**Hard Non-Metallic**	**Siliceous**	**Agricultural**	**Specialty**
Chilled cast iron	Silicon carbide	Quartz	Corn cobs	Dry ice
Cast steel	Garnet	Flint	Black walnut shells	Ice
Malleable iron	Refractory slag	Sand	Pecan shells	Plastic beads
Crushed steel	Mineral slag	Silica	Peach pit shells	Soda
Cut steel wire			Almond shells	Sponge
Aluminum oxide			Apricot pit shells	Glass beads
			Rice hulls	
			Sugar	

Table 6 Anchor / Profile Pattern and Associated Abrasive

Anchor Pattern		Pressure Blast (1) or Centrifugal Wheel
mm	**mils**	
12.7	0.5	80/120-mesh silica sand, 100-mesh garnet,120-grit aluminum oxide, or G-200 iron or steel grit (2)
25.4	1.0	30/60-mesh silica sand, 80-mesh garnet, 100-grit aluminum oxide, or G-80 iron or steel grit
38.0	1.5	16/40-mesh silica sand, 36-mesh garnet, 50-grit aluminum oxide, or G-50 iron or steel grit
50.8	2.0	16/40-mesh silica sand, 36-mesh garnet, 36-grit aluminum oxide, or G-40 chilled iron or steel grit
63.5	2.5	16/40-mesh silica sand, 36-mesh garnet, 24-grit aluminum oxide, or G-25 iron or steel grit
76.2	3.0	8/20-mesh silica sand, 16-mesh garnet, 16-grit aluminum oxide, or G-16 chilled iron or steel grit

Notes:

(1) Pressure blast: normally around 620 kPa (90 psi) nozzle pressure 60 cm (2 ft.) from surface.
Steel shot is not normally recommended when a sharp anchor pattern is required; its round shot peens the surface.
It should be remembered that abrasive sizes vary and a close inspection of size tolerance (measure to approximately +/-10 percent) should be maintained, especially where abrasives are reclaimed and reused. Reclaimed abrasives should be angular, not rounded, and free from oil, grease, iron oxide, etc.
These recommended abrasive sizes apply only to mild steel. As the hardness and type of metals vary, so will the anchor pattern.

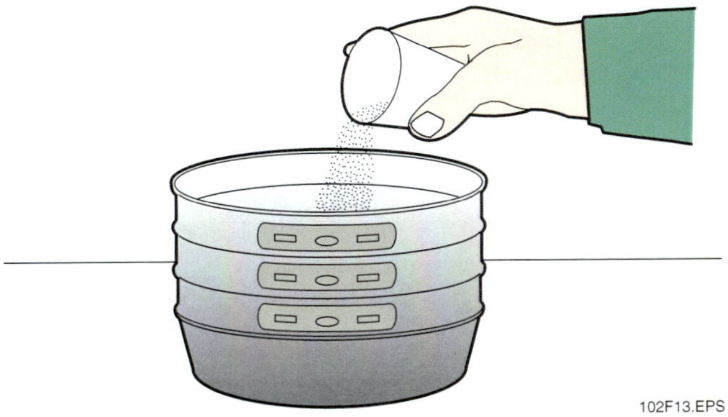

102F13.EPS

Figure 13 ◆ Typical selected stack of 8" screen sieves.

12.3.1 SSPC-AB 3, Ferrous Metallic Abrasive

This specification covers previously unused abrasive material to remove rust, mill scale, paint, or surface coating systems by blast cleaning. It does not address cleanliness of recycled abrasive processed through field or shop abrasive blast cleaning units. Requirements for recycled ferrous metallic abrasive are covered in SSPC-AB 2.

12.3.2 SSPC-AB 2, Cleanliness of Recycled Ferrous Metallic Abrasives

This specification covers the requirements for cleanliness of recycled ferrous metallic blast cleaning abrasives used for the removal of coatings, paints, scale, rust, and other foreign matter from steel or other surfaces.

Requirements are given for laboratory and field testing of recycled ferrous metallic abrasives work mix. Recycled ferrous metallic abrasives are intended for use in the field or for shop abrasive blast cleaning of steel or other surfaces.

12.4.0 Abrasive Additives

Sometimes additives are put into abrasives to improve their performance. There are additives that help reduce the amount of dust generated during blasting operations. There are also other additives that are added to abrasives to make them more environmentally acceptable for the disposal of heavy metals, such as lead.

12.5.0 Abrasive Storage

Depending on the type of abrasive being used, it is important to store the abrasives off the ground and in a dry environment. Wet abrasives can clump, and metallic abrasives can cause rust. Any contamination of the abrasive can also contaminate the surfaces being blasted.

12.6.0 Types of Dry Metallic Abrasives

12.6.1 Steel Grit

Steel grits are abrasives with irregular shapes. They are effective in cutting away surface deposits or imperfections. Steel grit is expensive and is generally used only in recycling systems and when less expensive abrasives are unavailable.

12.6.2 Steel Shot

Shot is spherical in shape. It can be produced accidentally as a by-product or intentionally for blast cleaning. Steel shot is good for heavy brittle deposits. Because many coating manufacturers require an angular profile to achieve good coating adhesion, shot may be mixed with steel grit to achieve benefits associated with each medium.

12.6.3 Cast Iron

Cast iron is the hardest metallic abrasive. It is used for removing scale and other hard surface deposits. Hardness should not be confused with toughness. These abrasives have a high initial cost, but can be recycled.

12.6.4 Malleable Iron

Malleable iron is a relatively hard metallic abrasive. It is used for removing scale and other hard deposits.

12.7.0 Hard Nonmetallic Abrasives (Oxides, Natural, or Synthetic)

Hard nonmetallic abrasives can be used indoors and outdoors, under various atmospheric conditions. They have a breakdown rate greater than any of the metallic abrasives. Although they have a low initial cost, they have a high loss rate during blasting. Reclaiming is critical due to the costs of using silicon carbide and aluminum oxide. The hard nonmetallic abrasives are free of the silica hazard present with siliceous abrasives. Types of hard nonmetallic abrasives include:

- *Garnet (oxide)* – A fast-cutting abrasive with a long life. It is commonly used for blasting cabinets and blasting rooms and has a diamond shape.

- *Silicon carbide (oxide)* – Expensive and fast-cutting. Silicon carbide is also used for grinding wheels.
- *Aluminum oxide* – The mineral corundum is natural aluminum oxide, and emery, ruby, and sapphire are impure crystalline varieties. Aluminum oxide can be manufactured from bauxite. It is fast-cutting, durable, uniform, and expensive.

12.8.0 Free-Silica-Containing Abrasives

Siliceous abrasives, like synthetic abrasives, can be used indoors or out, under a variety of atmospheric conditions. The potential to release free silica is a limiting factor because of the health hazards associated with breathing dust containing free silica. Types of free-silica-containing abrasives include:

- *Sand*
- *Flint/chert* – Flint (sometimes called chert) is a crystalline form of native silica or quartz
- *Quartz* – A sharp, fast cutting, and expensive abrasive

12.9.0 Slag Abrasives

Types of slag or conglomerate abrasives include:

- *Refractory slag* – Manufactured from by-products of burning coal, refining copper, and nickel. It is fast-cutting with medium durability.
- *Mineral slag* – SSPC-AB 1, Mineral and Slag Abrasives, is a specification that defines the requirements for selecting and evaluating mineral and slag abrasives used for blast cleaning steel and other surfaces for painting and other purposes. This specification mainly covers abrasives intended for one-time use without recycling. Reclaimed materials must be re-tested before reuse.

12.10.0 Specialty Abrasives

Specialty abrasives include:

- *Agricultural abrasives* – Agricultural abrasives are commonly used when the dust associated with other abrasives may be harmful to sensitive equipment. Agricultural abrasives may be used when the goal of the operation is simply to clean the surface as opposed to creating a surface profile.
- *Dry ice (carbon dioxide)* – This is frozen carbon dioxide. It does not change the surface profile of metal substrates. It produces little or no dust and usually leaves the substrate dry and cold.
- *Plastic beads* – These are small plastic beads (about the size of the holes in a button) used to remove coatings with a minimum of surface profile change to the substrate. Plastic beads are almost always used in recovery (recycling) systems. They are used extensively to strip coatings from aircraft.
- *Sodium bicarbonate (soda)* – Usually used in a water slurry to control dust and driven by compressed air. It is used to remove surface contamination and existing coatings with minimum substrate profile change. The substrate must be rinsed with fresh water as a final step.
- *Sponge* – Synthetic sponge particles are also used to remove surface contaminants. They are propelled by compressed air; the multiple impacts of the sponge particles clean the surface by absorbing contaminants. Sponge particles are also available with abrasives encased in the particle itself. These abrasive-embedded particles not only clean contaminants from the substrate, but are capable of removing coatings and creating a surface profile. While the material works less quickly than regular abrasives, dust levels in the work area are significantly reduced.

12.11.0 Recycling Abrasives

Recycling is usually done so that expensive abrasives, such as high-grade metallics, can be reused. It may also be done to reclaim less-expensive abrasives, such as sand, if shipping costs and/or availability are a factor. Most metallic abrasives, such as iron and steel shot and grit, and expensive abrasives, such as glass beads, may be recycled.

Care must be taken when recycled abrasives are used. While blasting a surface, the abrasive begins to fracture as it contacts the surface. Additionally, whatever contaminant that may remain on the surface mixes in with the abrasive. If you blast off oils or salts and recycle the abrasives, you can contaminate the substrate being cleaned.

To ensure that the abrasive has not broken up too much, samples can be run through a screen or a cyclone separator. It is even possible to wash the recycled abrasives to remove fine particles and salts. Abrasives must be dried before they are reused. If contamination cannot be removed from the recycled abrasive, it must be disposed of.

13.0.0 ♦ DRY ABRASIVE CLEANING STANDARDS

We have previously mentioned the NACE/SSPC blast standards that cover abrasive blasting and waterjetting. Since the majority of painting involves one of these methods, it is important to be familiar with the standards.

13.1.0 NACE No. 1/SSPC-CP 5, White Metal Blast Cleaning

This standard requires that after cleaning, the surface, when viewed without magnification, will be free of all visible oil, grease, dust, dirt, mill scale, rust, coating, oxides, corrosion products, and other foreign matter.

13.2.0 NACE No. 2/SSPC-SP 10, Near-White Metal Blast Cleaning

This standard requires that after cleaning, the surface, when viewed without magnification, will be free of all visible oil, grease, dust, dirt, mill scale, rust, coating, oxides, corrosion products, and other foreign matter, except that random staining may be present on not more than 5 percent of any unit area of the surface. The unit area is defined as a 3 × 3 ft area or an area that is 76 × 76 mm. The staining may consist of light shadows, slight streaks, or minor discolorations caused by stains of rust, stains of mill scale, or stains of previously applied coating.

13.3.0 NACE No. 3/SSPC-SP 6, Commercial Blast Cleaning

This standard requires that after cleaning, the surface (when viewed without magnification) will be free of all visible oil, grease, dust, dirt, mill scale, rust, coating, oxides, corrosion products, and other foreign matter. Random staining may be present on not more than 33 percent of the surface given any unit of area measurement.

The staining may consist of light shadows, slight streaks, or minor discolorations caused by stains of rust, stains of mill scale, or stains of previously applied coating.

13.4.0 NACE No. 4/SSPC-SP 7, Brush-Off Blast Cleaning

This standard requires that after cleaning, the surface (when viewed without magnification) will be free of all visible oil, grease, dirt, dust, loose mill scale, loose rust, and loose coating. Tightly adherent mill scale, rust, and coating that pass the dull putty knife test may remain.

13.5.0 NACE No. 8/SSPC-SP 14, Industrial Blast Cleaning

Industrial blast cleaning provides a greater degree of cleaning than brush-off blast cleaning (NACE No. 4/SSPC-SP 7), but less than commercial blast cleaning (NACE No. 3/SSPC-SP 6). The difference between industrial and brush-off is that the objective of a brush-off blast is to leave as much of the old coat as possible. The purpose of the industrial blast is to remove most of the coating. The industrial blast allows defined mill scale, coating, and rust to remain on less than 10 percent of the surface. It also allows defined stains to remain on all surfaces.

Mill scale can be detected by applying an acidic solution of copper sulfate to a blast-cleaned surface. The area will appear as a bright copper color on clean steel, and black when mill scale is present.

14.0.0 ♦ DRY ABRASIVE BLASTING TECHNIQUES

In spite of the common term *sand blasting*, sand is no longer commonly used as an abrasive because it contains free silica. Workers exposed to free silica dust may develop silicosis, a disease of the lungs. Some naturally occurring forms of sand are silica-free and may be used. Silica-free sand does not release silica in a harmful form during blasting. Sand may occasionally be used in the open-air work sites. In these cases, the operators and other personnel must be protected from the dust. Sand is used often in wet abrasive operations.

Correct application techniques are essential to make sure the substrate to be cleaned meets the required standards. It takes practice for the operator to maintain consistency and maximize efficiency. However, the basic processes that should be followed include:

- Maintain proper air pressure; normally 90 to 100 psi (620 to 690 kPa).
- Ensure the orifice size is appropriate for the desired results.
- For most surfaces, maintain a nozzle angle of 80 to 90 degrees to the surface being blasted.
- Maintain consistency in the speed with which you move the nozzle, as well as the distance of the nozzle from the surface.

- The speed of nozzle travel should be rapid enough to clean adequately without causing unnecessary dwell time.
- Each succeeding pass must overlap the preceding one and expose clean metal without any discolored patches.

The optimal distance the blast nozzle should be held from the surface is:

- ¼ in (6.3 mm) nozzle, optimal distance 14 in (36 cm)
- ⅜ in (9.5 mm) nozzle, optimal distance 22 in (56 cm)
- ½ in (12.7 mm) nozzle, optimal distance 28 in (71 cm)

Monitor the flow of abrasives to the nozzle. To adjust the flow of abrasives to the nozzle, use the abrasive flow valve on the blast pot. Adjust the position of the valve a small amount and then test the flow at the nozzle. If you hear a high-pitched whistling noise from the nozzle, there is not enough abrasive flow. If there is an irregular pulsating sound, there is too much abrasive.

Surfaces with heavy mill scale layers, thermal-sprayed zinc, aluminum coatings, initially need to be blasted at a shallower angle (e.g., 45 degrees). This technique removes the existing coating, but produces a reduced surface profile due to the angle of impact. Final blasting at right angles to the surface is required to achieve the correct surface profile.

To prevent injury to the operator, in case of an accident, use the deadman valve appropriately. Hold the deadman valve in a closed position to start and operate the blasting system. When the valve is opened, the blast machine is shut down, thus safeguarding the operator.

14.1.0 Protecting Areas and Equipment from Dry Abrasive Blasting

When blast cleaning is planned, there are generally some areas that will not be blast cleaned: some surfaces are already painted, or there is sensitive equipment in the area that would be harmed. These surfaces need to be protected by masking them with suitable templates of metal, rubber, or tough masking tape firmly secured to the surface.

Masking materials should be in place before any blast cleaning begins. They should be checked on a regular basis to ensure that they are providing good protection. When blasting and painting have been completed, carefully remove masking materials.

14.2.0 Considerations Before and During Blasting

Ambient environmental conditions can have an effect on the abrasive blasting process as well as the surfaces to be cleaned. Ambient conditions include the following:

- Air temperature (and substrate temperature)
- Relative humidity
- Dew point temperature
- Environmental exposure (for example, marine and industrial)

When a steel surface is much colder than the surrounding air, moisture may condense on the surface that is blasted, causing the surface to flash rust. The substrate temperature can be checked using a steel surface thermometer. Flash rusting can also happen in high-humidity (moisture) conditions. Final dry abrasive blasting should not be performed in wet or damp conditions (that is, during rainfall or when the relative humidity is very high, generally greater than 90 percent). When flash rust is a problem during dry abrasive blasting, it may be necessary to dehumidify the area during surface preparation.

14.3.0 ISO 8502-3:2000 Dust Tape Test

This test determines the quantity and size of dust particles on cleaned surfaces prepared for painting. This test must conform to the recommendations of BS EN ISO 8502-3:2000, either as a pass/fail test or as a permanent record of the presence of dust.

1. Remove a piece of tape about 200 mm (8 in) long, holding it by the ends only.
2. Place the sticky side of the tape onto the test surface. Use your thumb to make sure it sticks.
3. Grasp the tape at one end and pull it away from the test surface.

4. Check the quantity of dust on the tape. Compare the dust on the tape with the samples at the bottom of the comparator display board. Record the dust rating number that matches most closely.
5. Determine the size of the dust particles on the tape.
6. Repeat the test on at least three areas of the surface being investigated.

See *Table 7* for dust particle classes and descriptions.

15.0.0 ◆ WET ABRASIVE BLASTING

In wet abrasive blasting, potable or low-chloride water is added to conventional abrasive blasting nozzles via an adapter (retrofit water curtain device). The water ring attachment is connected to a low-pressure water source. The flow of water is controlled by a small valve on the attachment. In use, the water stream wets the abrasive stream outside and in front of the nozzle discharge, which helps keep dust to a minimum. This cleaning process may be used when dust is objectionable. The surface profile is similar to that achieved with dry abrasive blasting, except the surface is wet after blasting. The advantages of wet abrasive blasting are:

- The reduction of dust. This is particularly important when blast cleaning takes place in a public environment, such as when city buildings are blast cleaned. For this reason, equipment was created to provide a shroud of water around a normal flow of abrasive carried in compressed air.
- The ability to remove soluble salts and other contaminants from the surface.
- Hazardous components of the waste (including dust emissions) are reduced. Up to 75 percent of dust cannot escape from the water stream and is less likely to enter the immediate surrounding environment.

The disadvantages of wet abrasive blasting include:

- The problem of removing the used abrasive (in its wet form).
- The need for water removal (which can be contaminated with paint chips and surface contamination).
- The necessity of using an inhibitor in the water to prevent surface rusting when wet abrasives are used to clean steel.
- Clean-up time is longer, because of the difficulty of removing wet abrasive sludge.

15.1.0 Rust Inhibitors

Rust-inhibiting chemicals may be added to the water to stop the formation of rust on steel during the wet abrasive blast operation.

Some products use film-forming inhibitors that prevent rusting by depositing a thin film of corrosion inhibitors.

Others use volatile amines to prevent rusting. The amines temporarily coat the surface of the steel, keeping the water away from the steel until the water evaporates. Then, the amines evaporate, leaving a rust-free surface. These amines will also remove salts and other contaminants from the surface.

Table 7 Dust Particle Rating

Class	Description of Dust Particles
0	Particles not visible under 10X magnification
1	Particles visible under 10X magnification but not with normal or corrected vision (usually particles less than 50 m in diameter)
2	Particles just visible with normal or corrected vision (usually particles between 50 m 100 m in diameter)
3	Particles clearly visible with normal or corrected vision (particles up to 0.5 mm in diameter)
4	Particles between 0.5 mm and 2.5 mm in diameter
5	Particles larger than 2.5 mm in diameter

16.0.0 ◆ SAFETY AND ABRASIVE BLASTING

With abrasives and equipment under pressure, safety is very important. Abrasives and air leave the nozzle at speeds above 644 kph (400 mph). This is about half the speed of a shotgun blast. It can shoot abrasives far away from the operation. The entire system (including hoses, operator, and workpiece) must be grounded to prevent injury from electrical shock. Electrical grounding is especially important when the operator is working at heights (when shock may lead to a fall) or when blasting is done in a hazardous environment.

For the safety and comfort of the blast cleaning operator, it is essential that good-quality work clothing be worn. Usually, this would include:

- Safety boots with steel insert toe cap
- Coveralls
- Strong leather gloves with long cuffs
- Clean, cooled air-fed blasting helmet, incorporating a replaceable visor and leather cape
- Hearing protection

All participants in the surface preparation process should use common sense in detecting potential hazards. Any lighting, scaffolding, or equipment malfunctions presenting safety hazards should be reported to the appropriate person. Knowledgeable and responsible workers always ensure the safety of staging or scaffolding before using them for work or for inspection. Never get in the path of an abrasive blast. Follow all safety and health rules as established by the safety engineer or the person responsible for safety on the job.

Abrasive blast cleaning at high pressure is a dangerous operation. It is essential to use good safety practices to protect operators, spectators, and other personnel. Some considerations are as follows:

- Only the operator should be allowed within the area of the blasting operation.
- Warning notices and barricades should be posted at appropriate points and especially at blind corners.
- A look-out (or pot-man) should be on the alert.
- All equipment should be tested for proper operation before use.
- A deadman (remote control) valve, which allows the machine to be controlled at the nozzle by the operator, should be part of this unit.

17.0.0 ◆ WATER CLEANING AND WATERJETTING

Waterjetting is the use of a high-pressure water stream to clean surfaces. It is also used to strip existing coatings. With waterjetting, respiratory requirements may be more flexible than with other methods of surface preparation, but waterjetting has its own set of hazards. Even water under low pressure is powerful enough to scrape skin. Water under high pressure can be lethal.

The commercial waterjet unit can be skid-, trailer-, or truck-mounted. It usually consists of pumps, hoses, a prime mover (diesel, electric, etc.), and various tools such as guns, nozzles, and lances.

17.1.0 Waterjetting Standard, NACE No. 5/SSPC-SP 12

NACE No. 5/SSPC-SP 12 defines waterjetting as the use of standard jetting discharged from a nozzle at pressures above 70 MPa (10,000 psi) to produce a surface for coating or inspection. NACE No. 5 defines water cleaning as the use of pressurized water discharged from a nozzle to remove unwanted matter from a surface. The following four specific definitions are provided in the standard:

- *Low-Pressure Water Cleaning (LP WC)* – Water cleaning performed at pressures less than 34 MPa (5,000 psig). This is also called power washing or pressure washing.
- *High-Pressure Water Cleaning (HP WC)* – Water cleaning performed at pressures from 34 to 70 MPa (5,000 to 10,000 psig).
- *High-Pressure Waterjetting (HP WJ)* – Waterjetting performed at pressures from 70 to 210 MPa (10,000 to 30,000 psig).
- *Ultrahigh-Pressure Waterjetting (UHP WJ)* – Waterjetting performed at pressures above 210 MPa (30,000 psig).

NACE No. 5/SSPC-SP 12 defines three parameters of surface cleanliness, four levels of visible contaminants, four levels of flash rust, and three levels of nonvisible contaminants. The levels of visible contaminants are:

- *WJ-1, Clean to Bare Substrate* – A WJ-1 surface shall be cleaned to a finish which, when viewed without magnification, is free of all visible rust, dirt, previous coatings, mill scale, and foreign matter. Discoloration of the surface may be present.

- *WJ-2, Very Thorough or Substantial Cleaning* – A WJ-2 surface shall be cleaned to a matte (dull, mottled) finish which, when viewed without magnification, is free of all visible oil, grease, dirt, and rust except for randomly dispersed stains of rust, tightly adherent thin coatings, and other tightly adherent foreign matter. The staining or tightly adherent matter is limited to a maximum of 5 percent of the surface.
- *WJ-3, Thorough Cleaning* – A WJ-3 surface shall be cleaned to a matte (dull, mottled) finish which, when viewed without magnification, is free of all visible oil, grease, dirt, and rust except for randomly dispersed stains of rust, tightly adherent thin coatings, and other tightly adherent foreign matter. The staining or tightly adherent matter is limited to a maximum of 33 percent of the surface.
- *WJ-4, Light Cleaning* – A WJ-4 surface shall be cleaned to a finish which, when viewed without magnification, is free of all visible oil, grease, dirt, dust, loose mill scale, loose rust, and loose coating. Any residual material shall be tightly adherent.

The levels of flash rust defined in NACE No. 5/SSPC-SP 12 are:

- *No Flash Rust* – A steel surface which, when viewed without magnification, exhibits no visible flash rust.
- *Light (L)* – A surface which, when viewed without magnification, exhibits small quantities of a yellow-brown rust layer through which the steel substrate may be observed. The rust or discoloration may be evenly distributed or present in patches, but it is tightly adherent and not easily removed by lightly wiping with a cloth.
- *Moderate (M)* – A surface which, when viewed without magnification, exhibits a layer of yellow-brown rust that obscures the original steel surface. The rust layer may be evenly distributed or present in patches, but it is reasonably well adherent and leaves light marks on a cloth that is lightly wiped over the surface.
- *Heavy (H)* – A surface which, when viewed without magnification, exhibits a layer of heavy red-brown rust that hides the initial surface condition completely. The rust may be evenly distributed or present in patches, but the rust is loosely adherent, easily comes off, and leaves significant marks on a cloth that is lightly wiped over the surface.

NACE No. 5/SSPC-SP 12 limits its definition of nonvisible contaminant levels to water-soluble chlorides, iron-soluble salts, and sulfates. It is important to note that other nonvisible contaminants (e.g., organic contaminants) may be present. Greases and oils, for example, may be detected by water-break tests or black-light inspection. Other ionic species may be detected by measuring surface conductivity. In any case, the following levels of nonvisible surface contaminants are defined:

- *NV-1* – An NV-1 surface shall be free of detectable levels of soluble contaminants, as verified by field or laboratory analysis using reliable, reproducible test methods.
- *NV-2* – An NV-2 surface shall have less than 7 $\mu g/cm^2$ (0.0007 $grain/in^2$) of chloride contaminants, less than 10 $\mu g/cm^2$ (0.001 $grain/in^2$) of soluble ferrous ion levels, or less than 17 µg/cm2 (0.0017 $grain/in^2$) of sulfate contaminants as verified by field or laboratory analysis using reliable, reproducible test methods.
- *NV-3* – An NV-3 surface shall have less than 50 $\mu g/cm^2$ (0.005 $grain/in^2$) of chloride or sulfate contaminants as verified by field or laboratory analysis using reliable, reproducible test methods.

NACE, SSPC, and other societies have agreed to use the term *waterjetting* to describe the cleaning process where water alone is the cleaning medium. The term *water blast* has been discussed previously and is reserved to describe any cleaning process in which some type of abrasive will be incorporated with water to form the cleaning medium. Because of the high pressures required when waterjetting is performed, safe practice dictates use of great care in controlling the waterjet nozzles. A person struck by high-velocity water at short range could be seriously injured.

The advantages of high-pressure waterjetting include:

- The use of water as a cleaning material, because water suitable for water cleaning is generally available in inexpensive large quantities
- Removal of all rust, coatings, and contaminants
- Lack of contamination of surrounding areas because there are no abrasive particles
- Lack of dust and spark hazards

17.2.0 Waterjetting Equipment

Equipment used for waterjetting includes:

- High-pressure water pump attached to a motor of suitable size
- High-pressure hose
- Nozzle

In waterjetting, as in wet abrasive blasting, a rust inhibitor is sometimes added to the water to help prevent rusting of the cleaned surface before a coating can be applied. If an inhibitor is used, it is important to ensure that the coating is compatible with the inhibitor.

The waterjet unit consists of pumps, hoses, and a prime mover (diesel, electric, etc.). It also includes various tools such as guns, nozzles, lances, etc. The (high-pressure) hose, hose connections, and all other equipment, including the nozzle control valve, lance, and nozzle are included. The unit consists of:

- The hose connectors are fitted with a whiplock. A whiplock is a short length of cable or wire looped over each end of two hoses connected by a coupling.
- The section of hose next to the gun is fitted with a hose shroud. The shroud is a short length of heavy-duty hose placed over the high-pressure hose.
- Waterjets are produced by orifices or tips that come in different forms. Round jets are most commonly used. Other shapes are available.
- Typical water flow rates are 1 to 14 gallons per minute (gpm) (3.8 to 53 Lpm). Water can be propelled through a single jet, a fan jet, or multiple rotating jets.
- The tools can be handheld or mounted on a robot. When operators are working with pressures of 4,000 psi (281 kg/cm^2) or higher, the cleaning team consists of a nozzle operator, pump operator, and other workers, as needed.
- As with abrasive blasting at the same volume of water, a larger orifice size will result in lower pressure, and a smaller orifice size in higher pressure.

Depending on the size and scope of the project, other operators or workers may be required to assist in handling the jetting gun. This is especially true if the gun has more than one jetting extension or if the hose must be fed to the workpiece. To decrease operator fatigue and ensure a safe operation, workers can alternate the nozzle operator position at designated intervals.

17.3.0 Water Cleaning Preparation

Before starting the job, the surface preparation team should ensure the following:

- The work area is properly barricaded and appropriate warning signs are installed.
- Electrical equipment is properly covered and protected from the water.
- Electrical connections are not allowed to sit in water.
- One member of the team should be responsible for checking that all necessary permits are in place. This includes any confined space permits, as well as those that may be required by local, state, and federal agencies.
- All fittings and hoses are in good condition (not worn or damaged). They must be properly pressure rated for the working pressure to be used.
- Nozzles are open and free of obstructions.
- The complete hook-up is flushed clean and air removed from the system before the nozzle is installed.
- The dump system and all control systems are operational.
- All relevant moving equipment, such as conveyors, mixers, etc., is mechanically or electrically disabled. The appropriate lockout devices must be in place.

17.4.0 Waterjetting Techniques

The most common waterjetting technique is shotgunning. When using a shotgun technique to clean large surface areas, be aware of where the tip is pointed. The effectiveness of the cleaning depends on how close the tip is to the surface being cleaned. With a fan jet, the wider the area being covered in a sweep, the less force available.

17.5.0 Safety in Waterjetting

The following safety rules apply whenever waterjetting operations take place:

- As with a firearm, never point the nozzle of the lance at another person. Always be aware of where the lance is pointed.
- Do not point the nozzle at yourself. Remember, the water that comes out of the nozzle is capable of removing skin. If it is close enough, it may cut your flesh. Even with protective cloth-

ing, the jet will cut away the armor more quickly than is expected.

- Do not use the lance to clean your shoes or your face shield.
- Always be on the lookout for hazards.
- If the water does not cut off when the safety valve is released, stop work immediately. Shut off pressure at the pump, drain the system, and have the valve repaired or replaced.
- Never try to repair the system while it is pressurized. Always stop the operation, shut down the pump, and release pressure before repairing leaks or failures. Never leave the pump unattended while it is running.
- If you must work in an enclosed space, engine exhaust must be vented out of the area. The carbon monoxide in the exhaust can overwhelm workers in a matter of minutes.
- In addition to the standard equipment, all workers in the area of a waterjetting operation must wear waterproof suits, head protection, and protective goggles or a full-face shield (*Figure 14* and *Figure 15*).
- The sound level of a waterjetting operation frequently runs in excess of 90 decibels. Hearing protectors or earplugs are required.
- Take special precautions if removing a toxic substance.

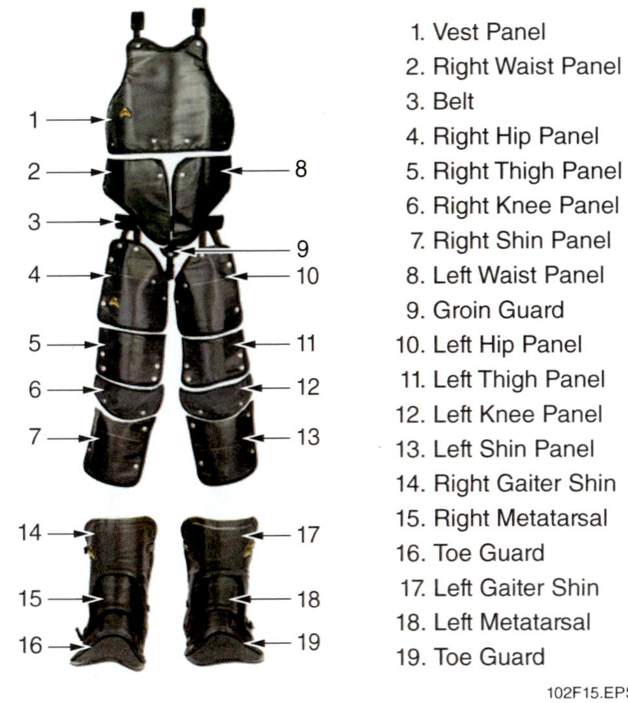

Figure 15 ◆ Armor.

- All blast accidents must be reported immediately. Even when the blast does not visibly cut the skin, any contact can force water, air bubbles, and toxins under the skin.

Figure 14 ◆ Waterproof clothing.

17.6.0 Waterjetting Environmental Issues

While waterjetting is a relatively clean technology and uses water efficiently, it can produce a lot of waste water. On many sites, clean water may be channeled to storm drain systems. In some places, water is recovered and recycled, using either filtration systems or a series of settling tanks, or both. Be sure to check local codes and company requirements affecting discharge of cleaning water before setting up equipment for a project.

It is very important that the waterjetting technician knows what is to be blasted. It may be toxic. If so, the project must capture waste water. Many times, the waterjetting technician will be working to remove potentially toxic materials from a surface. The materials may be blasted loose and either thrown into the air as tiny particles or as vapors mixed with water. The technician must know what materials may be released. Proper personal protective equipment and environment protection must be used at all times.

18.0.0 ◆ CHEMICAL STRIPPING

Sometimes conditions exist that make blasting undesirable. If a surface has been previously painted, it probably already has a surface profile. So, if the coating can be removed, the existing profile can be used. Depending on the coatings type, there are chemicals on the market that will dissolve or soften many coatings (*Figure 16*). The chemicals are applied and the coating can then be mechanically stripped off. Some of the problems with this method are that many strippers pose environmental problems and the strippers themselves become part of the waste. There are some environmentally friendly strippers on the market, but they do not work on all coatings.

102F16.EPS

Figure 16 ◆ Example of a paint removal operation using a stripper and a scraper.

19.0.0 ◆ SURFACE PREPARATION OF MASONRY AND CONCRETE

Concrete is the most widely used material in construction. It is made by mixing cement, which is a powdered mixture of lime and other minerals, some type of **aggregate**, and water. Concrete needs to cure over time (traditionally 28 days) to create a hard, durable surface.

Cured concrete can be finished in a variety of ways. *Figure 17* shows how a stiff-bristle broom is used for a rough finish. *Figure 18* shows how a smooth finish is achieved with a hand trowel and bull float.

102F17.EPS

Figure 17 ◆ Brooming concrete.

102F18.EPS

Figure 18 ◆ Concrete finishing with a bullfloat and trowel.

Depending on its application, concrete may or may not require coating. Concrete is a very **porous** material. It readily allows moisture to pass through it. This can speed corrosion of any metal reinforcing material in the concrete. In cold climates, moisture in the concrete expands as it freezes. This will break apart the concrete from internal pressure. If concrete is coated, it will last longer because exposure to moisture is limited.

When concrete is to be coated, it requires special preparation so the coating will adhere to the surface. These preparations are covered in detail in NACE No. 6/SSPC-SP 13, Joint Surface Preparation Standard for Surface Preparation of Concrete.

19.1.0 Methods of Surface Preparation of Concrete

Surface preparation of concrete is generally performed by one of the following methods:

- Abrasive blast cleaning blast cleaning will provide a rough, irregular surface (profile). Abrasive blasting opens holes and voids so they can be sealed.
- Wet abrasive blast cleaning
- Waterjetting
- Hand and power tool cleaning

Hand and power tool cleaning of concrete surfaces is time consuming, labor-intensive, and costly. Hand or power tools can be used to remove loose, powdery, weak concrete at the surface. This is slow work and does not open air pockets as well as abrasive blasting. However, it may be the only way to clean the edges of the concrete.

19.2.0 New Concrete

To prepare concrete for coating, a number of operations may be listed in the project specification, but the first step is always inspection. The purpose of the inspection is to spot problem areas in the surface of the concrete. These problem areas may include:

- Pits and voids
- **Efflorescence**, which is a fluffy white deposit on the surface
- Projections
- Porosity
- Moisture content
- Oils or grease
- Visible residues of dirt, chemical salts, or other foreign matter that can cause coatings to fail
- Ice or ice crystals on the surface, which require special attention when coating outdoors in very cold weather
- Water on the surface

19.3.0 Old Concrete

Chemical contaminants, oil, and grease can be removed by either steam cleaning or cleaning with chemicals or detergents. Sometimes, it is necessary to blast concrete clean with water, abrasives, or both. When severe contamination exists, it may not be possible to remove it. Instead, it might be necessary to replace the existing concrete. Coatings can only protect good concrete. Trying to coat damaged or contaminated concrete will not extend its life.

19.4.0 Water Blasting and Wet Abrasive Blasting

Lower-pressure water cleaning at pressures of 3,000 to 4,500 psi (246 to 316 kg/cm^2) is often used on poured concrete surfaces. This range of pressures does not usually open voids and pockets, and water cleaning will not achieve as good a surface profile on concrete as abrasive blasting. Wet abrasive blasting also may be used.

The advantages of waterjetting and wet abrasive blasting include:

- Speed
- Good surface profile (wet abrasive blasting)
- Washing away of surface dust
- Suppressing of abrasive and concrete particles in air

19.5.0 Acid Etching

Acid etching uses a diluted solution of hydrochloric acid to remove **laitance** and roughen the concrete surface. It is applied to a clean concrete surface and allowed to sit on the surface until it stops bubbling (usually about 5 to 10 minutes). Then, the surface is washed to remove the acid salts. Sometimes, several etchings are needed to achieve the required cleanliness and surface profile.

All acids are toxic and corrosive. Never let them touch your skin or clothing. Acids will eat holes in any fabric and can burn skin and eyes. Goggles, a face shield, rubber gloves, and rubber boots should be used whenever you are using acid solutions.

19.6.0 Concrete Repair

Concrete repair may be done on freshly poured concrete before surface preparation, or it may be done after cleaning and surface preparation. If it is done after surface preparation, make sure that the repaired areas are suitable for coating. Cement, synthetic putties, or grout can be used to repair services and voids, depending on the project specifications.

19.7.0 Measure the Moisture Content

Once the surface has been prepared, you will need to measure the moisture level of the concrete. The type of coating to be applied determines the desired moisture content. A moisture level that is too high during coating can lead to coating failure and promote corrosion of metal in the concrete.

The moisture level in concrete may be measured using a moisture meter (*Figure 19*). Moisture meters come in two types: with pins and without pins. To use a meter with pins requires drilling holes in the concrete to receive the pins. Generally, it is easier to use pinless meters for concrete. Moisture levels can also be assessed by the plastic sheet method or a calcium chloride test.

19.8.0 Concrete Surface Roughness Standards

An organization known as International Concrete Repair Institute (ICRI) is one of the standards groups for the concrete industry. They publish standards that define the approximate surface roughness of concrete.

Coating thickness varies depending on intended use, surface roughness, and profile. The International Concrete Repair Institute (ICRI) has developed a standard for concrete surface profile (CSP) ranging from one (smoothest) to nine (roughest).

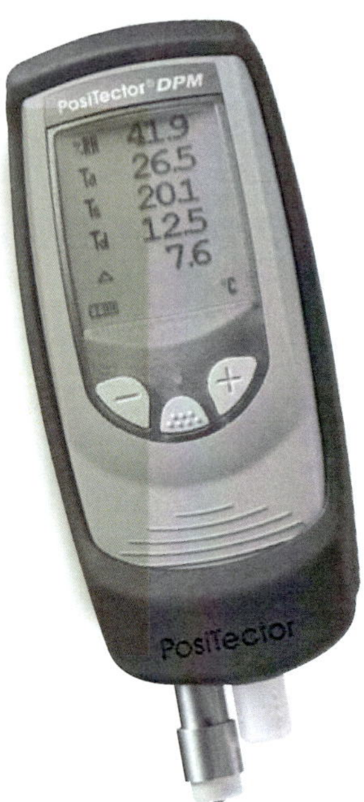

102F19.EPS

Figure 19 ◆ Moisture meter.

20.0.0 ◆ SAFETY AND DEBRIS MANAGEMENT

During surface preparation operations, you may be working with materials that are hazardous to your health or to the environment. This is not only because of chemicals being used, but also due to debris from surfaces. Things like silica from concrete and lead, arsenic, and cadmium from old coatings can affect your long-term health. In these cases, you must take special precautions.

CAUTION

If lead is present in the blast debris, the surface cannot be blown down. The contractor must follow all regulations regarding lead removal.

20.1.0 Protecting Yourself

- Wear the appropriate protective equipment.
- Be certain to avoid bringing your work home with you in the form of dust particles. This means you need to change from your work clothing into clean clothing as soon as possible. Do not launder your work clothing with other clothing.
- Paint chips should never be released into the environment, even if they are not considered hazardous. To protect the environment, debris must be collected and must be disposed of in an approved manner.
- Dust should be collected as it is generated. Vacuum devices are attached to equipment so paint chips and dust particles are collected immediately. This allows workers to work in a clean, virtually dust-free environment. Some devices are so effective that measured amounts of hazardous materials are so low that respirators are not needed.
- Debris management is more than collecting paint chips and dust. It is also controlling the amount of abrasives used. A more expensive abrasive that does not break down easily may be more cost-effective in the long-term.
- Debris management includes the collection of hazardous solvents as well as abrasives that have been used during surface preparation.

This module has presented a broad view of surface preparation as it pertains to the cleaning of steel and concrete substrates. It is important to become familiar with the different standards that apply to surface preparation. Always remember that conducting your work in safe and environmentally responsible manner is a key part of your job.

Review Questions

1. During surface preparation, you will *not* _____.
   a. apply coatings
   b. repair surface
   c. use power tools
   d. work with chemicals

2. Anchor profile refers to surface _____.
   a. pattern
   b. material
   c. thickness
   d. appearance

3. Coatings adhere best to a surface that is ____.
   a. smooth
   b. rough
   c. wet
   d. polished

4. Mill scale is _____.
   a. a fabrication imperfection
   b. rust that must be removed
   c. formed on hot steel as it cools
   d. the gross weight of the steel

5. While preparing a steel surface for coating, you notice a large amount of weld spatter in one area. The best thing to do is _____.
   a. ignore it because it won't affect the coating
   b. use a power grinder to remove it
   c. scrub it with an abrasive pad
   d. use epoxy to smooth it

6. Galvanized steel has been coated with _____.
   a. paint
   b. galvalum
   c. zinc
   d. metal

7. The surface corrosion on weathering steel must be removed to prevent further corrosion.
   a. True
   b. False

8. Aluminum _____ to be coated because its surface will _____.
   a. needs; oxidize
   b. doesn't need; oxidize
   c. needs; rust
   d. doesn't need; rust

9. Acid etching most commonly uses a diluted solution of _____.
   a. lye
   b. carbonic acid
   c. hydrochloric acid
   d. diatomaceous acid

10. When roughening the surface of concrete by acid etching, you must leave the etching solution on the surface _____.
    a. for exactly 5 minutes
    b. a minimum of 16 hours
    c. until it stops bubbling
    d. until it starts bubbling

11. After washing the surface with a detergent, _____.
    a. rinse it with clean water
    b. abrasive blast the surface
    c. wipe it dry with clean rags
    d. apply a coating of solvent

12. You can remove _____ with solvents.
    a. weld spatter
    b. mill scale
    c. some coatings
    d. sodium chloride

13. Which of the following statements is true?
    a. Acids can burn skin but alkalines cannot.
    b. Acid and alkalines can burn skin.
    c. Alkalines can burn skin but acids cannot.
    d. Acid and alkalines cannot burn skin.

14. To dispose of used acid and alkaline cleaners, _____.
    a. wash them down the drain
    b. pour them onto the ground
    c. collect them for processing
    d. let them dry on the surface

Review Questions

15. When cleaning surfaces with hand tools, wear eye protection _____.
    a. only when the material safety data sheet requires it
    b. whenever there are hand tools in use in the work area
    c. whenever Eye Protection Required signs are posted
    d. only when your employer requires them

16. One disadvantage of using power tools on steel surfaces is that they may _____.
    a. roughen the surface excessively
    b. use too much power
    c. polish the surface
    d. be too heavy

17. Power wire brushes are often used to _____.
    a. sweep up debris
    b. polish a surface
    c. smooth rough welds
    d. apply coatings

18. To achieve a deep surface profile using a power sander, _____.
    a. apply a lot of pressure to the tool
    b. move the tool in a circular motion
    c. use a coarse grit sanding disc
    d. use a fine grit sanding disc

19. A power tool that can remove a sound metal and leave burrs is a _____.
    a. wire brush
    b. chipping hammer
    c. disc sander
    d. grinding wheel

20. Which of the following is *not* true about dry abrasive blasting?
    a. It leaves a distinctive anchor pattern.
    b. It is good for removing mill scale.
    c. It leaves a good surface profile.
    d. It takes moisture out of the air.

21. You can test for mill scale on the surface of steel by applying _____, which turns clean steel a bright _____ color.
    a. copper sulfate; red
    b. ferrous sulfate; red
    c. ferrous sulfate; coppery
    d. copper sulfate; coppery

22. Sand is not used very often today for abrasive blasting because it _____.
    a. is too messy
    b. is in short supply
    c. can be a health hazard
    d is a non-renewable resource

23. It is necessary to recycle _____.
    a. crushed slag, ceramic grit, and iron shot
    b. glass beads, crushed slag, and ceramic grit
    c. iron shot, glass beads, and crushed slag
    d. ceramic grit, iron shot, and glass beads

24. It is very important that abrasives be kept _____.
    a. clean and free of moisture and oil
    b. clean and free of moisture
    c. free of moisture and oil
    d. clean

25. Which of the following components is *not* part of a basic dry abrasive system?
    a. Air compressor
    b. Water shroud
    c. Blast pot
    d. Hose

Summary

Surface preparation can be the most time consuming and expensive part of a coating operation. During surface preparation, you will use many different types of tools, chemicals, and equipment to clean and repair surfaces. Often, there will be varying degrees of cleanliness, depending on the surface and the coating. The goal of surface preparation is very simple and always the same: to remove grease, oils, salts, and other contaminates that will prevent coatings from adhering to a surface. Doing a good job during surface preparation does not guarantee a coating will tightly adhere to a surface and provide a long-lived, durable finish. Doing a bad job at surface preparation guarantees the coating will fail.

Notes

Trade Terms Introduced in This Module

Aggregate: Crushed rock or gravel that is screened to size for use in concrete.

Alloy: Often associated with metal. It indicates that another element has been added to a molten metal (such as carbon added to steel).

Efflorescence: A fluffy white deposit on the surface of concrete that is caused by moisture that carries salt in it rising to the surface.

Emulsify: To dissolve a fat or oil, usually with a detergent or other chemical, so that it can be washed away with water.

Flash rust: A type of rust that occurs after a steel surface has been exposed to moisture, either humidity in the air, or water.

Laitance: The accumulation of fine particles (usually salt) on the surface of freshly poured concrete. It is caused by water rising out of the concrete.

Mill scale: The black or dark blue coating on steel that is sometimes scaly in appearance. It is a result of hot steel cooling in the presence of oxygen.

pH: The alkalinity or acidity of a substance.

Porous: Indicates a material that has many small holes that allows liquid (usually water) to pass.

Saponify: Turning to soap. For example, when added to grease, some caustics turn the grease into a soap that can be washed away with water.

Stripe coat: A coat of paint the covers the edges of something, often used in overcoating operations to paint over the edges of the old coating before the final coating is applied.

Traps: Devices used to capture something. A water trap traps water.

Water-miscible: Indicates that a substance can be mixed with water.

Weathering: To expose the unprotected surface to weather conditions.

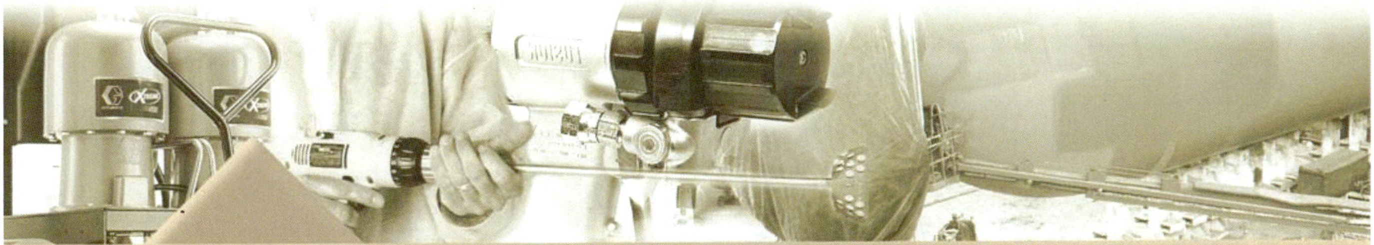

Appendix A

Request for Information (RFI)

Date:	**Request for Information**	RFI number:

Information Request

Project ______________________ Contractor ______________ Project Type ____________

Project Name ____________________ Purchase Order No. __________

Requested By ____________________

Information Requested:

______________________________ ______________________ ______________

Representative Signature Contact Number Date

Response Needed Date ________________

RESPONSE

Action: Per conversation with __________________ Dept. ____________________

No change

Repair

Reject

Other

______________________________ ______________________ ______________

Engineer Contact Number Date

Resolved

______________________________ __________

Representative Signature Date

Profiles in Success

John Troeger
Project Manager
Basic Industries, Inc.
Baytown, TX

How did you choose a career in the industrial coatings field?
While I was attending Texas Tech University I sold commercial/industrial protective coatings. Once I started in the business I followed one opportunity after another.

What types of training have you been through?
I have been through the NACE CIP program.

What kinds of work have you done in your career?
Industrial construction management including maintenance painting, insulation, and scaffolding as well as turnarounds.

Tell us about your present job and what you like about it.
I currently manage Basic Industries, Inc. sites in Texas, Illinois, Ohio, Iowa, and Maryland. Basic Industries provides specialty services contractors for the petrochemical, utilities, nuclear, pulp and paper, and refining industries. As a specialty contractor we do lead and asbestos abatement, coatings, scaffolding and shoring, tower and vessel maintenance, and general construction (IDIQ) for our clients.

What factors have contributed most to your success?
Be visible and accessible to your customers. Availability is just as important as ability. Develop a good relationship with your customers. Good relationships are built on communication and trust. I contact my customers frequently to let them know how the work is progressing. When they have concerns I make sure to follow up right away. By being responsive to my customers' needs I developed a good reputation and a good share of repeat business.

What advice would you give to those new to the field?
If you are willing to work hard it will pay off. Be reliable and hard working and you will be appreciated. Give your company and your customers your best every day. Treat them as well as you want them to treat you. A can do attitude goes a long way when it comes time for promotions and new opportunities.

Additional Resources and References

Additional Resources

This module is intended to be a thorough resource for task training. The following reference works are suggested for further study. These are optional materials for continued education rather than for task training.

Abrasive Blasting, Final Report, Prepared by Midwest Research Institute (MRI) for the Office of Air Quality Planning and Standards (OAQPS), U.S. Environmental Protection Agency (EPA), under Contract No. 68-D2-0159, Work Assignment Nos. 2-01 and 4-02. September 1997.

Coating Inspector Program Level 1© NACE International, 2003, January 2007.

Coating Inspector Program Level 2© NACE International, 2004, January 2007.

Coating and Color Manual COMDTINST M10360.3C COMTNOTE 10360 9 June 2007.

Maintenance Coating of Weathering Steel: Field Evaluation and Guidelines. Final Report, FHWA-RD-92-055, March 1992.

New Concepts for Coating Protection of Steel Structures – Stp 841: A Symposium by D.M. Berger, R.F. Wint, Steel Structures Painting Council (U.S.), ASTM Committee D-1 on Paint and Related Coatings and Material.

Solvent Waste Management, Inc. website http://www.solventwasher.com/index.htm.

Technical Advisory Uncoated Weathering Steel in Structures T 5140.22. U.S. Department of Transportation Federal Highway Administration, October 3, 1989.

Tips on Portable Shot Blasting, Heather Bayne, Journal of Protective Coatings & Linings, May 2007.

Figure Credits

Painting and Decorating Contractors of America, 102F01, 102F02, 102F08, 102F11, 102F13, 102F16

NACE International, 102F04, 102F05, 102F07, 102T01-101T07

Guide for the Visual Examination of Welds, AWS B1:2000, Figure 35. Reproduced with permission from the American Welding Society (AWS) Miami, FL, 102F03

Topaz Publications, Inc., 102F06, 102F14

Mike O'Brien, 102F09, 102F10

Elcometer, Inc., 102F12
800-521-0635, www.elcometer.com

Photo provided by TurtleSkin© Water Armor/Warwick Mills, Inc., 102F15

Baker Concrete Construction, Inc., 102F17, 102F18 (bottom)

Bon Tool Company, 102F018 (top)

DeFelsco Corporation, 102F19

NCCER CURRICULA — USER UPDATE

NCCER makes every effort to keep its textbooks up-to-date and free of technical errors. We appreciate your help in this process. If you find an error, a typographical mistake, or an inaccuracy in NCCER's curricula, please fill out this form (or a photocopy), or complete the online form at **www.nccer.org/olf**. Be sure to include the exact module ID number, page number, a detailed description, and your recommended correction. Your input will be brought to the attention of the Authoring Team. Thank you for your assistance.

Instructors – If you have an idea for improving this textbook, or have found that additional materials were necessary to teach this module effectively, please let us know so that we may present your suggestions to the Authoring Team.

NCCER Product Development and Revision

13614 Progress Blvd., Alachua, FL 32615

Email: curriculum@nccer.org

Online: www.nccer.org/olf

❑ Trainee Guide ❑ Lesson Plans ❑ Exam ❑ PowerPoints Other ________________________

Craft / Level: Copyright Date:

Module ID Number / Title:

Section Number(s):

Description:

Recommended Correction:

Your Name:

Address:

Email: Phone:

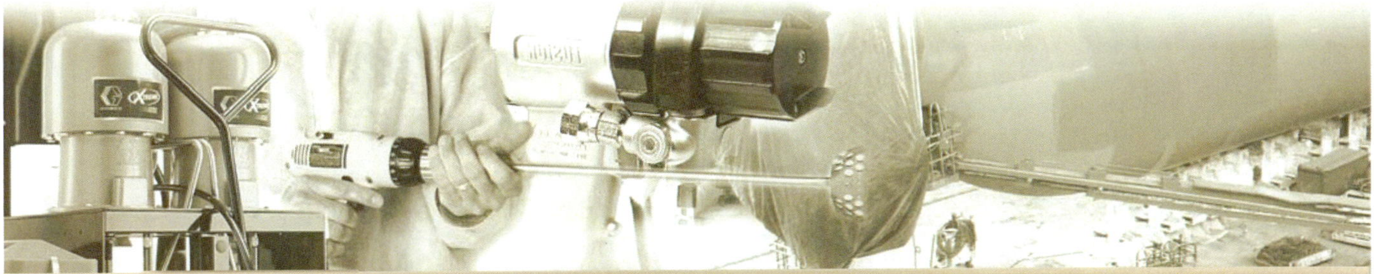

Industrial Coating and Lining Application Specialist Level One

69103-09

Industrial Coatings

69103-09

Industrial Coatings

Overview

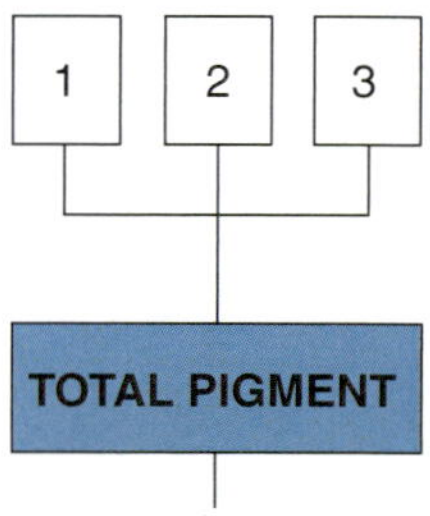

This module covers topics pertaining to basic and specialty coatings, coating composition, coating curing methods, and storage requirements. Other items covered include coating safety procedures, understanding material safety data sheets (MSDSs), and proper waste disposal.

Table of Contents

Figures

Tables

Objectives

When you have completed this module, you will be able to do the following:

1. State the purpose of each coating component.
2. List physical properties of paints.
3. List basic coating additives.
4. State the purpose for solvents (thinners) in cleaning.
5 State the difference between convertible and non-convertible coatings.
6. List curing mechanisms for coatings.
7. Identify conditions that must be considered before selecting a coating/lining.
8. Describe the coverage of coatings and learn to calculate wet- and dry-film thickness.
9. Locate and practice safety procedures listed on the material safety data sheet (MSDS).
10. List safety equipment and correct personal protective equipment (PPE) for use during coatings mixing operations.
11. Describe disposal techniques for hazardous and non-hazardous waste.

Trade Terms

Coalescence
Convertible coatings
Diluent
Intumescent materials
Lower explosive limit (LEL)
Non-convertible coatings
Polymerization
Short term exposure limit (STEL)
Subliming materials
Thermoplastic
Thermosetting materials
Threshold limit value (TLV)
Time weighted average (TWA)
Upper explosive limit (UEL)
Viscosity
Vitreous coatings
Volatile organic compound (VOC)

Required Trainee Materials

1. Pencil and paper
2. Appropriate personal protective equipment
3. Calculator

Prerequisites

Before you begin this module, it is recommended that you successfully complete *Core Curriculum*, Modules 00101-04 and 00106-04; and *Industrial Coating and Lining Application Specialist Level One*, Modules 69101-09 and 69102-09.

This course map shows all of the modules in the first level of the Industrial Coating and Lining Application Specialist curriculum. The suggested training order begins at the bottom and proceeds up. Skill levels increase as you advance on the course map. The local Training Program Sponsor may adjust the training order.

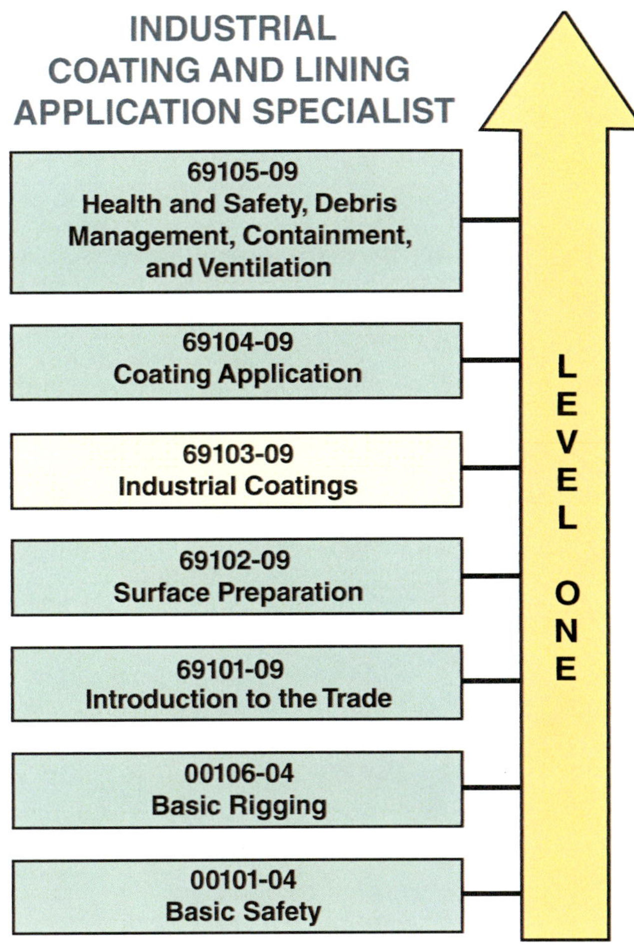

103CMAP.EPS

1.0.0 ◆ INTRODUCTION

Industrial coatings have many purposes, including:

- Waterproofing
- Sound deadening
- Heat transfer
- Fireproofing
- Aesthetics
- Corrosion control

When used to protect a surface from corrosion, the coating is usually a clear or colored film applied to a surface. There are three types of protective coatings:

- *Barrier coatings* – These coatings prevent bare surfaces from being exposed to the weather by stopping moisture vapor from touching steel surfaces, slowing the corrosion cycle. They also help to block the flow of electrons and ions to the surface, thereby preventing galvanic corrosion.
- *Inhibitive coatings* – These coatings work as a barrier coating; they are usually primers. The pigment in the primer reacts with moisture vapor that passes though the coating to slow corrosion.
- *Sacrificial coatings (also known as cathodic and galvanic coatings)* – Sacrificial coatings are used on steel. These coatings are rich in zinc, and they work by functioning as a more reactive substance that corrodes preferentially to the steel to which it is attached.

2.0.0 ◆ COATING MATERIAL PROPERTIES

There are many desirable traits for coatings. The most important is the ability to bond with the surface being coated. Other desired traits depend on a coating's use, but for most coatings, it is generally desired that they will be resistant to the following:

- Water
- Chemicals
- Impact
- Abrasion
- Extreme temperatures
- Hostile environment
- Cold flow

Coatings may also have the following properties:

- Low-moisture permeability
- Easy application to substrate
- Good surface bonding
- Tensile strength
- Cohesive strength
- Flexibility
- Dielectric strength

2.1.0 Components of Industrial Coatings

Conventional coatings have two major components:

- Pigment
- Vehicle (combination of resin and solvent)

Miscellaneous additional components can also be included to meet specified conditions.

2.1.1 Pigment

A pigment is a discrete particulate solid used to give specific protective or decorative qualities to the coating. It does not dissolve in the solvent component. Instead it remains solid individual particles in both the liquid coating and the solid film. Inert fillers such as chalk, talc, or mica (used to add bulk to the film) are also called pigments.

Some pigments also have protective properties such as the ones used in inhibitive primers. Other paints combine pigments with dyes. Not all paints contain pigment.

The functions of pigments (*Figure 1*) include the following:

- Adding color
- Hiding
- Extending the coating
- Serving as a barrier
- Serving as an inhibitor
- Serving as a sacrificial material

Table 1 shows metallic pigments used for corrosion resistance.

2.1.2 Solvent/Volatile Vehicle

A vehicle is the liquid base of the coating. Refer to *Figure 2*. The vehicle consists of:

- *Solvent* – usually referred to as the volatile portion of the vehicle

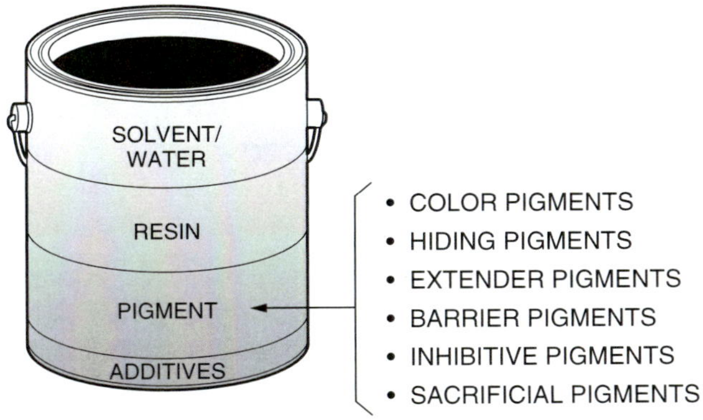

Figure 1 ◆ Functional categories of pigments used in coatings.

- *Binder (resin)* – nonvolatile portion of the vehicle, generally refers to the resin or resin blend that provides the film-forming portion
- *Required liquid additives*

The solvent is volatile (something that will readily evaporate) and does not become part of the paint film. It also controls flow and applica-

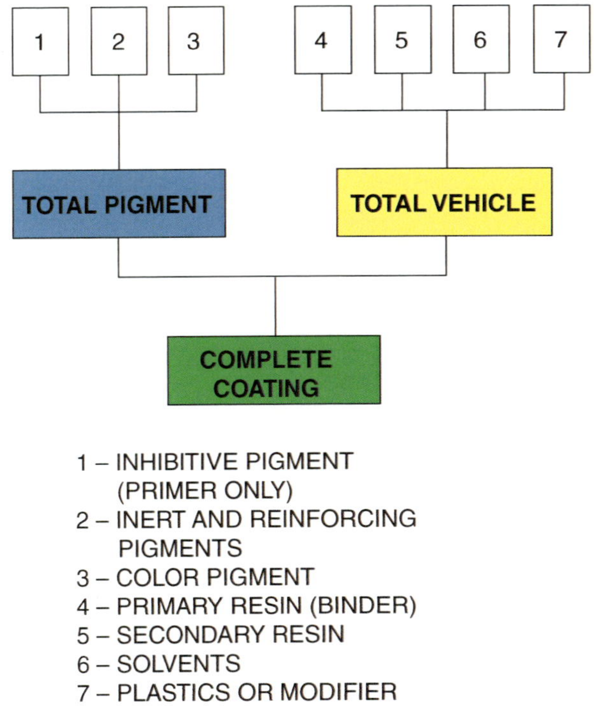

Figure 2 ◆ Complete coating.

Table 1 Metallic Pigments Used for Corrosion Resistance

Generic Type	Common Name	Resistance: Alkali	Acid	Water	Weather	Physical Characteristics
Aluminum	Aluminum flake	NR	NR	E	E	Creates single effect, protects binder, increases moisture upon transfer resistance.
Stainless	Stainless flake	E	E	E	G	Does not leaf as well as aluminum flakes. Reinforces binder without reducing chemical resistance.
Lead	Lead flake	E	E	E	E	Does not leaf as well as aluminum. Excellent chemical and water resistance.
Copper	Copper flake	NR	NR	G	F–G	Leafs well. Good copper color, chemical resistance only fair. Has good antiflaking properties.
Zinc	Zinc powder	NR	NR	E	E	Provides cathodic protection to steel. Reacts with inorganic vehicles to form hard adherent coating.
	Zinc flake	NR	NR	E	E	Provides some cathodic protection to steel. Reinforces some organic binders. May be used with zinc powder for reinforcing purposes.

Note: Leafing is the alignment of the metal flakes in the coating. A material leafs well when it remains parallel to the surface of the coating film and to the surface of the substrate.

E = Excellent G = Good F = Fair NR = Not Recommended

103T01.EPS

tion properties. The solvent affects the stability of the paint while in liquid state. Once the solvent has evaporated, the remaining paint is fixed to the surface. Solvents are added to coatings for the following reasons:

- To dissolve the base resin into a liquid; most resin binders are solids and would not be usable without solvents
- To reduce the **viscosity** (thickness) of the coating
- To improve the wetting of the substrate surface
- To slow evaporation in hot, dry conditions
- When adding a solvent to a coating, always use the one specified by the coating manufacturer. Using the wrong solvent can adversely affect the curing (drying), appearance, and durability of the coating.

2.2.0 Resins

The binder, or resin, is the actual film-forming component of paint. It is the only component that must be present. Other components are optional, depending on the desired properties of the coating.

Resins make a coating hard when it dries. They are the non-volatile vehicle in coatings because resins don't evaporate. Coatings are usually referred to by their principal resin—acrylic, for example. The only common exception to this is when coatings are named after the primary pigment such as inorganic zinc primers. Resins are both natural and synthetic, and because they are usually solid they need to be dissolved in a solvent. Sometimes a number of different resins are mixed into a coating to give the coating very specific characteristics. The following are the primary resins:

- Acrylic
- Alkyd
- Asphalt
- Chlorinated rubber
- Coal tar
- Drying oil
- Epoxy
- Phenolic
- Polyvinylchloride (PVC)
- Silicone
- Urethane
- Vinyl
- Soloxirane
- Bitumin

CASE HISTORY

Corrosion Kills

In 2001, there was an explosion at a Delaware refinery. Workers were repairing a catwalk above a sulfuric acid storage tank when a spark ignited flammable vapors in the tanks. The tank collapsed, and one worker was killed while eight others were injured. The tank had holes in it due to corrosion.

Source: *The U.S. Chemical Safety and Hazard Investigation Board web site.*

2.3.0 Additives

Paint can have a wide variety of additives that are usually added in very small amounts. Still, these additives have a very significant effect on the product, such as controlling surface tension, freezing properties, foaming, skinning, etc. Some additives are catalysts, thickeners, stabilizers, emulsifiers, texturizers, adhesion promoters, UV stabilizers, flatteners (de-glossing agents), and biocides. The following are additive groups:

- *Accelerators* – Decrease the drying time; commonly used with oil paints that would normally dry slowly without an accelerator additive.
- *Plasticizers* – Make the resin tough and flexible; they are dissolved by and into resins, so they are considered permanent solvents. Plasticizers can also increase the coating coverage area because they make coatings spread easily, but this is not their primary task.
- *Flow agents* – Improve the flow of coatings so that they go on smoothly; make coatings easier to apply evenly, helping to reduce application flaws.
- *Emulsifiers* – Used to help incompatible materials to mix. This helps to prevent coatings from settling during storage.
- *Ultraviolet light (UV) absorbers* – Protect coatings from sunlight and prevent deterioration of coatings exposed to the sun's rays. Acrylic polyurethanes have excellent color retention so they may be used on exterior surfaces.

3.0.0 ♦ CURING MECHANISMS

The curing mechanism is the way a coating goes from a liquid to a solid state. It is the most important reaction that takes place in the formation of a coating film. For a liquid coating to form a protective coat on a surface, the liquid portion must cure. This leaves the solids (resins and pigments) that must adhere tightly to the surface. This defines a resin's suitability as a binder.

There are several curing mechanisms for coatings. These curing mechanisms can be grouped into two broad types known as **non-convertible coatings** and **convertible coatings** (*Table 2*).

3.1.0 Non-Convertible Coatings (Thermoplastic)

Non-convertible coatings do not go through a chemical change when they dry. Because of this, they can be re-dissolved by their original solvent, or softened with heat and re-hardened by cooling. Non-convertible coatings are also referred to as **thermoplastic** and are defined as a coating that is formed by solvent or water evaporation or by a phase change such as melting and solidifying. The drying process, also called curing process, occurs through the evaporation of the solvent or **diluents**. Thermoplastic curing processes are evaporation and **coalescence**. Coalescence occurs when the particles of resin in a thermoplastic coating pull together to form a hard coating. Cooling and plating curing mechanisms are used with metallic and **vitreous coatings**. *Table 3* shows types of coatings classified by their curing mechanism.

3.2.0 Convertible Coatings (Thermoset)

Convertible coatings go through a chemical change as they dry. Therefore they cannot be re-dissolved by their original solvent, or softened with heat and re-hardened by cooling. Convertible coatings are also referred to as thermoset and are defined as a coating that is formed as a result of a chemical crosslinking reaction (oxidation, **polymerization**, chemical additive reaction, heat, or a combination of these.).

Polymerization occurs when two or more resin molecules combine to form a single, more complex molecule. This is a chemical reaction in which a compound is formed when many similar chemical groups called monomers join. Polymerization is the most common curing method of today's high performance coatings. The primary characteristics of the monomers in polymerization are stability and the ability to link together chemically.

Table 2 Curing Mechanisms

Non-Convertible Coatings (Thermoplastic)	Convertible Coatings (Thermoset)
Solvent-evaporation cure	Oxygen induced
Coalescence cure	Chemically induced
Fusion cure	Heat induced
Respirator	Moisture induced
	Carbon dioxide cure
	Ultraviolet cure

Table 3 Types of Coatings by Curing Mechanism

Convertible Oxidation Cure	Convertible Co-Reaction Cure
Alkyd	Epoxy
Drying oil	Coal tar epoxy
Epoxy ester	Zinc-rich epoxy
Phenolic (oil-modified)	Epoxy emulsion
Silicone alkyd	Rubber sheet lining
Urethane (oil modified)	Phenolic (epoxy-modified)
	Urethane (two-component)
	Inorganic zinc, post-cure type 1-A (waterbase)
Convertible Condensation Cure	**Convertible Hydration Cure**
Vinyl Wash Primer	Cementitious Materials
Furan	Inorganic zinc, self-cure type 1-C (solvent reducible)
Phenolic (highbake)	Urethane (moisture-cure)
Inorganic zinc, self-cure type 1-B (water reducible)	
Non-Convertible Evaporation (Lacquer Dry)	**Non-Convertible Coalescence (Latex Dry)**
Asphalt/coal tar	Acrylic/latex emulsion
Chlorinated rubber	
Polyethylene	
Vinyl	
Vinyl acetate	
Cooling and Plating Vitreous	**Cooling and Plating Metallic**
Porcelain/glass enamels	Cementation
Ceramics	Cladding/electroplating
	Condensation of metal vapor
	Metal spraying
	Hot dipping

103T03.EPS

4.0.0 ◆ SERVICE CONDITIONS

Coatings are selected based on the service conditions of the material being coated. For example, a marine craft, a potable water tank, and the lining of a pipe carrying crude oil would all require different coatings.

Deciding which coating is needed depends on many factors. Industrial coatings can provide protection against corrosion, thermal fatigue, and oxidation. They also reduce heat transfer, and increase or decrease friction. Some service condition factors are:

- Water
- Chemicals
- Solvents
- Sunlight
- Alcohols
- Salts
- Acids
- Scratches
- Food
- Alkali

Many times, coatings are expected to perform multiple functions. Some relevant questions to ask are the following:

- What is to be painted?
- Why does it need to be painted?
  - Corrosion
  - Abrasion
  - Chemical attack
  - Chalking
  - Weathering
- When does it need to be painted?
- Who will paint it?
- How will it be painted?
- What are the environmental concerns?
  - Temperature
  - Location
  - Indoor or outdoor
  - VOC regulations
  - Air quality regulations
- What is the substrate?
  - Steel
  - Concrete
  - Fiberglass
  - Wood
  - Existing coating
  - Rust
- Where will it be painted?
  - In the field
  - Steel fabrication shop
  - OEM
  - Restricted work site

5.0.0 ◆ CONCRETE COATINGS

Concrete is the most widely used material in construction. NACE International reports that concrete covers more surface area than all other materials. Concrete is sometimes applied to steel for corrosion protection and fireproofing, and it provides the most protection against fresh-water and weather conditions.

Concrete presents a special challenge in selecting coatings. Concrete is a very porous material, so coatings easily penetrate the surface. However, concrete easily absorbs moisture and chemicals that can prevent the coating from bonding to the concrete. In addition, concrete can reach a pH level of 12 to 13. This high alkaline level makes it hard for coatings to stick to concrete. Hardeners or curing compounds are frequently applied to the surface of concrete. However, they can prevent coatings from adhering to the surface, so they need to be removed before any coatings are applied.

Because of its porous nature, concrete can be difficult to coat. For example, even though you can't see it, air will bubble out of the concrete pores as the temperature increases. If a coating is applied to the concrete during the cooler part of the day, you will most likely see air bubbles and pinholes on the coating surface as the temperature increases.

Some coatings to be applied over a concrete surface may have good chemical resistance. Vinyl and epoxy resins are good choices for coating concrete, and both will protect the surface from other reactive materials. While vinyl coatings may have broader chemical resistance than epoxies, the epoxies stick better. Epoxy coatings are often used on concrete that will be submerged in water.

A key reason for coating concrete is to prevent water from absorbing into it. Without waterproofing, the concrete surface can become damp, leading to efflorescence on the surface. Efflorescence is a white crystalline deposit that is caused by salts in the concrete rising to the surface with the moisture. In cold weather, water in concrete can freeze and expand, which can shatter it. Using a waterproof coating on concrete will reduce the amount of water it absorbs, and help to protect it from freeze damage.

Coatings that do not provide waterproofing can peel or blister when moisture is passed through them. Waterproof coatings prevent the water or moisture from passing through the concrete, so they are essential for underground or underwater structures.

If the concrete absorbs chloride, sulfate, or oxygen, reinforcing steel in the concrete can corrode. Coatings that are resistant to these elements are a practical answer for the protection of concrete and embedded reinforcing steel.

Pipes in wastewater systems and treatment plants are usually constructed with concrete. The sewage running through these pipes often contains hydrogen sulfide, which is easily absorbed by the concrete. Hydrogen sulfide oxidizes into sulfuric acid which attacks calcium in the concrete, causing it to weaken. Concrete surfaces exposed to sewage need to be thoroughly coated, because bacteria and liquid sewage will pass through any gap or defect in the coating.

6.0.0 ◆ SPECIALTY COATINGS

There are a number of specialty coatings available today. Three specialty coatings will be discussed in this module. They are powder, pipeline, and fireproofing coatings.

Powder coatings are also called non-liquid coatings. These coatings are used when a fast-curing finish is needed without producing any **volatile organic compound (VOC)** vapors.

Pipeline coatings are used to line the inside of pipes to protect them from the material flowing through them, and to coat the outside of pipes to protect them from the environment.

Fireproofing coatings resist fire for a specified period of time.

Resistant materials are divided into two types: coatings (exterior) and linings (interior). A resistant coating is applied to the exterior of a structure or surface to help protect it from corrosion due to weathering, condensation, fumes, dust, splash, or spray (but will not necessarily protect it from immersion in any liquid or chemical).

A resistant lining is applied to the interior of pipes, tanks, containers, or equipment that is in direct contact with liquids, chemicals, or food products. A lining must prevent corrosion by the product, but the lining itself must not contaminate the product. The most important function of a resistant lining may be preventing product contamination.

6.1.0 Powder Coatings

Powder coatings are non-liquid and do not contain a solvent. If applied electrostatically, they are charged electrically and sprayed onto a surface that is electrically grounded. The difference in electrical charge makes powder particles stick to the surface. This allows the thickness of a powder coating to be much greater than many conventional liquid coatings. Most often, the surface is pre-heated before the coating is applied. Usually, just before the surface is coated it is heated at a high temperature so the coating melts, creating a durable finish when it cools.

The most common uses for powder coatings are pipelines, reinforcing bar (rebar), household appliances, and automotive exteriors. Some powder coatings are chemically resistant and may be used on pump impellers and agitation blades or other devices that are used with chemicals.

Powder coatings are similar to liquid coatings in their makeup. Powders generally contain some of the following:

- Pigment
- Curing agents (partially reacted solid resins)
- Wetting agents
- Flow-control agents
- Fillers and extenders
- Foam breakers and other additives

All powder coatings are dry particles that look like a fine powder, similar to flour or talc except it may be colored. Powders fall into the same two broad categories as previously discussed: thermoplastic and thermosetting. Recall, thermoplastic materials soften whenever they are heated and harden when cooled. Also recall, **thermosetting materials** harden when heated and stay hard when cooled. Thermoplastic coating materials include the following:

- Polyvinyl chloride (PVC)
- Polypropylene
- Kynar®
- Halar®
- Polyethylene
- Teflon® (polytetrafluoroethylene or PTFE)

Thermosetting coating materials include the following:

- Epoxy
- Urethane
- Polyester
- Acrylic
- Alkyd

6.2.0 Pipeline Coatings

A wide variety of materials are used in the construction of pipelines. Some examples are provided below:

- Carbon steel is widely used for pipe construction, and is used with crude oil, natural gas, and refined petroleum products. Carbon steel is often used as slurry and raw water lines, and as re-circulating water pipes in power plants.
- Glass reinforced plastic (GRP), also called fiberglass reinforced plastic (FRP), is a plastic that is reinforced with glass fibers. GRP is used for pipes that carry some chemicals, including acids, alkalis, and chemical wastes. It is primarily used in and around petrochemical complexes.
- Ductile iron is less brittle than regular cast iron, so it has greater durability when stressed. Ductile iron is very difficult to weld, so it is almost always joined with mechanical couplers or mechanical joints.
- Asbestos cement (AC) pipe is made from high-grade portland cement reinforced with asbestos fibers. This pipe is very dense, with a smooth interior surface that permits good fluid flow through the pipe.
- Plastics are widely used in the pipeline industry to carry natural gas, water wastes, and chemicals. Depending on the service and operating conditions, polyethylene (PE), polypropylene (PP), or polyvinyl chloride (PVC) pipe may be used.
- Pre-stressed concrete forms a thick-walled pipe with great strength. It is manufactured by applying a concrete coating to thin-wall steel pipe; a steel wire under controlled tension is spirally embedded into the concrete and then covered with more concrete. The pipe is then cured in a steam kiln.
- Steel-cylinder concrete pipe is formed by lining and coating a steel pipe with concrete, which is then air-cured or cured in a steam kiln.
- Both pre-stressed and steel-cylinder pipe are used to transport water (untreated and treated), domestic and industrial wastes, and treated sewage. They are frequently coated to provide chemical and corrosion resistance to the waste.

6.2.1 Pipeline Coatings Standards

Pipelines, whether above or below ground, are exposed to hostile environments (*Figure 3*). The purpose of coating the exterior of a pipe is to separate the pipe from its environment, as well as to provide cathodic protection. The coating should provide a barrier to moisture and electron flow. Providing a barrier to moisture and electron flow is a difficult task. Although many coatings are very resistant to current flow, as they break down they allow current flow, thus accelerating the corrosion process.

NACE International has a number of standards that discuss pipeline coatings. Further discussion of pipeline coating standards can be found through the NACE International web site at http://www.nace.org.

6.2.2 Pipeline Coatings Personal Protective Equipment

When using pipeline coatings, you must check the product's material safety data sheet (MSDS) to see what PPE you'll need to use when handling it. With coatings, you'll most often need some type of respirator and protective clothing, including protective gloves. You may also need to wear fireproof clothing. Always wear chemical eye protectors.

6.3.0 Fireproofing

Passive fireproofing materials can dramatically reduce fire damage to a structure. These materials work by slowing the transfer of heat from a fire to the structure by providing insulation between the fire and the structure. The term *fireproofing* is a bit misleading. These materials will burn given enough heat, but when used with active systems such as water sprinklers, they significantly slow the progress of a fire, which reduces fire damage.

103F03.EPS

Figure 3 ◆ Alaskan pipeline.

Most fireproofing materials are applied to structural steel, but some protect other construction materials in critical areas, such as reinforced concrete used in tunnels. It is important not to confuse these fireproofing materials with refractory materials. Refractory materials resist the heat from a fire but have poor insulating properties, so the heat is transferred to the surrounding structure.

6.3.1 Fire Protection Standards

Fire protection standards may be specified by local law. Materials are usually rated by the time they can delay the progress of a fire, usually in 15-minute increments. Often a temperature is specified with the time. In addition, NACE has a standard for corrosion under fireproofing material: NACE Standard RP0198, *The Control of Corrosion Under Thermal Insulation and Fireproofing Materials—A Systems Approach*.

6.3.2 Fireproofing Materials

A wide range of fireproofing materials is available. They include the following:

- *Cementitious products* – work by absorbing heat from the fire as water evaporates from the coating.
- ***Subliming materials*** – when heated, change from a solid to a gas without going through the liquid stage and absorb heat as they change.
- ***Intumescent materials*** – turn into foam and swell when exposed to heat, forming an insulative barrier between the fire and the structure.
- *Fibrous materials* – made into boards, sprayed on, or made into blankets; they are often used for fireproofing because some don't burn at lower temperatures. Ceramic fibers are a good choice because they don't burn in the first stages of a fire and some can withstand temperatures up to 2,100°F (1,150°C).

6.3.3 Fireproofing Personal Protective Equipment

You will generally need to wear a respirator, gloves, goggles, and protective clothing when working with fireproofing coatings, but always check the MSDS for the product that you are using.

7.0.0 ◆ COATING SAFETY

There are a number of hazards related to coatings. Information about specific hazards for a coating can be found on the product MSDS. A sample MSDS can be found in the *Appendix*.

The two main hazards that exist with coatings are fire and health. The flash point of a flammable liquid is the lowest temperature at which it can ignite when exposed to an ignition source. Sources of ignition include open flames, static electricity, heat from welding or grinding operations, lit cigarettes, sparks from tools, and other sources.

All coatings and solvents should be handled with care. The National Institute of Occupational Safety and Health (NIOSH) approves breathing apparatus and personal protective equipment for use when working with industrial coatings.

Specific terminology is used when describing the safety concerns of working with or around coatings and solvents. Understanding the following terms and definitions will help the applicator maintain safe working conditions:

- The explosive limit of a gas or a vapor is the limiting concentration that is needed for the gas to ignite and explode. There are two explosive limits for any gas or vapor: the **lower explosive limit (LEL)** and the **upper explosive limit (UEL)**. At concentrations in air below the LEL there is not enough fuel to burn. At concentrations above the UEL the vapor has displaced so much air that there is not enough oxygen to

INSIDE TRACK

Fireproofing Structural Members

A common method of fireproofing structural members is to apply the fireproofing material under pressure with a hose.

103SA01.EPS

permit burning. Between the LEL and the UEL there is a danger of vapor explosion.

- **Threshold limit value (TLV)** is the amount and concentration of an airborne vapor that can be safely inhaled by workers during a specified period.
- **Short term exposure limit (STEL)** is the highest concentration of an airborne vapor that workers can be exposed to over a stated period of time.
- **Time weighted average (TWA)** is the threshold limit of vapor concentration to which workers can be exposed for a specific work period, usually given in hours allowed to work per day or week.
- A volatile organic compound (VOC) is an organic chemical compound that vaporizes under normal conditions and enters the atmosphere as a gas. The term is often used in a regulatory context, such as found in the Environmental Protection (EPA) guidelines.

7.1.0 Personal Protection Equipment

A respirator is a mask worn over the mouth and nose to prevent the inhalation of fumes and vapor. Respirators are necessary.

Also, some sort of respiratory protection is dictated by regulations established in the United States, such the Occupational Safety and Health Administration (OSHA) and National Institute for Occupational Safety and Health (NIOSH).

Even though the concentration of flammable gas or vapor may be below the LEL, it may be far above the safe limit for breathing. The maximum allowable concentration (MAC) is the amount of gas or vapor that must not be exceeded when a worker is exposed to the hazard for an eight-hour workday. This concentration pertains to vapors, gases, mists, and solids. In the United States, MACs are published annually in the Archives of Industrial Hygiene. These same MACs are often adopted in other countries.

Aromatic solvents, such as xylene and toluene, are more hazardous than the aliphatic solvents, such as mineral spirits. Unfortunately, mineral spirits are rarely used in high-performance coatings, and hazardous solvents are still widely used.

Aromatic solvents are used almost exclusively in some synthetic paints, such as lacquer and vinyls, and in oleoresinous paints, such as phenolic varnish and some alkyds.

Allowable concentrations for the aliphatic solvents are greater than for the aromatic solvents,

THINK ABOUT IT — **HazCom**

The U.S. Department of Labor Occupational Safety and Health Administration requires employers to notify workers of hazardous materials at the workplace using a hazardous communication program called HazCom. Why do you suppose that is?

but in every case the MAC for breathing is far less than the LEL. This means that while an air space may be considered safe from fire or explosion, it might be extremely dangerous to people. Exposed workers should be supplied with fresh air masks connected to a source of clean, cool, filtered air.

The common practice of using low-flash-point solvents such as MEK or acetone as universal thinners can be extremely hazardous, particularly in enclosed spaces.

Other items of PPE include the following:

- Gloves impervious to the chemicals
- Eye protection – safety glasses
- Full face protection
- Body protection worn over work clothes

8.0.0 ◆ MIXING AND THINNING

Before mixing any coatings, inspect the coating to see if it appears to have been damaged by age or improper storage and handling. Products in cans that appear to have been opened should be examined carefully. The pigment in coatings can settle over time, so coatings that are prone to settling may require repeated mixing. Coatings that have separated or gelled should also be carefully examined. If in doubt, do not use the coating. Set it aside and call the manufacturer for information.

Ensure the coating is thoroughly mixed (*Figure 4*) before it is used. When using a hand agitator, place the agitator in the coating before you start the device to avoid paint spatter and waste. The specific effects for failing to mix a coating thoroughly vary between different types of coatings. Generally, a poorly mixed coating will not provide the desired protection. Additionally, failure to mix components in the proper proportions can cause inadequate curing or separation of components after application.

Some coatings, such as acrylic, vinyl, and enamels, are single-component coatings. They may require mixing to distribute all ingredients evenly, but beyond adding a thinner (per manufacturer specifications only) to decrease viscosity, they are ready to use from the container. Other coatings, like epoxy, polyester resins, and aliphatic urethane, are two-component coatings. Once a two-component coating is mixed, it must be used within the manufacturer's specified pot life. When using a two-component coating kit, never mix a partial kit. The recommended industry practice is to mix the entire kit even if you don't need it all. To reduce waste you should check to see if a smaller quantity kit is available.

Sometimes a coating needs to be thinned in order to execute proper application. In general, thinners should not be used unless necessary. Always consult the manufacturer's instructions and be certain to use the correct amount of thinner. Too much thinner can cause the dry-film thickness to be too thin, and may cause runs and sags. Too little thinner can result in a non-uniform film with poor integrity, pinholes, or poor appearance.

Often workers will need to mix an additive into a coating just before use. One such additive is methyl ethyl ketone peroxide (MEKP), which is a catalyst that is used with polyester resins such as vinyl ester resins to start the polymerization process. MEKP must be handled carefully because it is extremely irritating to the skin. When mixing MEKP into resin, start mixing the resin and then carefully pour the MEKP into the vortex. Continue mixing, being certain to scrape material from the sides of the container, until the components are thoroughly mixed.

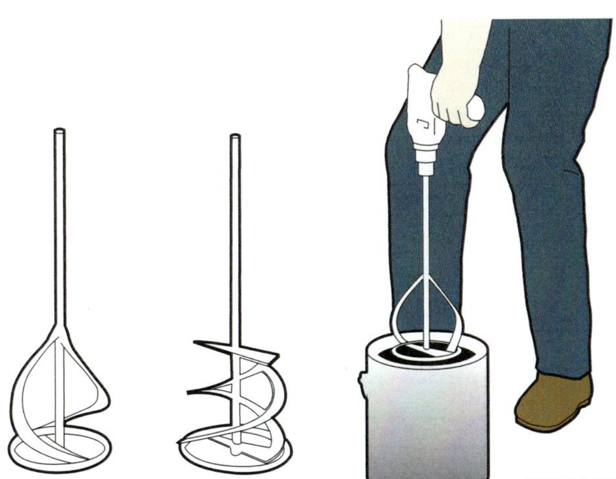

Figure 4 ◆ Mixing a coating with a portable drill-powered mixer.

8.1.0 Calculating Dry-Film Thickness (DFT)

To calculate the amount of coating necessary for a job, the worker must understand the relationship between wet-film and dry-film thickness. Dry-film thickness (DFT) is the usual specification criterion in the coating contract (specification). Wet-film thickness (WFT) measurements help to determine how much of a coating must be applied to reach the specified DFT. WFT measurements are made at the time each coat is applied so that errors in thickness can be properly corrected. WFT is only a guide. DFT is the defining measurement.

The dry-film/wet-film ration is based on the percentage of solids by volume in the coating. Manufacturers' data sheets often list solids by weight as well as by volume, but the volume measurement must be used to calculate WFT. The formula is as follows:

$$\frac{\text{DFT}}{\text{\% solids by volume}} = \text{WFT}$$

For example, the coating specification calls for each application to have a DFT of 4 to 6 mil (100 to 150 µm). The coating has 30 percent solids volume. First, convert 30 percent to decimal form as shown below:

$$30\% = \frac{30}{100} = 0.30$$

Next, place the known information into the formula as follows:

$$\text{WFT} = \frac{\text{4 mil}}{\text{0.30 solids by volume}} = \text{13.3 mils (.34 mm)}$$

$$\text{WFT} = \frac{\text{6 mil}}{\text{0.30 solids by volume}} = \text{20.3 mils (.52 mm)}$$

One gallon of fluid will cover 1,604 ft^2 (149 m^2) to 1 mil (25 µm) thickness when it is wet. When it is necessary to know how much coating you will need to cover a surface, divide 1,604 by the desired WFT. For example, using the above WFT of 20 mil (500 µm), calculate how much area one gallon of coating covers, as shown below:

$$\frac{1{,}604}{20} = 80.2 \text{ ft}^2 \text{ per gallon (7.45 square meters)}$$

To figure out how many gallons of coating are needed to coat a 1,200 square ft. (110 square m) area, use the equation below:

$$\frac{1{,}200 \text{ ft}^2}{80.2 \text{ ft}^2 \text{ per gallon}} = \text{14.96 gallons (56.6 liters)}$$

As shown in the example, the job needs between 10 and 15 gal (38 and 57 L) of coating. No matter how carefully work is performed, there will always be a small amount of waste, so ordering 15 gal (57 L) of coating is more realistic than ordering 10 gal (38 L).

If the coating is thinned, this must be taken into account. Thinner increases the total volume of the coating without increasing the amount of solids. This introduces a new variable called percentage of thinner by volume (% TBV). The formula to be used when thinner has been added is as follows:

$$\frac{\text{DFT } (1 + \%\text{ TBV})}{\%\text{ solids by volume}} = \text{WFT}$$

Occasionally it is necessary to thin a coating by a percentage of volume. In this case, put the percentage into the formula. For example, using the WFT of 4 mils with a solid of 30 percent thinned 25 percent (0.25), the WFT is calculated as follows:

$$\text{WFT} = \frac{4 \text{ mil } (1 + 0.25)}{0.30 \text{ solids by volume}} = 16.7 \text{ mils } (.424 \text{ mm})$$

Occasionally it is necessary to add a specific volume of thinner to each gallon of coating. If so, calculate the percentage of thinner before you calculate the WFT. For example, you need to add 1 pint (0.5 liter) of thinner to each gallon of coating. Calculate the percentage as follows:

$$1 \text{ gallon} = 8 \text{ pints, so } \frac{1 \text{ pint}}{8 \text{ pints}} = \frac{1}{8} = 0.125\%$$

8.2.0 Labeling and Sampling

Coatings that are not in their original container after they are mixed will need to be labeled. The employer will likely have a policy for this, so be sure to follow it. When in doubt, label the coating with all of its relevant information so anyone looking at the label will know exactly what the coating is and how it was mixed. Include information required by federal, state, and laws such as the following:

- Name of coating
- Manufacturer's name
- Batch number
- Date mixed
- Proportion and names of any thinners or additives
- Your name
- Identity code for tracking sample, if required

Coatings must be sampled during mixing. These samples will be retained in case they are needed later to determine why a coating failed, or to prove that a particular coating was used and correctly mixed. Samples of individual components of multicomponent systems are collected in clean, unused containers. They are labeled with the relevant details, including the following:

- Name of coating
- Manufacturer's name
- Batch number
- Date sample retained
- Inspector's name
- Identity code for tracking sample, if required

9.0.0 ◆ MSDS: MATERIAL SAFETY DATA SHEETS

The employer is required to have a MSDS present on any job site where any chemicals are present. The MSDS must be in a location that all employees have access to and be made available to all employees when requested.

All projects should start with a safety meeting and the MSDS should be part of this meeting. Many chemicals used for industrial painting are hazardous, toxic, flammable or carcinogenic. The best way to avoid getting hurt is to know what chemicals you are working with. The MSDS contains all the safety and toxicity information about the chemicals (including paints) that you will be using. See the example MSDS provided in the *Appendix*.

The government has mandated that this safety and toxicity information be disclosed on the MSDS. However, the order in which the information is presented is not mandated. MSDSs can look very different and have the relevant information in different places in the document.

9.1.0 Typical MSDS Sections

Section 1: Product and Company Identification

- Details of the manufacturer issuing the data sheet
- Emergency call-out information
- Who to call at different times

Section 2: Composition – Information on Ingredients

Section 2 identifies the material, and gives other registry numbers. These help emergency response crews that may not be familiar with the product name.

Section 3: Hazards Identification

Section 3 summarizes the major hazards associated with use of the chemical. The R and S codes in this section are explained in the text.

Section 4: First Aid Measures

Section 5: Fire Fighting Measures

Section 6: Accidental Release Measures

Section 6 outlines procedures to be followed in case of accidental release of the chemical, including methods to be used to clean up spills. Note that these measures are unlikely to be sufficiently detailed if the chemical is particularly hazardous, and local procedures should be drawn up to supplement safety information in the MSDS.

Section 7: Handling and Storage

Section 7 is a critical section, sometimes overlooked by those who use chemicals in a laboratory. It contains information about the possible formation of peroxides in storage, flammability, explosive risks, etc. Pay particular attention to the possible need for fire-proof storage cabinets, explosion-proof refrigerators, and the need to avoid storing incompatible chemicals in proximity with one another.

Section 8: Exposure Controls — Personal Protection

Section 8 provides information on regulatory standards for exposure to toxic materials. It details the maximum time that people can be exposed and usually describes the appropriate personal protective equipment.

Section 9: Physical and Chemical Properties

Section 9 describes the physical and chemical properties of the coating.

Section 10: Stability and Reactivity

Section 10 describes chemical incompatibilities and hazardous conditions to avoid.

Section 11: Toxicological Information

Section 11 outlines the risks to which the applicator may be exposed when using the particular chemical. The following sub-sections will likely be present:

- *Acute toxicity* – The acute toxicity gives an indication of the quantities of the chemical that may cause immediate damage to health if swallowed, inhaled or absorbed through the skin.
- *Lethal dose (LD)* – given this dose, fewer than 50% of test specimens die.
- *Lethal concentration (LC50)* – given this concentration, 50% of test specimens die. This section should be read particularly carefully, since the range of health effects may be broad, and may include carcinogenic or sensitizer effects.
- *Primary irritant effect/sensitization* – Chemical sensitization, for example by platinum compounds, is a potentially debilitating problem. Pay particular attention to any information which may suggest that the chemical is a sensitizer.
- *Subacute to chronic toxicity* – Details of the possible long-term effects of exposure to the chemical
- *Additional toxicological information* –This section contains important comments regarding the carcinogenicity. The acronyms such as IARC, NTP or OSHA refer to regulatory or health agencies.

Section 12: Ecological Information

Section 12 lists effects on the environment.

Section 13: Disposal Considerations

Section 13 deals with disposal. Safe chemical disposal is a must.

Section 14: Transport Information

There are usually United Nations (UN), Department of Transportation (DOT) and International Maritime Organization (IMO) hazard codes given in this section.

Section 15: Regulatory Information

Section 15 lists the hazard codes that indicate the principle hazards associated with the chemical and the precautions that should be taken when working with it. It may also contain any state or country-specific information.

Section 16: Other Information

Section 16 lists additional information, such as the name of the person preparing the data sheet, a list of references from which data have been drawn, and disclaimers.

10.0.0 ◆ WASTE MANAGEMENT

A waste stream is any waste material generated at a particular location. The waste may be hazardous or non-hazardous. By law, waste of any kind must be disposed of properly and proper disposal can be very expensive. The first step in waste management is to control it at the source.

Minimizing waste decreases costs in two ways. First, it saves money on materials. Second, it saves money on disposal fees. Minimizing waste is also good for the environment. There are other ways to control waste:

- Use only the amount of coating that is needed.
- Place recyclables in their proper containers.
- Never mix hazardous and non-hazardous waste.
- Use care to not release waste into the environment.

10.1.0 Non-Hazardous Waste Disposal

Non-hazardous waste will be disposed of in accordance with the employer's policy. These policies are usually created based on local laws about waste disposal, so be sure to follow company policy.

10.2.0 Hazardous Waste Disposal

Hazardous waste is dangerous, or potentially dangerous, to human health or the environment. Hazardous wastes can be in the form of liquids, solids, contained gases, or sludge. They can be the by-products of manufacturing processes, or discarded products, such as solvents. These products need to be disposed of in accordance with the Environmental Protection Agency (EPA) *Resource Conservation and Recovery Act (RCRA) 40 CFR 261* or with local law. Local law can be stricter than the RCRA but it cannot be more lenient. It is important to read the MSDS for the product to find out how to dispose of the product properly.

INSIDE TRACK

Exploding Garbage Truck

According to a *Savannah Morning News* article on February 16, 2007, a Savannah, Georgia sanitation truck exploded. Police initially called in the bomb squad, but later discovered the cause—common household chemicals that had mixed together in the truck's closed trash compactor. The Savannah-Chatham Metropolitan Police spokesperson was quoted as saying, "Combustible chemical household products caused the blast. Some of those things, when combined in a confined space and subject to shock and mixture, can create a blast."

The EPA has determined that a number of wastes are hazardous. These are defined in EPA document *Title 40, Protection of Environment Part 261—Identification and Listing of Hazardous Waste*. Other wastes can also be classified as hazardous. Some may waste may be toxic or spontaneously combustible, or have a low flash point.

It is very important to know what waste is hazardous and what waste is not. Hazardous waste must be placed in the appropriate container in accordance with your company's policy. This waste must be disposed of by a contractor authorized to dispose of it. Sometimes waste needs to be tested to determine its exact contents.

Should a hazardous material ever be spilled in an area, check the MSDS for cleanup instructions. If necessary, inform the EPA of the spill, depending on the material and how large of an area is involved. Some states may have stricter rules about handling hazardous materials and cleaning up spills. Follow established company rules and always consult a supervisor with any questions.

Review Questions

1. The term *sacrificial coatings* is most commonly applied to coatings that _____.
   a. are rich in zinc
   b. are rich in iron
   c. are used on concrete
   d. need a topcoat to prevent corrosion

2. Which of the following qualities is most important for coatings?
   a. Flexibility
   b. Good bonding ability
   c. Permeability
   d. Dielectric strength

3. Convertible coatings cure by _____.
   a. evaporation
   b. chemical reaction
   c. humidification
   d. chain reaction

4. Non-convertible coatings are also called _____.
   a. thermoset
   b. thermoplastic
   c. thermogenic
   d. thermostatic

5. The coating vehicle is the _____.
   a. solid portion of the coating
   b. vapor portion of the coating
   c. liquid portion of the coating
   d. colored portion of the coating

6. Resins are the portions of the coating that _____.
   a. make the coating harden when it cures
   b. evaporate as the coating is mixed
   c. thin the coating so it is easier to apply
   d. dissolve when exposed to the air

7. Acrylic is an example of a(n) _____.
   a. solvent
   b. pigment
   c. resin
   d. epoxy

8. Natural gas is most often transported through _____.
   a. plastic pipes
   b. carbon steel
   c. ductile iron
   d. concrete

9. Fireproofing coatings are designed so that they _____.
   a. prevent fires from starting
   b. slow the transfer of heat
   c. do not catch fire
   d. turn to liquid when heated

10. Fire protection intervals represent the time that a coating can protect from a fire at a certain temperature, and they are normally specified in _____ minute intervals.
    a. 10-
    b. 15-
    c. 30-
    d. 60-

11. Powder coatings are _____.
    a. non-liquid and do not contain a solvent
    b. non-liquid and contain a solvent
    c. liquid and do not contain a solvent
    d. liquid and contain a solvent

12. When exposed to heat, intumescent materials turn to _____.
    a. steam
    b. liquid
    c. foam
    d. gas

13. When used as fireproofing, ceramic fibers _____.
    a. do not burn during the initial stages of the fire.
    b. expand to twice their original size at 2,000°F (1,093.3°C)
    c. foam when exposed to heat
    d. turn into a gas when exposed to heat

14. Resins are _____.
    a. natural, but not synthetic
    b. synthetic, but not natural
    c. natural and synthetic
    d. neither natural or synthetic

Review Questions

15. When working with coatings, it is important to know that explosions can occur when vapor concentrations are _____.
    a. between the LEL and the UEL
    b. a ratio of the LEL and UEL
    c. above the LEL and the UEL
    d below the LEL and the UEL

16. A volatile organic compound (VOC) _____.
    a. is used as a fireproofing coating
    b. explodes when exposed to oxygen
    c. is organic and safe for the environment
    d vaporizes into the atmosphere

17. A waste stream is _____.
    a. unauthorized dumping of waste into a stream
    b. any waste generated at a particular location
    c. hazardous material from industrial work
    d. a polluted stream near an industrial area

18. When mixing paint, it is considered best practice to _____.
    a. mix as much as you can and store unused portions
    b. mix as little as you can and try to avoid waste
    c. use whatever thinner you have in the shop
    d. never mix any quantity less than 50 gallons

19. The pigment portion of a coating _____.
    a. provides color and protective properties
    b. is the liquid portion
    c. is the volatile portion
    d. is dissolved by the solvent

20. Concrete requires special coating consideration because it has _____.
    a. a very dense, nonporous surface
    b. a very porous surface
    c. a highly acidic chemical composition
    d. a neutral chemical composition

Summary

Protective coatings are clear or colored films applied to a surface to protect it from corrosion. Coatings are classified into two broad classes: convertible and non-convertible. Convertible coatings cure by a chemical reaction. Non-convertible coatings cure when the solvent evaporates, but there is no chemical reaction.

One of the jobs performed by an application specialist is mixing paint. Proper mixing is essential because an improperly mixed coating will not provide the desired protection. Failing to mix components in the proper proportions can cause faulty application, inadequate curing, or separation of components after application. This is a misuse of time and money and it is unprofessional.

Additionally, the application specialist must know how to handle hazardous and non-hazardous waste. Non-hazardous waste must be disposed of in accordance with approved policy. Hazardous waste needs to be disposed of in accordance with the *Environmental Protection Agency (EPA) Resource Conservation and Recovery Act (RCRA) 40 CFR 261* or with local law.

Notes

Trade Terms Introduced in This Module

Coalescence: Curing property of acrylic, epoxy, and latex-based coatings. The particles of resin pull together as the coating dries to form a hard coating.

Convertible coatings: Coatings that cure by a chemical reaction.

Diluent: Something added to a coating for the express purpose of diluting it. Solvent and water can be diluents when used to thin coatings.

Intumescent materials: Fireproof materials that turn into foam when exposed to heat.

Lower explosive limit (LEL): Lowest concentration of a gas or vapor in air capable of producing a flash of fire in presence of an ignition source (arc, flame, heat). Concentrations lower than LEL are too low to burn.

Non-convertible coatings: Air-drying coatings that dry by solvent evaporation with no change in the chemical nature of the binder. The coating remains soluble in the original solvent after drying.

Polymerization: A chemical reaction that joins molecules to create long chains.

Short term exposure limit (STEL): The highest concentration of an airborne material that workers may be exposed over a prescribed period of time.

Subliming materials: Fireproof materials that change when heated from a solid directly to a gas without going through a liquid phase.

Thermoplastic: Characteristic of a coating that is pliable when cool and hardens as it is heated.

Thermosetting materials: Convertible coatings where heat causes the coating to harden by chemical reaction.

Threshold limit value (TLV): The level it is believed that a worker can be exposed to chemical substances day after day for a working lifetime without adverse health effects.

Time weighted average (TWA): Limit of amount of time a worker can be exposed to a material over a certain period, usually given in hours per day or week.

Upper explosive limit (UEL): The highest amount of vapor concentration in the air that can explode. Vapor concentrations above the UEL displace so much oxygen that there is not enough oxygen in the air to explode.

Viscosity: The thickness of a fluid.

Vitreous coatings: Coatings that have the appearance of glass.

Volatile organic compounds (VOC): Organic chemical compounds that have high enough vapor pressures under normal conditions to significantly vaporize and enter the atmosphere.

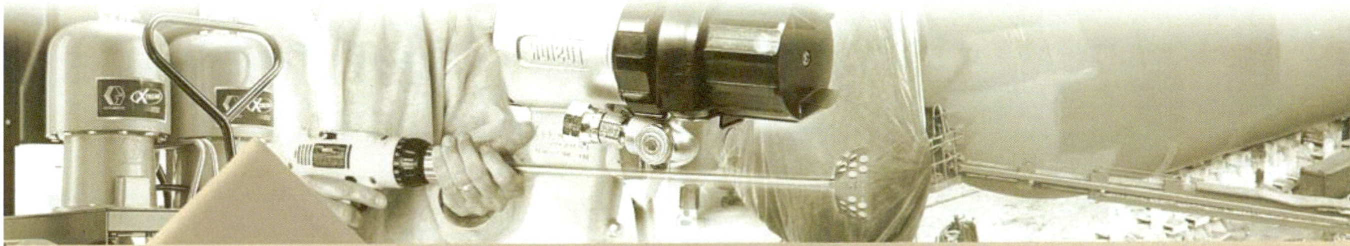

Appendix A

Material Safety Data Sheet

MATERIAL SAFETY DATA SHEET
R7K145
04 00
==
Section 1 — PRODUCT AND COMPANY IDENTIFICATION

PRODUCT NUMBER HMIS CODES
Health 2*
R7K145 Flammability 3
Reactivity 0
PRODUCT NAME
Reducer No. 145
MANUFACTURER'S NAME EMERGENCY TELEPHONE NO.
THE SHERWIN–WILLIAMS COMPANY (216) 566–2917
101 Prospect Avenue N.W.
Cleveland, OH 44115
DATE OF PREPARATION INFORMATION TELEPHONE NO.
02-APR-07 (216) 566-2902
==
Section 2 — COMPOSITION/INFORMATION ON INGREDIENTS
% by WT CAS No. INGREDIENT UNITS VAPOR PRESSURE
----
7 100–41–4 Ethylbenzene
ACGIH TLV 100 ppm 7.1 mm
ACGIH TLV 125 ppm STEL
OSHA PEL 100 ppm
OSHA PEL 125 ppm STEL
41 1330–20–7 Xylene
ACGIH TLV 100 ppm 5.9 mm
ACGIH TLV 150 ppm STEL
OSHA PEL 100 ppm
OSHA PEL 150 ppm STEL
51 107–98–2 1–Methoxy–2–propanol
ACGIH TLV 100 ppm 10.9 mm
ACGIH TLV 150 ppm STEL
OSHA PEL 100 ppm
OSHA PEL 150 ppm STEL
==
Section 3 — HAZARDS IDENTIFICATION

ROUTES OF EXPOSURE
INHALATION of vapor or spray mist.
EYE or SKIN contact with the product, vapor or spray mist.
Contains alcohols and acetates which can be absorbed through the skin.
EFFECTS OF OVEREXPOSURE
EYES: Irritation.
SKIN: Prolonged or repeated exposure may cause irritation.
INHALATION: Irritation of the upper respiratory system.
May cause nervous system depression. Extreme overexposure may result in unconsciousness and possibly death.
SIGNS AND SYMPTOMS OF OVEREXPOSURE
Headache, dizziness, nausea, and loss of coordination are indications of excessive exposure to vapors or spray mists.
Redness and itching or burning sensation may indicate eye or excessive skin exposure.
R7K145 page 2
==
MEDICAL CONDITIONS AGGRAVATED BY EXPOSURE

R7K145 page 2

==

MEDICAL CONDITIONS AGGRAVATED BY EXPOSURE

None generally recognized.

CANCER INFORMATION

For complete discussion of toxicology data refer to Section 11.

==

Section 4 — FIRST AID MEASURES

EYES: Flush eyes with large amounts of water for 15 minutes.

Get medical attention.

SKIN: Wash affected area thoroughly with soap and water.

Remove contaminated clothing and launder before reuse.

INHALATION: If affected, remove from exposure. Restore breathing. Keep warm and quiet.

INGESTION: Do not induce vomiting. Get medical attention immediately.

==

Section 5 — FIRE FIGHTING MEASURES

FLASH POINT LEL UEL

80 F PMCC 1.0 13.7

FLAMMABILITY CLASSIFICATION

RED LABEL — Flammable, Flash below 100°F (38°C)

EXTINGUISHING MEDIA

Carbon Dioxide, Dry Chemical, Foam

UNUSUAL FIRE AND EXPLOSION HAZARDS

Closed containers may explode when exposed to extreme heat.

Application to hot surfaces requires special precautions.

During emergency conditions overexposure to decomposition products may cause a health hazard. Symptoms may not be immediately apparent. Obtain medical attention.

SPECIAL FIRE FIGHTING PROCEDURES

Full protective equipment including self-contained breathing apparatus should be used.

Water spray may be ineffective. If water is used, fog nozzles are preferable. Water may be used to cool closed containers to prevent pressure buildup and possible auto ignition or explosion when exposed to extreme heat.

==

Section 6 — ACCIDENTAL RELEASE MEASURES

STEPS TO BE TAKEN IN CASE MATERIAL IS RELEASED OR SPILLED

Remove all sources of ignition. Ventilate the area.

Remove with inert absorbent.

==

Section 7 — HANDLING AND STORAGE

STORAGE CATEGORY

DOL Storage Class IC

R7K145 page 3

==

PRECAUTIONS TO BE TAKEN IN HANDLING AND STORAGE

Contents are FLAMMABLE. Keep away from heat, sparks, and open flame. During use and until all vapors are gone: Keep area ventilated – Do not smoke – Extinguish all flames, pilot lights, and heaters ? Turn off stoves, electric tools and appliances, and any other sources of ignition.

Consult NFPA Code. Use approved Bonding and Grounding procedures.

Keep container closed when not in use. Transfer only to approve containers with complete and appropriate labeling. Do not take internally. Keep out of the reach of children.

==

Section 8 — EXPOSURE CONTROLS/PERSONAL PROTECTION

PRECAUTIONS TO BE TAKEN IN USE

Use only with adequate ventilation.

Avoid contact with skin and eyes. Avoid breathing vapor and spray mist.

Wash hands after using.

VENTILATION

Local exhaust preferable. General exhaust acceptable if the exposure to materials in Section 2 are maintained below applicable exposure limits.

Refer to OSHA Standards 1910.94, 1910.107, 1910.108.

RESPIRATORY PROTECTION

If personal exposure cannot be controlled below applicable limits by ventilation, wear a properly fitted organic vapor/particulate respirator approved by NIOSH/MSHA for protection against materials in Section 2.

PROTECTIVE GLOVES

Wear gloves which are recommended by glove supplier for protection against materials in Section 2.

EYE PROTECTION

Wear safety spectacles with imperforated side shields.

OTHER PROTECTIVE EQUIPMENT

Use of barrier cream on exposed skin is recommended.

OTHER PRECAUTIONS

Intentional misuse by deliberately concentrating and inhaling the contents can be harmful or fatal.

==

Section 9 — PHYSICAL AND CHEMICAL PROPERTIES

PRODUCT WEIGHT 7.41 lb/gal 887 g/l

SPECIFIC GRAVITY 0.89

BOILING POINT 248 – 292°F 120 – 144°C

MELTING POINT Not Available

VOLATILE VOLUME 100 %

EVAPORATION RATE Slower than ether

VAPOR DENSITY Heavier than air

SOLUBILITY IN WATER N.A.

VOLATILE ORGANIC COMPOUNDS (VOC Theoretical – As Packaged)

7.40 lb/gal 887 g/l Less Water and Federally Exempt Solvents

7.40 lb/gal 887 g/l Emitted VOC

==

Section 10 — STABILITY AND REACTIVITY

STABILITY ?? Stable
CONDITIONS TO AVOID
None known.
INCOMPATIBILITY
None known.
HAZARDOUS DECOMPOSITION PRODUCTS
By fire: Carbon Dioxide, Carbon Monoxide
HAZARDOUS POLYMERIZATION
Will not occur

==

Section 11 — TOXICOLOGICAL INFORMATION

CHRONIC HEALTH HAZARDS
Ethylbenzene is classified by IARC as possibly carcinogenic to humans (2B) based on inadequate evidence in humans and sufficient evidence in laboratory animals. Lifetime inhalation exposure of rats and mice to high ethylbenzene concentrations resulted in increases in certain types of cancer, including kidney tumors in rats and lung and liver tumors in mice.
These effects were not observed in animals exposed to lower concentrations.
There is no evidence that ethylbenzene causes cancer in humans.
Prolonged overexposure to solvent ingredients in Section 2 may cause adverse effects to the liver, urinary and reproductive systems.
Reports have associated repeated and prolonged overexposure to solvents with permanent brain and nervous system damage.

TOXICOLOGY DATA
CAS No. Ingredient Name

100–41–4 Ethylbenzene
LC50 RAT 4HR Not Available
LD50 RAT 3500 mg/kg
1330?20–7 Xylene
LC50 RAT 4HR 5000 ppm
LD50 RAT 4300 mg/kg
107?98–2 1–Methoxy–2–propanol
LC50 RAT 4HR Not Available
LD50 RAT 6600. mg/kg

==

Section 12 — ECOLOGICAL INFORMATION

ECOTOXICOLOGICAL INFORMATION
No data available.

==

Section 13 — DISPOSAL CONSIDERATIONS

WASTE DISPOSAL METHOD

Waste from this product may be hazardous as defined under the Resource Conservation and Recovery Act (RCRA) 40 CFR 261.

Waste must be tested for ignitability to determine the applicable EPA hazardous waste numbers.

Incinerate in approved facility. Do not incinerate closed container. Dispose of in accordance with Federal, State/Provincial, and Local regulations regarding pollution.

==

Section 14 — TRANSPORT INFORMATION

US Ground (DOT)

1 Gallon and Less may be Classed as CONSUMER COMMODITY, ORM?D

Larger Containers are Regulated as:

UN1263, PAINT RELATED MATERIAL, 3, PG III, (ERG#128)

DOT (Dept of Transportation) Hazardous Substances & Reportable Quantities

Ethyl benzene 1,000 lb RQ

Xylenes (isomers and mixture) 100 lb RQ

Bulk Containers may be Shipped as (check reportable quantities):

RQ, UN1263, PAINT RELATED MATERIAL, 3, PG III, (XYLENES (ISOMERS AND MIXTURE)), (ERG#128) Canada (TDG)

UN1263, PAINT RELATED MATERIAL, CLASS 3, PG III, LIMITED QUANTITY, (ERG#128) IMO

UN1263, PAINT RELATED MATERIAL, CLASS 3, PG III, (27 C c.c.), EmS F–E, S–E

==

Section 15 — REGULATORY INFORMATION

SARA 313 (40 CFR 372.65C) SUPPLIER NOTIFICATION

CAS No. CHEMICAL/COMPOUND % by WT % Element

100?41–4 Ethylbenzene 7

1330–20–7 Xylene 41

CALIFORNIA PROPOSITION 65

WARNING: This product contains chemicals known to the State of California to cause cancer and birth defects or other reproductive harm.

TSCA CERTIFICATION

All chemicals in this product are listed, or are exempt from listing, on the TSCA Inventory.

==

Section 16 — OTHER INFORMATION

This product has been classified in accordance with the hazard criteria of the Canadian Controlled Products Regulations (CPR) and the MSDS contains all of the information required by the CPR. The above information pertains to this product as currently formulated, and is based on the information available at this time. Addition of reducers or other additives to this product may substantially alter the composition and hazards of the product. Since conditions of use are outside our control, we make no warranties, express or implied, and assume no liability in connection with any use of this information.

Additional Resources and References

Additional Resources

This module is intended to be a thorough resource for task training. The following reference works are suggested for further study. These are optional materials for continued education rather than for task training.

Waste Stream Measurement and Analysis © Copyright 1995, 2007 by the California Integrated Waste Management Board. http://www. ciwmb.ca.gov/LGCentral/WasteStream.

Coating Inspector Program Level 1 © NACE International, 2003, January 2007.

Coating Inspector Program Level 2 © NACE International, 2004, January 2007.

Protective Coatings and Linings Course II © NACE International, 1998, January 2005.

Control of Pipeline Corrosion, 2nd Edition, A.W. Peobody, edited by R.L. Bianchetti, © NACE International, 1967, 2001.

Introduction to Treatment, Storage and Disposal Facilities (40 CFR Parts 264/265, Subpart A-E), United States Environmental Protection Agency, September 2005, Solid Waste and Emergency Response (5305W)EPA530-K-05-017.

Savannahnow.com, copyrighted by Morris Publishing Group, LLC, "Cleaning Solvents Mess Up Trash Truck" by Megan Matteucci, February 16, 2007.

Preventing Occupational Fatalities in Confined Spaces, NIOSH ALERT:January 1986, DHHS (NIOSH) Publication No. 86-110.

Figure Credits

Painting and Decorating Contractors of America, 103F01, 103F04

NACE International, 103T01 - 103T03

Photo courtesy of Alyeska Pipeline Service Company, 103F03

Topaz Publications, Inc., 103SA01

Sherwin–Williams Co., Appendix

NCCER CURRICULA — USER UPDATE

NCCER makes every effort to keep its textbooks up-to-date and free of technical errors. We appreciate your help in this process. If you find an error, a typographical mistake, or an inaccuracy in NCCER's curricula, please fill out this form (or a photocopy), or complete the online form at **www.nccer.org/olf**. Be sure to include the exact module ID number, page number, a detailed description, and your recommended correction. Your input will be brought to the attention of the Authoring Team. Thank you for your assistance.

Instructors – If you have an idea for improving this textbook, or have found that additional materials were necessary to teach this module effectively, please let us know so that we may present your suggestions to the Authoring Team.

NCCER Product Development and Revision

13614 Progress Blvd., Alachua, FL 32615

Email: curriculum@nccer.org
Online: www.nccer.org/olf

❑ Trainee Guide ❑ Lesson Plans ❑ Exam ❑ PowerPoints Other ______________

Craft / Level: Copyright Date:

Module ID Number / Title:

Section Number(s):

Description:

Recommended Correction:

Your Name:

Address:

Email: Phone:

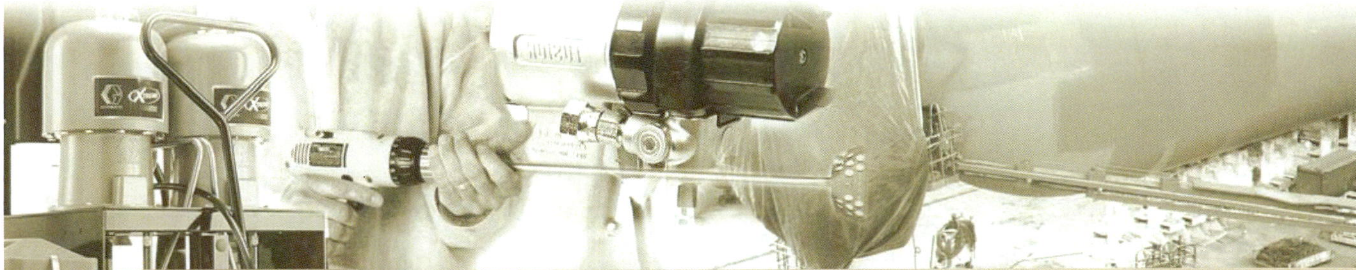

69104-09

Coating Application

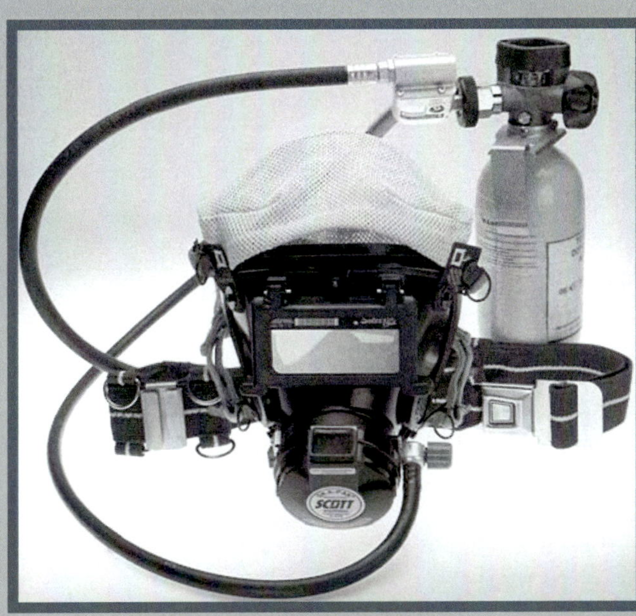

69104-09

Coating Application

Overview

After learning about corrosion, roles in the coating profession, surface preparation, and coatings in previous modules, applicator trainees must learn the basics of coating application. This module describes the information found in coating specifications and product data sheets. It covers what protective covering to use, such as masking and sheeting, and how to use them, as well as the equipment and techniques used to mix various types of coatings. Also included are descriptions of equipment and ways to apply coatings manually using brushes, rollers, and mitts, as well as by spraying methods such as conventional air spray, HVLP, airless, and air-assisted airless spray.

This module also describes what environmental conditions are necessary when applying coatings, and what quality control and inspections are required during coating application. Another important issue covered is coating application failures and how to remedy them. Specialty spray systems and their application methods are also briefly covered.

Table of Contents

Figures

Tables

Objectives

When you have completed this module, you should be able to:

1. Locate information about application equipment and other things on a coating specification and a product data sheet (PDS).
2. Mix one or more of the following using proper PPE:
   - Single-component coatings
   - Multi-component liquid coatings
   - Coatings with a powder component,

   Consider the importance of mixing time and the condition and appearance of the mixed material.
3. Describe induction time and pot life in relationship to material temperature.
4. Define *viscosity*, list the methods of viscosity control, and reduce the viscosity of coatings.
5. Install protective coverings and discuss the safe use of basic application equipment (setup, operation, adjustment, cleanup) on simple structures; list the strengths, weaknesses, and limitations of each piece; and identify basic application equipment problems with:
   - Brush, roller, mitt
   - Trowel, squeegee
   - Spray (air, airless, air-assisted airless, HVLP)
6. Demonstrate correct application techniques with:
   - Brush, roller, mitt
   - Trowel, squeegee
   - Spray (air, airless, air-assisted airless, HVLP)
7. Describe and demonstrate striping procedures.
8. Recognize the importance of the following during coating application operations:
   - Impact of environmental conditions on application and drying
   - Achieving correct thickness of each coating layer
   - Achieving a pinhole/holiday-free lining
9. Identify application defects during and after coating.
10. Describe how to calculate wet film thickness (WFT) from DFT, how to measure WFT, and explain how to calculate percentage of thinners by volume.
11. Recognize the importance of achieving a cured and a solvent-resistant film.

Trade Terms

Arc
Boxing
Confined space
Cutting in
Feathering
Flammable
Ground
Mitting
Orifice
Overspray
Personal protective equipment (PPE)
Pot life
Sweat-in time
Transfer efficiency

Prerequisites

Before you begin this module, it is recommended that you successfully complete *Core Curriculum*, Modules 00101-04 and 00106-04; and *Industrial Coating and Lining Application Specialist Level One*, Modules 69101-09 through 69103-09.

This course map shows all of the modules in the first level of the Industrial Coating and Lining Application Specialist curriculum. The suggested training order begins at the bottom and proceeds up. Skill levels increase as you advance on the course map. The local Training Program Sponsor may adjust the training order.

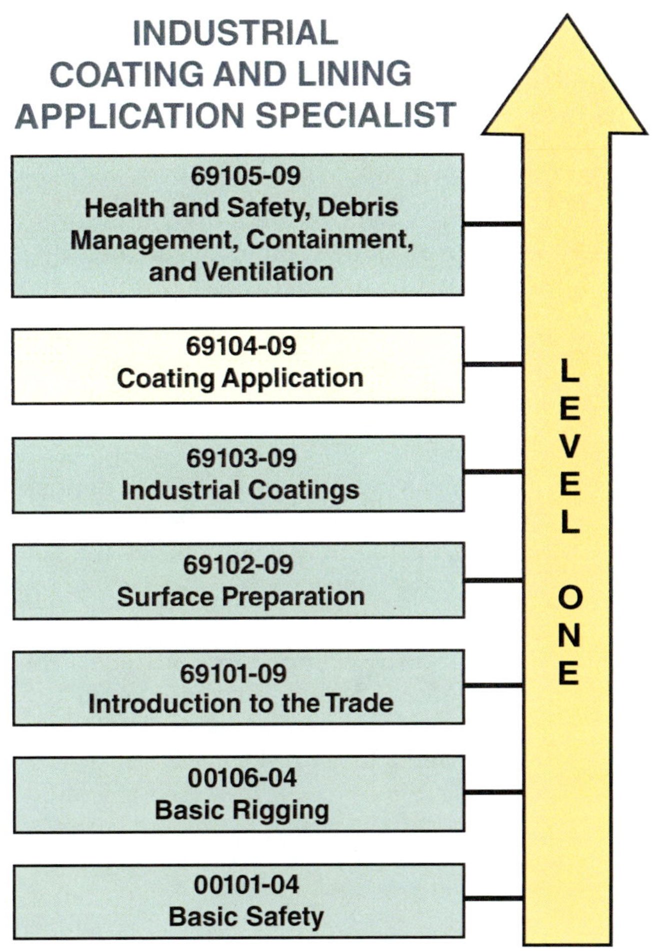

104CMAP.EPS

1.0.0 ◆ INTRODUCTION

Many methods of applying a coating have been developed. The most common are manual methods that include brush or roller application, and spray application methods. Other specialized methods include production methods like thermal spraying or hot dipping.

2.0.0 ◆ MANUAL METHODS

Brushing and rolling are the easiest and most commonly used manual coating application techniques. **Mitting**, sometimes called palming, is an application method in which the worker uses a hand mitt to apply the coating. This method is used for cables and other hard-to-cover items where appearance is not critical. Concrete, mortars, epoxy, polyester, furan, and other materials may be applied as heavy coatings by hand troweling. Organic materials might be reinforced with glass cloth, and inorganic materials might use wire mesh.

3.0.0 ◆ SPRAY METHODS

The following are several methods to spray coatings on surfaces:

- Conventional air spray uses an air stream to atomize the coating (convert the material into very fine particles or droplets) and propel it onto the workpiece.
- Airless spray pressurizes the coating and forces it through the tip of the gun to atomize it.
- Air-assisted airless spray can be used to apply a coating that has an even finer finish.
- Electrostatic spray uses electrically-charged particles of coating aimed though the air toward a surface that has a charge opposite to that of the particles. This way, the coating is attracted to the surface and will tend to deposit with little **overspray**. It is mainly used on production lines, although it is sometimes used in the field.

Thermal spray feeds powders or wires into a suitable flame and the coating sprays toward the surface being coated. The material coalesces (individual tiny particles brought together to form a whole) and solidifies to form the coating.

4.0.0 ◆ PRODUCTION METHODS

Production facilities use various methods to apply coatings. These include hot dipping, where the workpiece is dipped into a bath of molten (melted) metal. When molten zinc is the material applied, the process is called hot-dip galvanizing. In the fluidized bed method, powder coating is suspended in a tank through which air is pumped so that the particles behave like a liquid. Workpieces to be coated are preheated and run through the bed. Their hot surface melts the powder and fuses the coating. As the piece cools, a uniform plastic sheathing is formed. In the powder spray method, parts to be coated are heated to the appropriate temperature before or after a fusible coating powder is sprayed onto the surface. Powder spray is a production line technique that can produce very uniform coatings. Both the hot dipping and powder spray methods can use electrostatic spray to transfer the coating materials to the workpieces.

5.0.0 ◆ TYPICAL COATING SPECIFICATION

Application specialist trainees need to know that most industrial projects use formal contractual documents to list, in detail, exactly what all the work requirements are. A coating specification is an example of a commonly used contractual document. The coating specification tells a contractor (applicator) what to do and where to do it. Applicators need to be familiar with the work requirements defined in a coating specification because they are legally bound by that contract to follow all of them. A good coating specification should contain all of the sections listed below, with each containing the specific information and criteria noted.

5.1.0 General

This section may only show the owner's name and address, the job number, and any designated representative.

5.2.0 Terms and Definitions

Here words and terms are defined to make sure everyone understands exactly what they mean. Terms defining the contractor, owner, specifier, coating supplier, and inspector may also be included.

5.3.0 Reference Standards and Codes

This section lists the standards and/or codes referred to in other sections of the specification.

5.4.0 Environmental Regulations

Any local, state, or national air, land, or sea environmental regulations that apply to the job should be listed in this section.

5.5.0 Scope

This section describes what, where, and when the work is to be done. It also describes what the contractor is required to supply. There may also be a coatings schedule that lists each item to be coated, as well as areas that not to be coated and must be protected from coatings.

5.6.0 Safety

This section states that work is to be done in a safe manner and in accordance with all federal, state, and local safety codes that apply. Also listed here are any other safety requirements, like wearing hard hats, or specially colored protective clothing that identifies contractors and other work site personnel. This section often identifies access and permit badges for specific areas.

5.7.0 Pre-Job Conference

A good specification includes this section and requires a conference between the owner, coating application contractor, coating supplier, inspector, and others to review standards and work requirements for the project. This should take place before the project starts.

5.8.0 Surface Preparation

This section states the methods and standards of sub-surface preparation to be used on the project. These include pre-inspection for fabrication defects, pre-cleaning requirements, and cleaning operations. Each method stated should refer to the specific standard that applies to that method. In addition, information about abrasives, equipment techniques, and restrictions for surface cleaning may be included.

INSIDE TRACK — Coating Specifications

Coating specifications are normally tailored to meet the requirements for a particular job, and they come in many forms and in various degrees of quality and clarity. However, specifications often are copied from a previous job without regard to their accuracy, practicability, and adaptability to the current job, or with any consideration of technological advances.

5.9.0 Coating Materials

This section usually identifies the coating to be used, by manufacturer and type, and indicates both the required storage conditions and how to handle the material.

5.10.0 Sample Coatings

This section may require coating samples to be taken. Information on the number of samples along with labeling and storage procedures will be specified.

5.11.0 Coating Schedule

This section is a specific list of exactly what items are to be coated. It may also identify each area to be protected, covered, and not coated.

5.12.0 Workmanship

This section should contain a general statement stating that all work is to be performed according to the specification, the coating manufacturer's current instructions, and by skilled workers in a safe and workmanlike manner. To prevent misunderstanding of what that means, the specification should indicate exactly which standard to follow, for instance, SSPC-PA 1 or another standard.

5.13.0 Application

This section defines the approved methods of coating application such as brush, roller, or spray. A good specification states the minimum and maximum dry film thickness (DFT) of each coat and the resulting total minimum and maximum DFT of all the coats together. If stripe coating is required, it should state the areas to be stripe coated and the appropriate procedure to follow. (Stripe coating is a preliminary coating applied to odd-shaped objects, edges, corners, rivets, bolts, etc., that are within the area to be coated.)

5.14.0 Work Schedule

This section lists the specific order in which coatings are to be applied including pre-inspection of the substrate, pre-cleaning, surface repair, surface preparation, coating application, inspection intervals, repair/remedial coating work, as well as records and reports. This section should also identify access points to the project and working hours of the contractor along with any other known restrictions and limitations regarding the work schedule.

5.15.0 Repairs and Remedial Coating Work

This section identifies criteria and procedures for repair work.

5.16.0 Inspection

Inspections cover numerous aspects of any job. These include:

- Monitoring ambient climate conditions throughout the job
- Pre-inspection
- Pre-cleaning inspection
- Surface preparation inspection
- Coating material inspection
- Storage
- Identification
- Mixing
- Thinning
- Application inspection
- Wet film thickness (WFT)
- Dry film thickness (DFT)
- Recoat times
- Coating defect inspection
- Documentation inspection

This section may contain when, where, and how many measurements are required, the pass/fail criteria for each measurement, and the inspection tools to use. The section may also give guidelines for inspection reports, and a very detailed description of the inspector's responsibilities and authority. It may also show the client's organization chart to show the chain of command and the inspector's position.

A good specification will include a complete Inspection and Testing Plan in the inspection section. This plan identifies all the inspection points and states if they are witness, verification, or hold points. This document would list which section of the specification applies to each inspection step and what the pass/fail criteria are for each step.

5.17.0 Documentation

This section describes all the record keeping and reports that will be required as a result of the inspection process.

6.0.0 ◆ ENVIRONMENTAL CONDITIONS

Environmental conditions must be considered not only as you plan and begin a coating project, but also throughout the entire project. The conditions to take into consideration follow.

- Air temperature (ambient temperature) is simply the air temperature where the work is being performed. It is important to remember that the conditions required for applying the coating often need to stay the same even through the cure. If the coating requires 24 hours to cure, then relative humidity and dew point need to be the same until the coating is fully cured.
- Surface temperature is the temperature of the surface being painted.
- Relative humidity is the amount of moisture the air can hold at the ambient temperature. It is important because the amount of moisture in the air can affect the cure of some coatings.
- Dew point is the temperature at which water will condense out of the air. This is critical because when the surface temperature reaches the dew point, water can condense on the surface. This can cause the substrate to flash rust, and create adhesion and curing problems. When coatings that contain amines are used, condensation can cause an amine blush that affects appearance and can prevent an overcoat from adhering to the surface.
- Wind speed has to be considered in any outdoor operation because dust and dirt can be blown onto the wet coating and, if spraying, overspray can be blown around, even when wind speeds are as low as three to five miles per hour (4.8 to 8 kph).
- The temperature of the coating itself needs to be checked with a probe temperature gauge (*Figure 1*).

It is very important to take measurements as close as possible to where the work is being done and immediately before work begins. If work is being done in the sun, measurements should be taken in the sun. If work is being done in shade, measurements should be taken in the shade. If working in a tank, take measurements inside the tank.

It is important to measure environmental conditions because they can have a direct effect on induction time, **pot life**, and cure time. Generally, higher temperatures will shorten all three of these. There are instruments available to measure each environmental factor, for example, sling psychrometers (*Figure 2*) to measure humidity, and gauges to measure air and surface temperature.

Also available are some relatively inexpensive all-in-one electronic instruments like the one shown in *Figure 3*. This particular instrument measures air temperature, humidity, surface temperature, dew point temperature and wet bulb temperature. It also can display any three simultaneously.

In general, most coatings should be applied at less than 85% relative humidity and when the dew point is at least 5°F (3°C) less than the surface temperature. Some manufacturers specify on their label or PDS whether or not additives such as defoamers for rolling, flow control agents, and accelerators or retarders to modify drying and curing time can be used with their product to compensate for environmental conditions.

7.0.0 ♦ COATING MATERIAL PREPARATION

Applicators must be aware that good stewardship of coating materials is a necessity for any project. Stewardship means that coating materials need to be properly organized and properly stored. This can include product rotation, a controlled storage environment that is protected from the elements, and shelf life monitoring.

8.0.0 ♦ PRODUCT DATA SHEETS

The first step is to check the coating manufacturer's product data sheet (PDS), which the manufacturer supplies for each of its products. The PDS

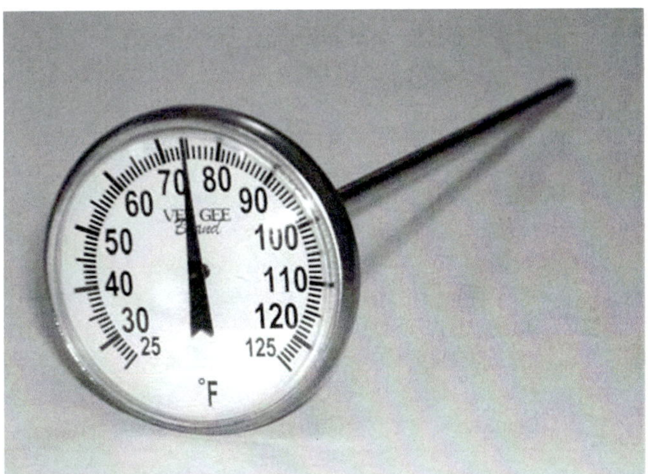

104F01.EPS

Figure 1 ♦ Probe temperature gauge.

104F03.EPS

Figure 3 ♦ Digital psychrometer and infrared temperature instrument.

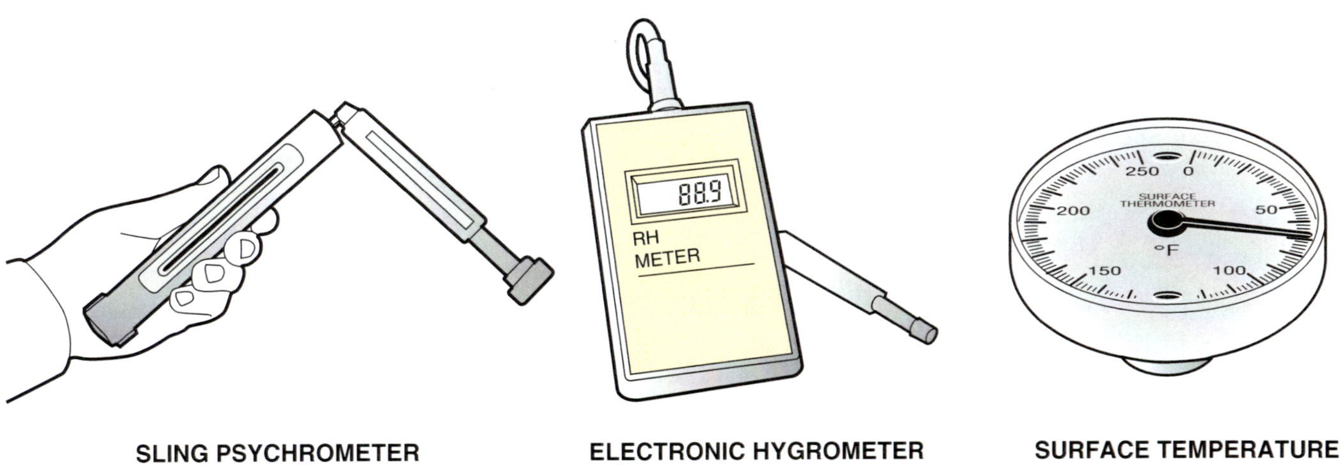

104F02.EPS

Figure 2 ♦ Devices used to measure relative humidity, dew point, and surface temperature.

should be read by all applicators. This sheet contains specific information including:

- Product description
- Product characteristics (spreading rates, drying time, WFT, DFT, etc.)
- Recommended uses
- Recommended application systems (primer and finish coat combinations)
- Safety references or precautions
- Surface preparation
- Coloring information
- Application criteria such as temperature and humidity requirements

The way the information is arranged in a PDS varies from manufacturer to manufacturer, but the information it contains is generally similar. *Appendix B* is a typical PDS from one manufacturer. This PDS can be used as a reference when a PDS is mentioned in other sections of this module. Applicators should always cross-check the latest PDS data for coating materials and solvents against the requirements in the project coating specification.

If there are any conflicts, they should be resolved before the project starts. If you notice a conflict between the PDS and specification, the specification is what to follow. If you see a conflict between the PDS and specification, you should immediately call it to the attention of your supervisor.

Once the surfaces to be coated are properly cleaned, and protective sheeting for areas not to be coated are in place, the next step is to mix the coating material properly. The coating material must be mixed and, if necessary, thinned or reduced according to the manufacturer's instructions. This is to make sure that the coating is uniform in color and texture and can be applied smoothly and evenly.

9.0.0 ◆ PERSONAL PROTECTIVE EQUIPMENT

Most industrial applications use some hazardous solvents or coating components. The MSDS for a coating identifies the particular hazards it presents and the proper procedures to follow if someone is exposed to them. In general, the following **personal protective equipment (PPE)** may be required depending on what the coating or solvent is made of and the environment in which the coating is being mixed or applied:

- Safety goggles (*Figure 4*)
- Solvent-proof rubber or vinyl gloves (*Figure 5*)
- Organic respirators along with adequate ventilation
- Supplied-air respirators to use in **confined spaces** when hazardous materials that can be inhaled are present (*Figure 6*)
- Full body suits with supplied-air hoods to wear when hazardous materials that can be inhaled are present or come into contact with the skin (*Figure 7*)

In certain cases, hard hats, high-top steel-toed shoes, hearing protection, body covering closed at the wrist and neck, or solvent-resistant skin cream may be required. In confined areas, equipment may be needed to monitor **flammable** and/or hazardous solvent levels to prevent fires, explosions, or too little oxygen.

10.0.0 ◆ MIXING GUIDELINES

It is important to understand the guidelines for mixing and thinning various types of coating materials, including single-component coatings, plural-component liquid coatings, and liquid coatings with a powder component.

Before mixing any coating, inspect it to see if it has been damaged by age or improper storage and handling. If cans have been opened before, examine the contents carefully. The pigment in coatings can settle over time, so coatings that tend to settle may need repeated mixing. Carefully examine coatings that have separated or gelled. If in doubt, don't use the coating. Set it aside and call the manufacturer for information.

Figure 4 ◆ Use proper eye and/or face protection for the job at hand.

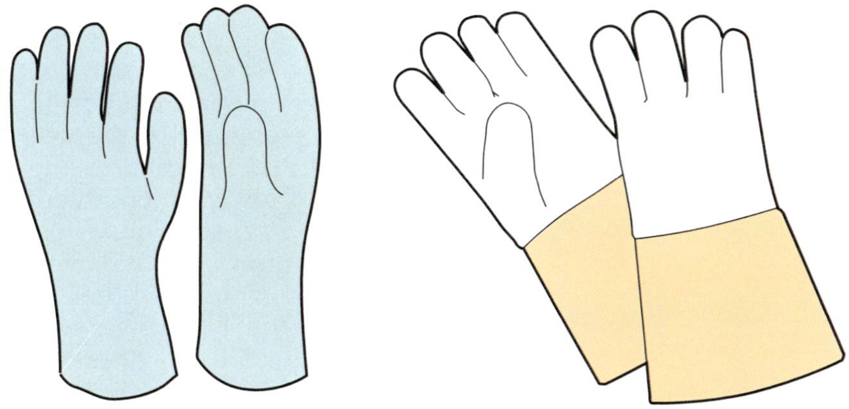

Figure 5 ◆ Use the proper type of glove for the job.

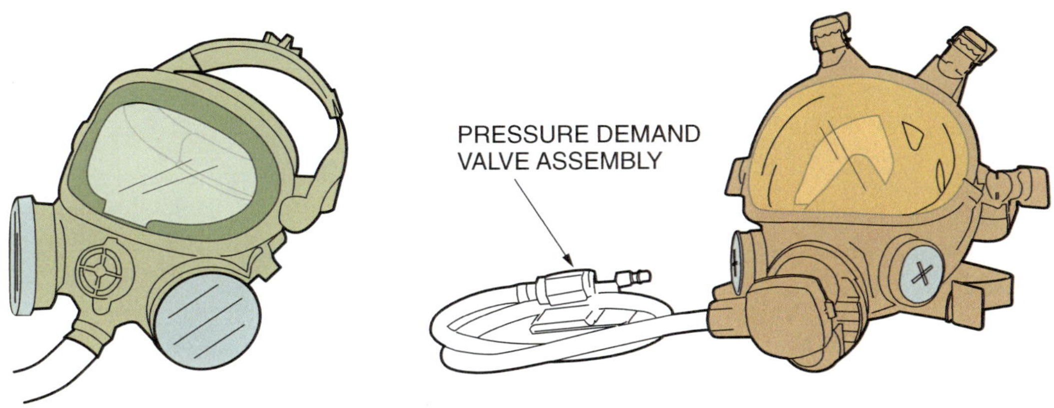

Figure 6 ◆ Supplied-air respirators.

WARNING!

Power mixing can present a potential explosion hazard. Always use non-sparking mixer motors to prevent ignition of any flammable or explosive solvent vapors.

Make sure the coating is completely mixed before use. When using a hand agitator, place the agitator in the coating before you start the device to avoid paint spatter and waste. The specific results from failure to mix completely will vary between different types of coatings, but generally, a poorly mixed coating will not protect properly. If the components are not mixed in the correct proportions, the coating may not cure the way it should and the components might separate after the coating is applied.

Figure 7 ◆ Full body suit with supplied-air hood.

10.1.0 Coating Terms Essential to Mixing

- *Viscosity* – Viscosity is the resistance to flow. This is a property of fluids based on their thickness and stickiness. Water has a lower viscosity than maple syrup, and maple syrup has a lower viscosity then peanut butter. You may be familiar with viscosity in relation to motor oils. The easier it flows, the lower the viscosity.
- *Induction time (**sweat-in time**)* – Catalyzed coatings such as epoxies require contact between the components to cure the coatings. When coatings are in the vessel (the pot) that holds them prior to application, there is more contact between components than when they are applied in a thin film on the surface. This is the induction time. Not all two-component coatings require an induction time, but the Product Data Sheet (PDS) will list the time needed if induction is required. Always consult the PDS before mixing. It is possible to leave coatings in the pot too long, so always follow the PDS induction time instructions.
- *Pot life* – Once the two parts of the coating are mixed, the coatings begin to cure. As the coatings cure they start to harden, becoming thicker and less viscous with time. They will reach a point at which they can no longer be applied. For a coating to be applied, it must be at less than the maximum viscosity stated by the manufacturer. The time it takes a coating to reach the exact viscosity is called the pot life. Some applicators think it is acceptable to add additional thinner once the pot life is exceeded, but this is not at all true. Although adding thinner when the coating is past the pot life may reduce the viscosity, the physical and chemical properties of the coating are changed and it will not do the job it should.
- *Temperature* – Temperature affects both induction time and pot life. When the temperature of a liquid gets hotter, the molecules in the liquids move around faster. The faster the molecules move, the more quickly they bump into each other. This speeds up the reaction. The product data sheet will state induction times and pot life based on temperature. At higher temperatures, the induction time and pot life are both shorter.

10.2.0 Single-Component Coatings

Many types of tools and equipment can be used for mixing single-component coatings. *Figure 8* shows two mixing tools. Always follow the manufacturer's instructions. All air bubbles should be allowed to disappear before the coating is used. For single-component coatings that contain pigments, the amount of mixing required depends on the amount of pigment that has settled (pigment separated from the liquid resin and solvent). How much pigment settles depends on the coating type, formula, storage time, and storage temperature. To avoid problems when mixing a coating, or to keep the coating or coating components from breaking down, always use fresh stock since it isn't known how long the older stock has been stored. Before mixing a partially used container of coating, remove any skin that may have formed on top of the liquid.

If coating is mixed in a mechanical shaker, some may foam after the shaking and should sit until the foam disappears.

> **NOTE**
>
> When hand power mixing any coating, always insert the mixing tool into the coating container before starting the mixing motor. When mixing is completed, stop the mixing motor before withdrawing the mixing tool. The mixing tool can be cleaned by operating it in a can of the coating solvent for a short time.

When a hand-powered mixer is used:

- Mix at a slow rate to help prevent trapping air in the mix and/or foaming the coating; when

104F08.EPS

Figure 8 ◆ Mixing tools.

the mixing rate is correct there will be a small whirlpool around the shaft of the mixer.

- After mixing, the coating should look smooth and have a uniform color.
- Make sure that there is no settled pigment at the bottom of the container.
- Allow the coating to sit for a short time so that any air bubbles can rise to the surface.

Boxing is a technique used to mix two or more containers of a coating. To box a coating, pour about half of the first container into a clean container. Then, pour about half of the second container into the first container, then remix everything in the first container. This process should be repeated as needed for any additional containers. Finally, fill the last container with the coating that was poured from the first container and remix the last container.

The coating should be strained after mixing to remove any lumps or other contaminants. If the coating will be sprayed, it should be thinned if required by the PDS. Thinning and straining are covered in a later section.

10.3.0 Plural- (Multi-) Component Liquid Coating Systems

Plural-component liquid coatings have two or more components that must be mixed together before applying. Plural-component liquid coatings have a shorter pot life. The coatings begin to cure in a specific amount of time when a converter, hardener, or activator is mixed into the coating. These agents cause a chemical reaction that starts a curing process that cannot be slowed or stopped. There is limited amount of time to thin and apply the coating.

Be sure to mix only the amount of coating that can be thinned and/or applied within the coating's pot life. When the end of the pot life is reached, the mixed coating must be discarded and all the tools and equipment must be cleaned with an appropriate solvent.

Some plural-component coatings also need an induction (sweat-in) time to allow the catalyst, converter, hardener, or activator to react chemically and reach a turning point before they are thinned and/or applied.

The induction time and pot life can be affected by the ambient temperature at the time the coating is mixed. Humidity has less of an effect. High temperatures shorten the time and cold temperatures extend it. The manufacturer's PDS will state the specific induction time and pot life at various temperatures. Using the wrong pot life can result in a defective coating application, wasted material, and damaged application equipment.

When mixing a plural-component coating:

- First mix each individual component by itself with a power mixer.
- Next, power mix the components together in the correct ratio to start the chemical curing reaction. Mix only long enough to provide uniform color and consistency since the induction and pot life clock is running.
- Follow the manufacturer's label instructions or PDS for the proper mixture ratios, mixing time, formulas, application data, pot life, and, thinning and induction time.
- Pay close attention to which component is added to what other component, because different manufacturers specify and label the components differently.
- If you cannot use the original container for mixing, use a separate clean metal container.
- Do not use plastic containers.
- Do not use the catalyst, converter, hardener, or activator containers as mixing containers.
- If only part of each component will be used, be careful to not contaminate the original component containers when each component is added to the mixing container.
- It is a good practice to mix small rather than large quantities of a multi-component coating in a big container. Chemical reactions between the components may cause large amounts of heat. The more coating that is mixed, the more heat is generated. This can result in temperatures that may be a safety hazard.

WARNING!

Never mechanically shake a metallic powder-filled coating. Shaking can cause the container to explode.

10.4.0 Liquid Coatings with a Powder Component

Liquid coatings with a powder component are multi-component coatings in which one component is a metallic dust (powder) such as zinc. Follow the PDS, label instructions, and the MSDS carefully to mix or thin any of the components.

Step 1 Each non-powder component is usually mixed first by itself using a shaker or power mixer.

Step 2 The metallic dust (powder) is then slowly added through a sifter or sieve to one of the liquid components and slowly power mixed into the liquid.

Step 3 When all of the powder has been added and mixed for a short time, the coating should be strained to remove any lumps.

Step 4 If an activator component is needed, it is then added and mixed.

Step 5 During application, zinc-rich coatings must be constantly and slowly mixed to prevent settling. If an activator has also been used, the coating must be applied and all the tools and equipment must be cleaned before the pot life expires.

When the powder (metallic dust) is handled or mixed into the coating, the applicator must wear the appropriate respirator to keep from breathing in the dust, which is considered a hazard. Closely follow the instructions (on the PDS, label instructions, and the MSDS) for mixing components or thinning that may be needed for spraying.

10.5.0 Mixing Equipment

A variety of tools can be used to mix coating materials properly. Which ones are used depends on the coating type and the amount of material to be mixed.

Power mixing is commonly used for larger projects or applications involving large quantities of a coating.

Portable or semi-portable mechanical shakers for use with metal or plastic containers allow mixing without opening the containers, as shown in *Figure 9*. The electrically powered 1 gal (4 L) unit is useful for maintenance departments and smaller projects. Similarly sized units powered by an air motor are also available. The 1 to 5 gal (4 to 19 L) unit is electrically powered and can be either floor mounted or mounted on a trailer, heavy cart, or van. Both units are equipped with an adjustable timer switch to set the mixing time.

Drum mixers with stirring paddles or spirals powered by variable-speed air motors are used for 55 gal (200 L) drums of coating material (*Figure 10*). They continuously mix coating material that is pumped from the drums and applied during coating operations.

All mixing tools and equipment must be promptly cleaned after use with the proper solvents specified in the coating manufacturer's PDS or on the container label.

10.6.0 Straining and Thinning Guidelines

After coatings have been mixed, it is good practice to filter the coating through a strainer to remove any lumps, hardened coating chips, or coating skin remnants. Use a bag strainer inside

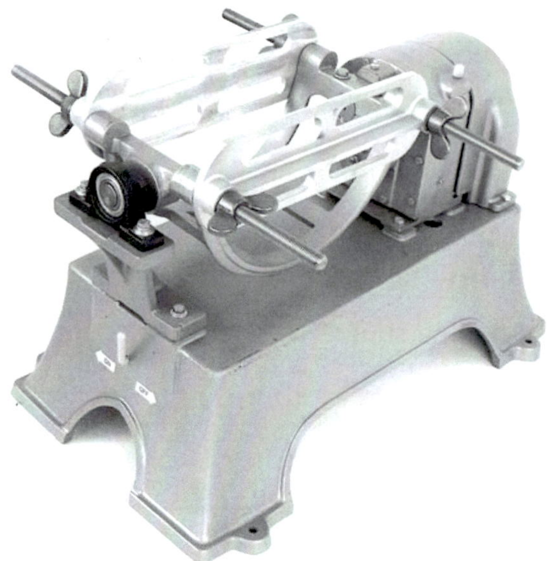

FIVE GALLON UNIT

ONE GALLON UNIT

104F09.EPS

Figure 9 ◆ Mechanical shakers.

a clean 5 gallon (19 liter) container as shown in *Figure 11*, and pour the coating to be strained into the container. Lift the bag up slowly to strain the coating. (The bag is cleanable and reusable.) Other strainers made of a cleanable fine metal mesh that can be placed over the top of a container are also available.

Sometimes a diluent (a substance used to dilute something) is added to the coating to thin it, but in general, thinners should not be used unless necessary. If you need to use one, consult the manufacturer's instructions and be certain to use the correct amount of thinner. Too much thinner can cause the dry-film thickness to be too thin, and may cause runs and sags. Too little thinner can cause a non-uniform film with poor integrity, pinholes, and/or poor appearance.

Coatings are normally supplied with the proper consistency for the most common methods of application. However, if thinning/reducing solvents are required to reduce the viscosity of the coating to use with spray equipment, measure the viscosity of the coating to prevent errors. Viscosity is a measure of the liquidity (thickness and stickiness) of a fluid. This measurement can be done manually by using various types of viscosimeters. *Figure 12* shows a shell cup viscosimeter.

The cup of the viscosimeter is dipped into the coating and held for a short period until the cup reaches the temperature of the coating. It is then removed completely from the coating. The time it takes for the coating to drain out of a calibrated (specifically sized) hole at the bottom of the cup is measured in seconds from the time the cup breaks the surface until the draining stops. Find the time it took to drain on the temperature graphs to determine the viscosity of the coating in shell cup seconds. There are other viscosimeters similar to shell cups, but they aren't as accurate because they dribble when the last of the coating drains from the cup. Shell cups make measuring the drain time more exact. Cups with different sized holes are available to cover a wide range of viscosities.

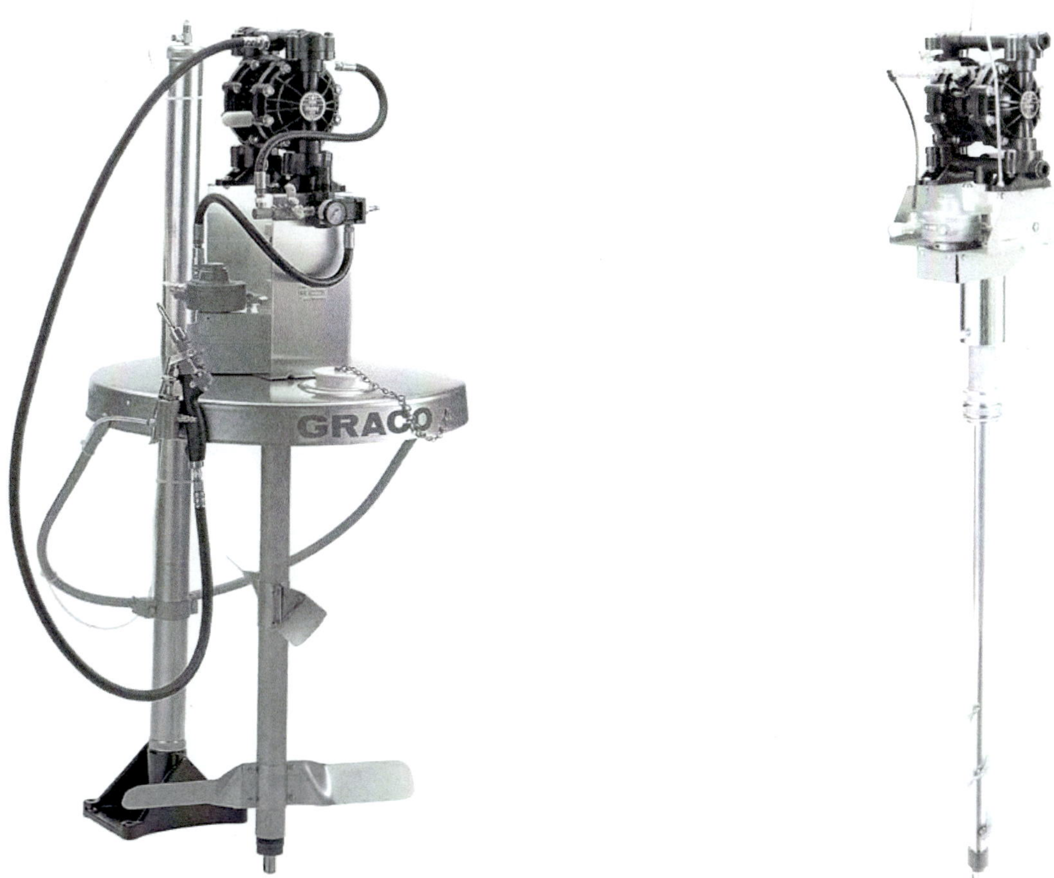

STAND-MOUNTED OPEN DRUM PADDLE MIXER WITH AIR-POWERED DRUM COVER LIFT AND PUMP

DRUM-MOUNTED SPIRAL MIXER FOR CLOSED DRUMS WITH AIR-POWERED MOTOR FOR SPIRAL AND PUMP

104F10.EPS

Figure 10 ◆ Air-motor powered drum mixers.

Because spray equipment used to apply a coating may need specific coating viscosities to produce the proper sprayed surface, the coating viscosity will most likely have to be checked. If the coating is not thin enough to be sprayed, the viscosity must be reduced by heating or thinning. It should then be remixed and tested again with the viscosimeter. Repeat this process until the proper viscosity is reached.

Spray coating systems (solvent or water-based) used in large-scale manufacturing operations have very specialized equipment. These have built-in, automatic equipment to measure the viscosity of the coating and to strain it as well. *Figure 13* shows typical viscosity sensors. A controller that can be programmed is used to operate the sensors along with coating temperature probes. The controller determines the viscosity based on the data from the sensors and temperature probes. It also controls valves to inject solvent into mixing barrels (that run constantly) if the viscosity needs to be changed to a preset value. The viscosity sensors can be mounted on both open and closed barrels (at atmospheric pressure), or in low pressure pipelines feeding the spray gun pumps.

The sensors operate at a timed interval (typically twice a minute) set at the controller. Each time they operate, a piston in a tube immersed in the coating is raised and a sample is drawn in under the piston. For the barrel sensors, the piston is released and allowed to fall by gravity to empty the sample back into the mixing barrel. For the pipeline sensors, compressed air is used to move the piston in both directions. In either case, the measurement phase is the time that is required to force the sample back into the barrel or pipeline. The controller calculates the viscosity measurement which is compared to a set-point value. Then, if necessary, the controller alters any solvent or water injection into the coating mixing barrels.

STRAINER BAG IN BUCKET

STRAINER BAGS

104F11.EPS

Figure 11 ◆ Coating strainers.

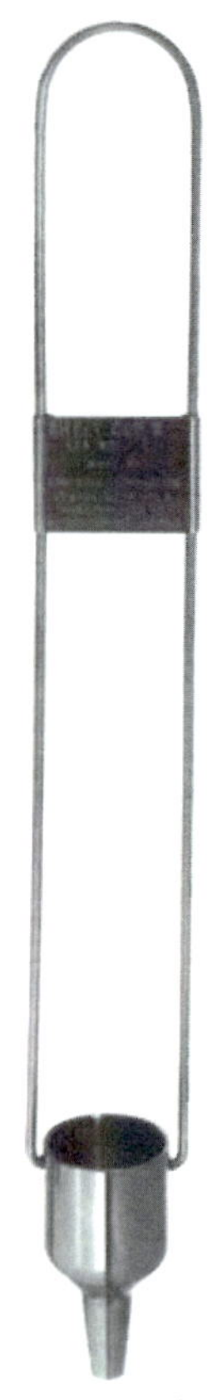

104F12.EPS

Figure 12 ◆ Shell cup viscosimeter.

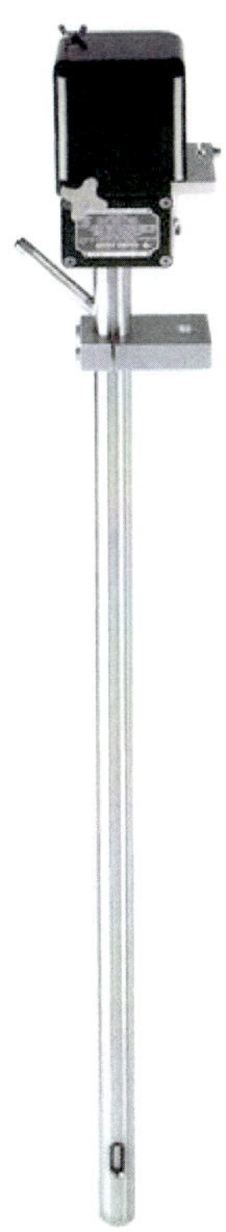

OPEN OR CLOSED BARREL-MOUNTED VISCOSITY SENSOR

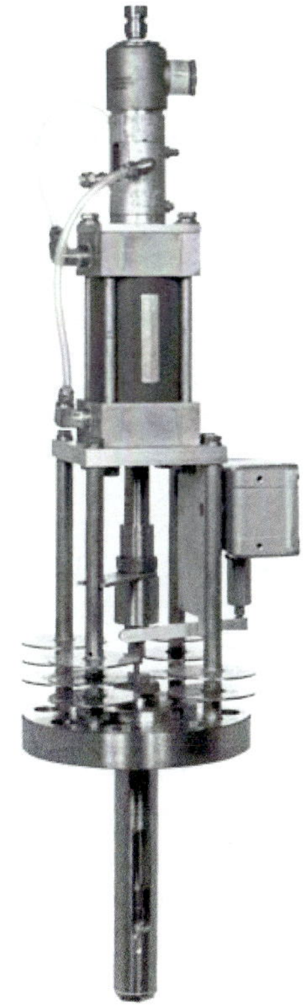

IN-LINE VISCOSITY SENSOR (MOUNTS TO PIPELINE TEE)

104F13.EPS

Figure 13 ◆ Control system components.

11.0.0 ◆ MEASURING WET FILM THICKNESS (WFT)

In order to ensure that the DFT (dry film thickness) of the finished coating meets the requirements in the job specification, the applicators must know the wet film thickness of the coating while it is being applied. Since often more than one layer of coating is applied, the applicator must measure the wet film thickness of each layer immediately after it is applied to be able to reach the required DFT.

The coating manufacturer's PDS usually specifies the maximum amount of thinner/reducer that can be added. The PDS will state that amount as a percentage by volume based on the solids volume of the coating. In any case, the least thinning possible should be done because thinning can affect the quality of the coating. In addition, thinning with solvents must be done according to the volatile organic compound (VOC) regulatory restrictions.

Adding thinner/reducer to the coating to make it easier to spray will increase the wet film thickness (WFT) needed to get the dry film thickness (DFT) that is asked for in the specification. For instance, if the volume of unreduced solids (percent solids) in a coating is a certain percentage, the DFT that results from the thinned coating can be calculated as shown in Equation 1. To figure the WFT needed for a thinned coating, insert the DFT number into Equation 2. The formula will give you the WFT that is needed to reach the correct DFT.

Equation 1:

$$\text{DFT} = \text{DFT} - \text{WFT}\left(\frac{\%\ \text{solids by volume}}{1 + \%\ \text{thinner by volume}}\right)$$

Equation 2:

$$\text{WFT} = \left(\frac{\text{DFT}\ (1 + \%\ \text{thinner by volume})}{\%\ \text{solids by volume}}\right)$$

11.1.0 Wet-Film Thickness Gauge

NOTE

For gauges that are not calibrated in both microns and mils, mils can be converted to microns or microns to mils by using the following equations:

$$\text{Mils} \times 25.4 = \text{microns}$$

$$\frac{\text{Microns}}{25.4} = \text{mils}$$

One of the most common tools used to measure DFT is the wet film thickness gauge. If the applicator knows the volume-solids content of the coating, he or she can calculate the WFT that is needed to achieve the required final DFT.

The most common instrument to measure WFT is the comb gauge. *Figure 14* shows a detailed diagram of a gauge. A comb gauge consists of a series of teeth cut to different heights, set between two outer teeth. The gauges are usually marked in both microns (μm) and mils. Gauges can be shaped as triangles, octagons, circles, squares, or rectangles.

Measure immediately after the coating is applied, and make sure it is done on a flat surface with no surface irregularities. To measure the WFT, push the gauge firmly into the wet paint film so that the outer teeth touch the substrate, if measuring a primer coat; touch the primer, if measuring the finish coat; or touch the previously dried coat for a job that requires several coats. Hold the gauge at a right angle to the surface. Remove the gauge and examine the teeth. The heads of some of the teeth will be coated with paint and others will be clean. The true WFT lies between the last tooth that is coated and the next (higher) tooth that is uncoated. The WFT is the measurement of the last wet or coated tooth on the gauge. *Figure 15* shows in detail how to use the gauge. In *Figure 16*, applicators are shown making WFT measurements. After making measurements and calculations, it may be necessary to recoat the area to prevent pinholes if the coating is not wet enough to allow reflow over the gauge marks.

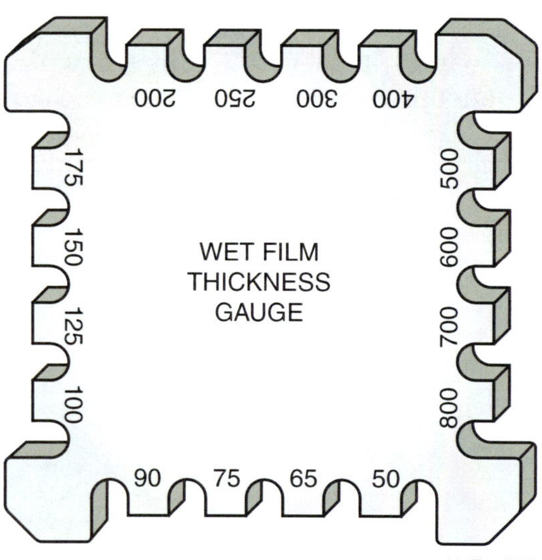

Figure 14 ◆ Typical WFT gauge.

11.2.0 How to Calculate WFT on a Basic Calculator

The following are step-by-step instructions to do a simple WFT calculation with most basic calculators. Some calculators have an equation mode that allows you to put numbers in exactly as they appear in the formula. Some simple calculators require you to perform calculations in groups.

The following procedure will not work for all calculators but will work for most.

To calculate what the coverage of a 55 percent solids coating at 2 mil (50 μm) thickness will be, use Equation 3:

Equation 3a:

$$TC = \frac{1{,}604 \times \%V.S.}{100 \times DFT}$$

When you put in the numbers it looks like Equation 3b:

Equation 3b:

$$TC = \frac{1{,}604 \times 55}{100 \times 2}$$

On a calculator, perform the following steps:

Step 1 Press 1640

Step 2 Press × (multiplication sign)

Step 3 Press 55

Step 4 Press / (division sign)

Step 5 Press 100

Step 6 Press / 2

To calculate the amount of coating you will need for a job, you need to understand the relationship between wet film and dry film thickness. Dry film thickness (DFT) is the criterion that is usually specified in the coating contract.

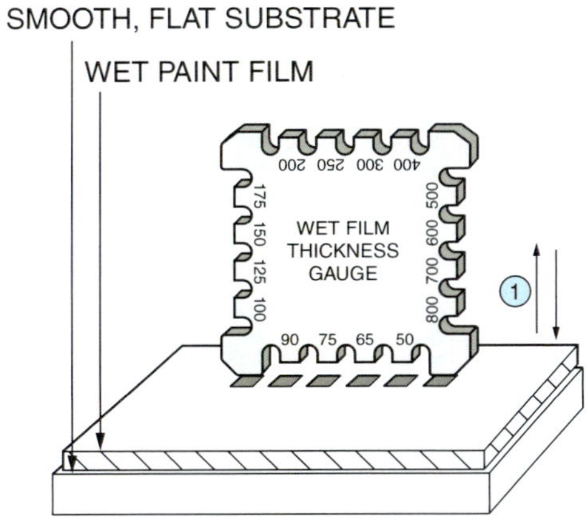

1. Begin with the lowest set of readings on the gauge. While holding the gauge perpendicular to the wet surface, gently press down onto the surface. The end notches should land onto the substrate or previously cured primer or finish. Remove from the surface without rocking the gauge. In this example, all of the notches (from 50-90 µm) are covered with the coating. Clean off all residues of the coating on the gauge with a soft cloth.

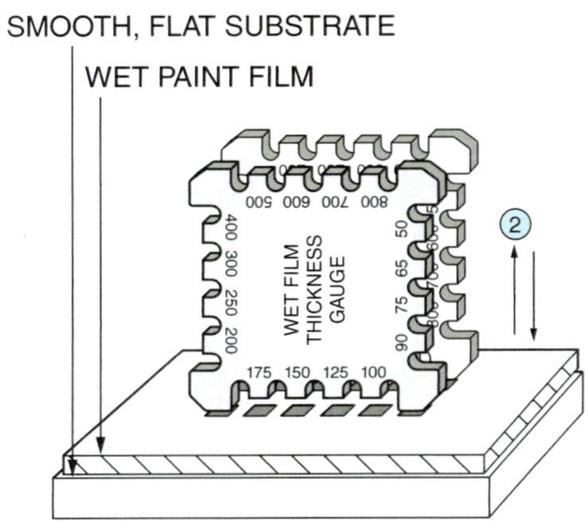

2. Rotate the gauge around to the next higher set of readings. Repeat the reading as per Step 1, taking the reading in close proximity to the first reading. Again, in this example, all of the notches (from 100 and 175 µm) are covered with the coating. Clean off all residues of the coating with a soft cloth.

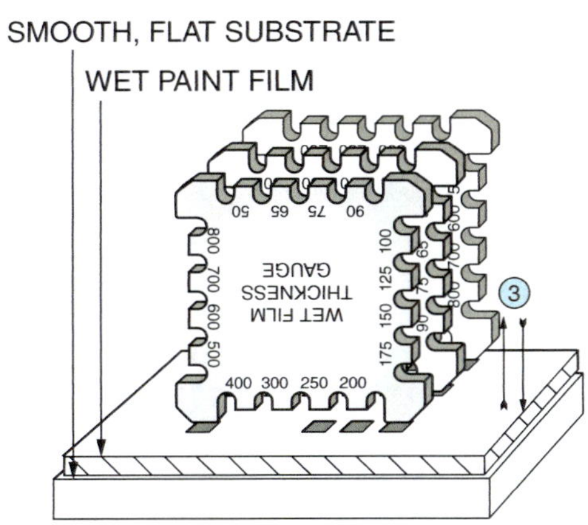

3. Rotate the gauge around to the next higher set of readings. Repeat the reading as per Step 1, taking the reading in close proximity to the first and second reading. In this example, the notches 200 and 250 µm are covered with the coating, as are the outside notches. The marks left on the coating surface show the same indents. This means that this reading is over 250 µm but under 300 µm. Clean off all residues of the coating with a soft cloth.

Repeat taking measurements in an area of around 1 yd^2 around the initial measurement. Average the result. This is the average WFT in this area. From this information, you should be able to determine whether the correct amount of paint is being applied to the substrate.

VIEW OF A WFT GAUGE INSERTED INTO AND THEN REMOVED FROM A WET PAINT FILM

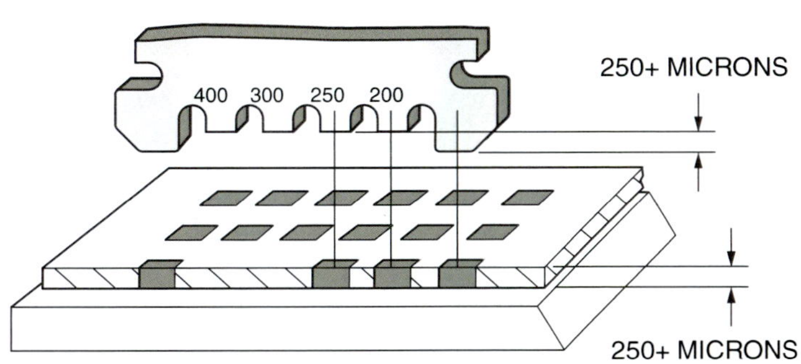

CROSS SECTIONAL VIEW OF A WFT GAUGE READING
THE RESULT OF THIS MEASUREMENT IS BETWEEN 250-300 MICRONS

104F15.EPS

Figure 15 ◆ Checking WFT with a gauge.

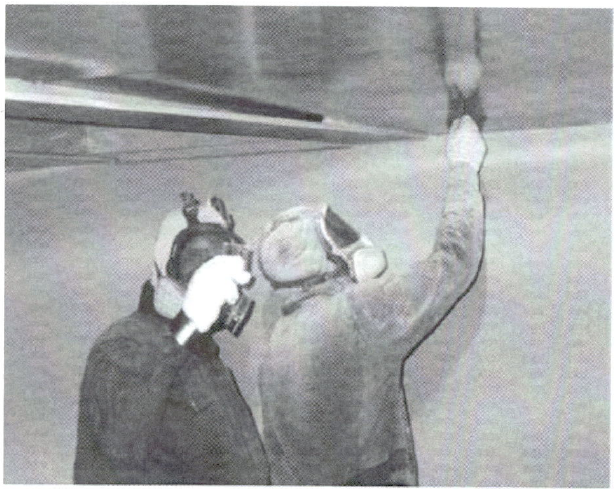

104F16.EPS

Figure 16 ◆ Wet film thickness measurement.

Wet film thickness (WFT) measurements help the applicator determine how much coating must be applied to reach the specified DFT. WFT measurements are made at the time each coat is applied so that errors in thickness can be corrected immediately. However, wet film thickness is only a guide. DFT is the defining measurement.

The dry film/wet film ratio is based on the percentage of solids by volume in the coating being used. Manufacturers' data sheets sometimes list solids by weight as well as solids by volume, but volume measurement must be used to calculate WFT. Use Equation 4:

Equation 4:

$$\frac{\text{DFT}}{\text{\% solids by volume}} = \text{WFT}$$

For example, the coating specification calls for each application to have a DFT of 4 to 6 mil (100 to 150 μm). The coating has 30 percent solids volume. First, convert 30 percent to decimal form as Equation 5 below shows:

Equation 5:

$$30\% = \frac{30}{100} = \text{WFT}$$

Next, put information you know into Equations 6 and 7 as follows:

Equation 6:

$$\text{WFT} = \frac{\text{4 mil}}{\text{0.30 solids by volume}} = \text{13.3 mils (.34 mm)}$$

Equation 7:

$$\text{WFT} = \frac{\text{6 mil}}{\text{0.30 solids by volume}} = \text{20.3 mils (.52 mm)}$$

One gallon of fluid will cover 1,604 ft^2 (149 m^2) to 1 mil (25 μm) thickness when it is wet. When you need to know how much coating you will need to cover a surface, divide 1,604 by the desired WFT. For example, using the above WFT of 20 mil (500 μm), calculate how much area one gallon of coating covers as shown in Equation 8:

Equation 8:

$$\frac{1{,}604}{20} = 80.2\ \text{ft}^2 \text{ per gallon (7.45 square meters)}$$

You can figure out how many gallons of coating are needed to coat a 1,200 sq ft (110 square m) area as Equation 9 shows:

Equation 9:

$$\frac{1{,}200\ \text{ft}^2}{80.2\ \text{ft}^2 \text{ per gallon}} = \text{14.96 gallons (56.6 liters)}$$

As shown in the example, you need between 10 and 15 gallons (38 and 57 liters) of coating. No matter how careful you are, there will always be a small amount of waste, so ordering 15 gallons (57 liters) of coating would be more realistic than ordering 10 gallons (38 liters).

If the coating is thinned, this must be taken into account. Thinner increases the total volume of the coating without increasing the amount of solids. This introduces a new variable called percentage of thinner by volume (% TBV). Use Equation 10 when thinner has been added:

Equation 10:

$$\frac{\text{DFT (1 + \%TBV)}}{\text{\% solids by volume}} = \text{WFT}$$

You may be told to thin a coating by a percentage of volume. In this case, you put the percentage into the formula. For example, using the WFT of 4 mil (100 μm) with a solid of 30 percent thinned 25 percent (0.25) the WFT is calculated as shown below in Equation 11:

Equation 11:

$$\text{WFT} = \frac{\text{4 mil (1 + 0.25)}}{\text{0.30 solids by volume}} = \text{13.3 mils (.34 mm)}$$

You may be told to add a specific volume of thinner to each gallon of coating. If so, calculate the percentage of thinner before you calculate the WFT. For example, you need to add 1 pint (0.5 liter) of thinner to each gallon of coating. Calculate the percentage as shown Equation 12:

Equation 12:

$$\text{1 gallon = 8 pints, so } \frac{\text{1 pint}}{\text{8 pints}} = \frac{1}{8} = 0.125\%$$

11.3.0 How WFT Affects the Application of a Common Coating System

As previously mentioned, one of the most critical measurements during a coatings project is dry film thickness, which is achieved by having the correct WFT. This has a direct relationship with the life of the coating system. It is important to recognize that each of the layers of a coating system work together to perform a specific function. One of the more common coating systems is made up of layers of inorganic zinc, epoxy, and urethane. By paying attention to the thickness of each layer (WFT and DFT), the coating life will be maximized and will keep the coating costs within budget. The specific function of each layer of the inorganic zinc, epoxy, and urethane coating system is detailed below.

Inorganic zinc provides corrosion protection for steel, and due to the nature of the product, is generally limited to 2 to 4 mil (50 to 100 μm). A wet film gauge cannot be used to measure wet film thickness of inorganic zincs. Because of this fact, only experienced applicators should apply inorganic zinc coatings. If it is applied too thinly, the coating system cannot provide the full protection it is designed to give. If it is applied too thickly, mud cracking can result.

Epoxy provides good film build and barrier protection. Because the zinc has a limiting thickness and the polyurethane topcoat is generally expensive, the epoxy is used as an inexpensive filler to achieve the film build needed. If the epoxy is applied too thinly, the coating will not reach the required film build. If polyurethane is added to make up the difference, the project will cost a great deal more. The epoxy may not cure properly since the thickness of the coat affects both the rate of cure as well as whether or not it will cure. If the epoxy is applied too thickly, not only can it have a bad effect on the cure, but it can cause solvent entrapment which can cause the whole coating system to fail.

Polyurethane is generally the most expensive component and provides both appearance and UV protection. The problems associated with polyurethane are the same as for epoxy. The other critical factor is the cost. Applying too little polyurethane can affect the appearance, and applying it too thickly can substantially increase the cost of the job.

12.0.0 ◆ PROTECTIVE COVERINGS

Areas and items that must be protected before coating begins will be described in the specification. A wide variety of masking tapes, films, sheeting, and canvas drop cloths are available. If at all possible do not create confined spaces when protective coverings are put in place.

12.1.0 Protective Sheeting and Drop Cloths

Protective sheeting can range from 0.35–10 mil (0.0089–0.25 mm) thick, depending on the job site requirements.

High-density thin film sheeting is normally used for large scale protection. Its thickness ranges from 0.35–1.5 mil (0.0089–0.038 mm) and is easily secured with masking tape. This sheeting is tough, light, and easy to handle. Paint cannot soak through it either. However, it will move in very slight air currents, so all loose edges must be held in place with masking tape. This material comes in a boxed roll (*Figure 17*), with widths up to 20 ft (6 m) (folded) and in lengths up to 1,000 ft (300 m). It is a dry film without any slip agent applied and sticks well to masking tape. Because of this, it can only be used once unless it is used as a loose drop cloth over objects. This type of film is also available in a 3.5 mil (0.9 mm) embossed form on 400 ft (122 m) rolls to use as drop cloths. The embossing helps prevent slipping on floors and helps contain paint spills.

Heavier grades of clear polyethylene sheeting, from 4–10 mil (0.01–0.025 mm), are also available. These are primarily used as environmental barriers and drop cloths. Most are normally coated with a slip agent that makes them easier to handle. Unfortunately, the slip agent prevents most masking tapes from sticking to it and makes it dangerous to use as floor coverings. Use heavy-duty cloth-backed tapes with extra adhesive to hold this type of sheeting.

Canvas, as well as other materials, is used for heavy-duty drop cloths. It is reusable and available in a variety of sizes. Canvas is especially suited to cover floors when it is crucial to prevent lad-

104F17.EPS

Figure 17 ◆ Boxed roll of high-density thin film sheeting.

ders from slipping and when painting equipment needs to be moved. *Figure 18* shows polyethylene and canvas drop cloths.

Plastic shrink wrap (similar to that used to wrap pallets for shipping) is used to wrap equipment that cannot be moved, piping, and in-line piping equipment. These often need to be protected from overspray when coating is done nearby. This method is also used in some industrial facilities.

12.2.0 Tapes and Taping Guidelines

Some crepe masking tapes (*Figure 19*) are available in widths from ¾ in (1.9 cm) to 4 in (10 cm). Some of the tapes are made to pull off easily after a certain number of hours or days, even if they have been in direct sunlight. There are also flat-back fine-line tapes that are used to make a fine, clean edge for striping when applicators need to apply different color coatings. Heavy-duty cloth-backed tapes are also available to join heavy plastic sheeting sections together. Other heavy-duty fabric-backed tapes include those with a tough rubber-like coating to protect items or edges during abrasive blasting. Some masking tape roll applicators (*Figure 20*) allow one-handed application of the tape when protective sheeting is fastened in place or an edge is masked.

12.2.1 Tape Application

Large film sheeting can be installed in one operation using a masking tape roll applicator. If the position of the tape edge is critical, use short pieces of tape placed near the permanent tape line, to hold the sheeting in place temporarily. Then use the roll applicator to apply an unbroken length of tape over the sheeting and the temporary tape pieces.

Do not stretch masking tape when applying. Unroll about 1½ in (45 cm) at a time and secure over 1 in (30 cm) intervals until the area being masked is completed. Make sure that the tape is pressed down firmly to prevent paint from seeping under the edge(s).

For sharp edge striping, use a flat-back fine-line tape. If additional large masking is needed for the areas adjacent to the fine edge, use crepe masking tape on top of the flat-back fine-line edging tape, but set back from the sharp edge to hold the protective sheeting in place. To make sure the crepe tape sticks to the edging striping tape, the applicator may need to slightly rough up the back of the edge striping tape with an abrasive pad. Make sure that both tapes are pressed down firmly, especially at the edge(s).

104F19.EPS

Figure 19 ♦ Crepe masking tapes.

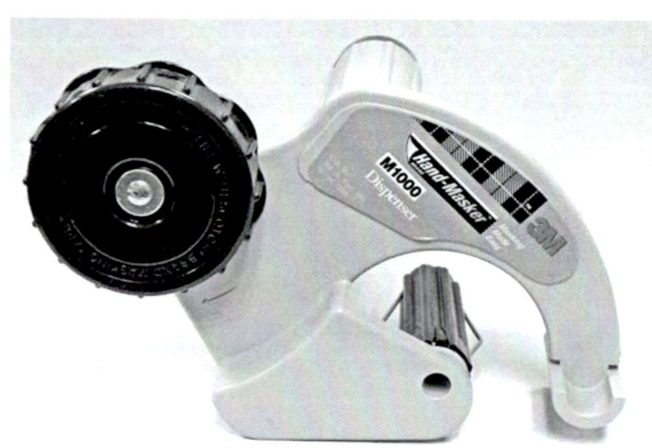

104F20.EPS

Figure 20 ♦ Masking tape applicator.

104F18.EPS

Figure 18 ♦ Polyethylene and canvas drop cloths.

Always make sure all surfaces to be taped and painted are free of dust, dirt, oil, or grease.

13.0.0 ◆ HAND APPLICATION OF COATINGS

Brushes, rollers, mitts, trowels, and squeegees are the most basic tools for applying a coating by hand. They can be used in circumstances when spray painting cannot be done.

A good coating application depends on choosing the correct tools and on the skill of the applicator. To get a project done on time, within budget, and with acceptable results, it is necessary to have qualified applicators. Some application methods, like brushwork for fine finishes, need a great deal of practice to do well, but using a roller is somewhat less demanding. These methods are outlined in the following sections and beginning applicators should master them.

Brushing and rolling are the easiest and most commonly used manual coating application techniques. Mitting (palming) is a technique in which the coating is applied using a hand mitt. This is used on cables and other hard-to-cover items when appearance is not critical. Hand trowels or squeegees are used to apply heavy coatings like concrete, mortars, epoxy, polyester, furan, and other materials. Organic materials may be reinforced with glass cloth, while wire mesh might be used for the inorganic materials.

13.1.0 Brushes

Various types and sizes of professional brushes are available along with a brush comb used to clean brushes (*Figure 21*). Brushes come in natural bristle (animal hair) or synthetic bristle. Some applicators prefer to use natural bristle with solvent-based coatings, but most synthetic bristle brushes can be used for both solvent- and water-based coatings. When selecting a brush, make sure it is constructed as shown in *Figure 22*, and has a tapered, feathered tip plus an adequate paint reservoir (the middle empty space between the sets of bristles).

13.1.1 Uses

Brushes are the best choice of applicators for stripe coating in difficult areas including edges, around bolts, corners, and in crevices. Brushes should be used on critical areas where spray could cause damage, including sensitive machinery, motors, instruments, and gauges that cannot be effectively masked with sheeting. They are also to be used for repairing holidays, and for small scattered areas.

13.1.2 Limitations

There are some limitations in using brushes for applying coatings. WFT may be uneven and several coats may be required to obtain a specified DFT, and application is slow. Also, surface quality may not be as visually pleasing as with other methods of application.

13.1.3 Problems

An inferior brush will shed bristles, not hold enough paint, and the metal ferrule can rust and become loose.

13.1.4 Brush Techniques

Before applying coatings with a brush, be sure to comb new brushes to remove loose bristles. To fully load a brush, dip the bristles about halfway into a partially filled coating container. To remove excess coating, tap the brush on the lip of the container. Unload both sides of the brush in an uncoated area above or below and in front of any wet previously coated area by touching the brush lightly in spots in the uncoated area (*Figure 23A*). When the brush is unloaded, use a back-and-forth movement (*Figure 23B*) to spread out the coating. Finish each area with long leveling brush strokes. Then, use the tip of the brush to smooth the appearance (*Figure 23C*).

To prevent ridges and lap marks, always stroke back into any previously coated wet area above or below and behind. To prevent bristle tip marks, lightly lift the brush off the surface at the end of a stroke as it approaches previous wet areas (called **feathering** or tipping off). See *Figure 24*. For large flat areas requiring multiple coats, each additional coat should be applied at right angles to the previous coat. When brush striping (**cutting in**) corners, use the wide portion of a 2 in (5 cm) brush held at an angle, to coat each side of the corner (*Figures 25* and *26*). Do not use the side of the brush directly in the corner. It will cause streaks and ridges and will shorten the life of the brush.

To cut in a corner using a different-colored coating on an adjoining surface, coat the corner as described above, but keep the new coating color ¼ in (0.64 cm) from the adjoining surface. Then use the wide portion of an unloaded brush (pushed against the surface at about a 45-degree angle) to create a slight bead of paint at the tip of the brush. Move the brush tip carefully across or down the corner, pushing the bead constantly up the tip of the brush to create the final cut-in (*Figures 27* and *28*). Finally, carefully long-stroke the corner with a lightly loaded brush, keeping slightly away from the bead.

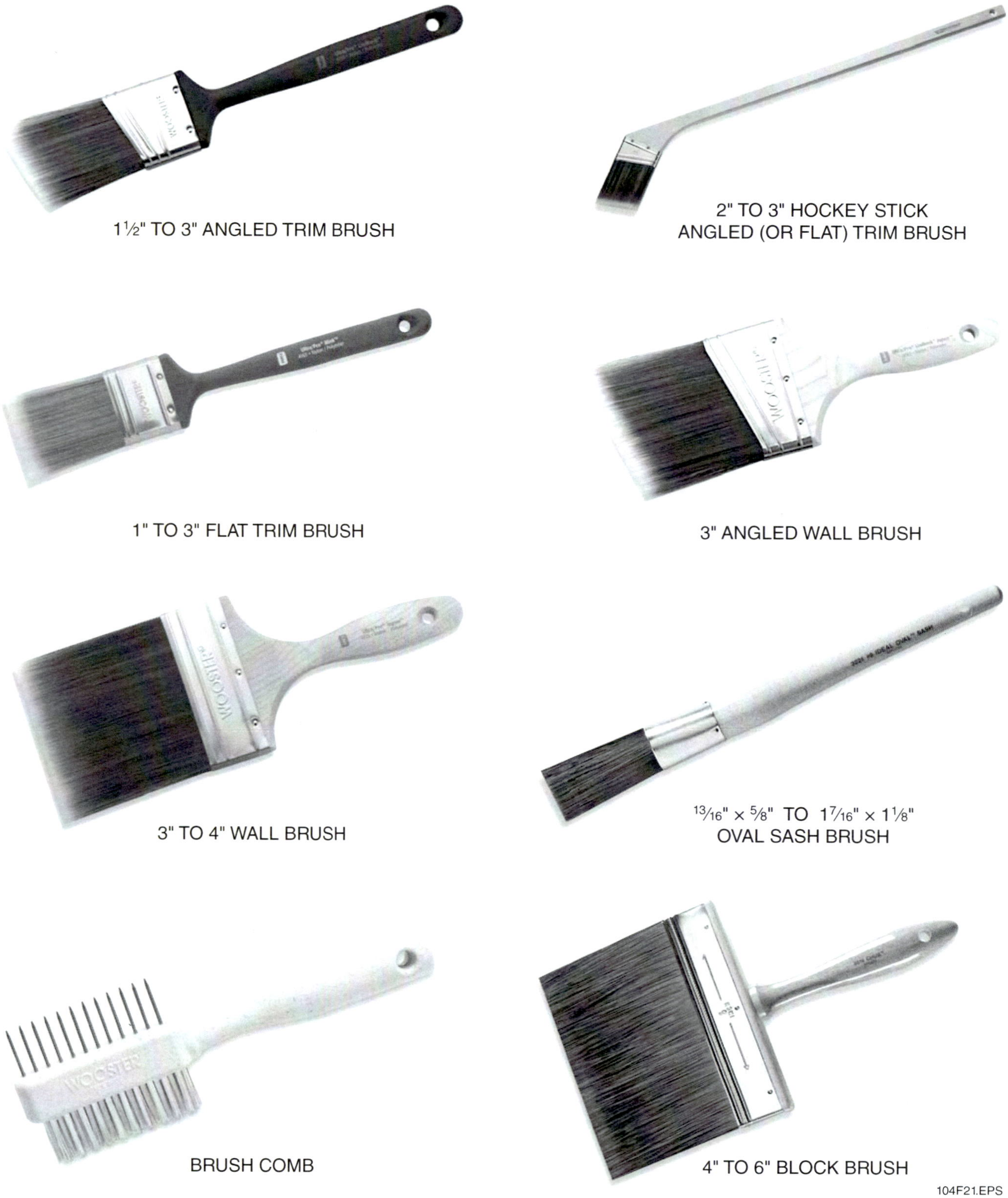

Figure 21 ◆ Professional brushes.

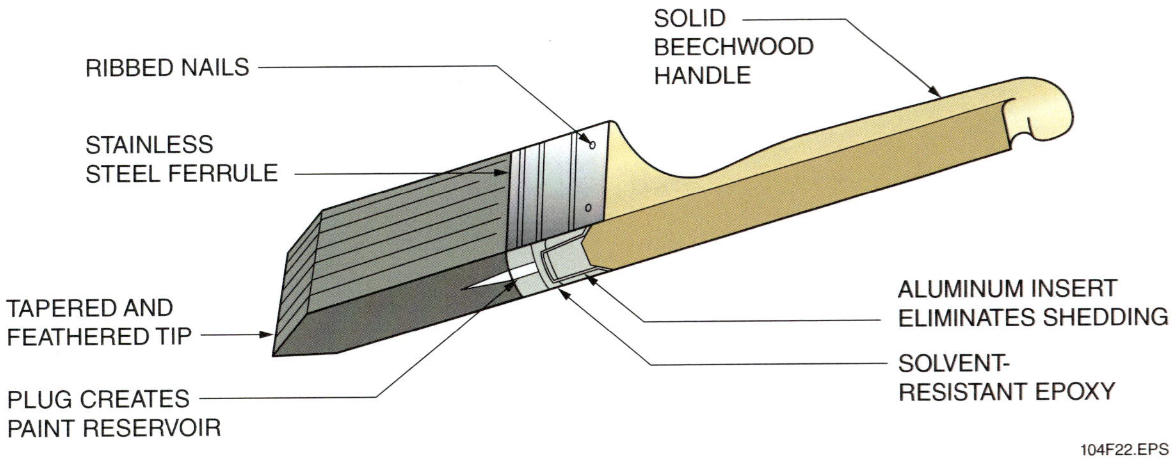

Figure 22 ◆ Professional brush construction.

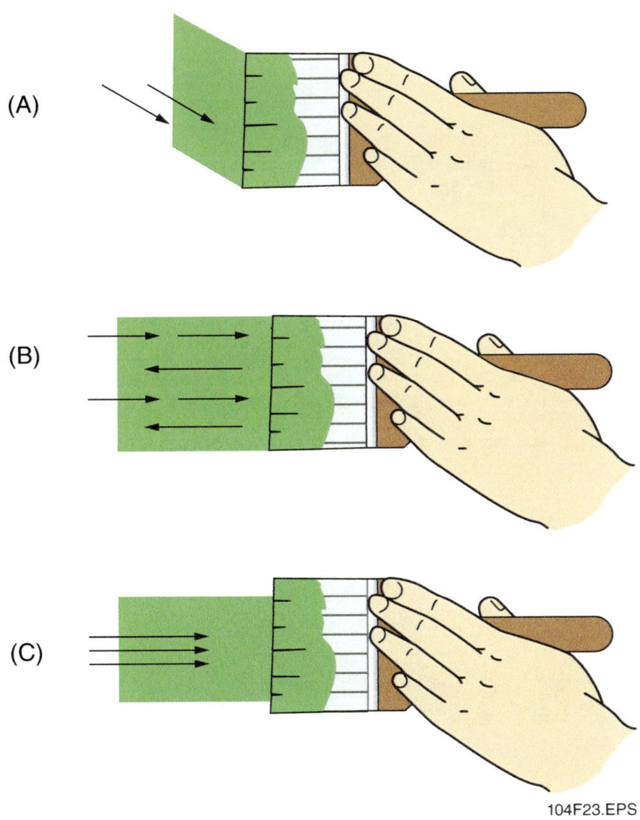

Figure 23 ◆ Applying coating to a surface.

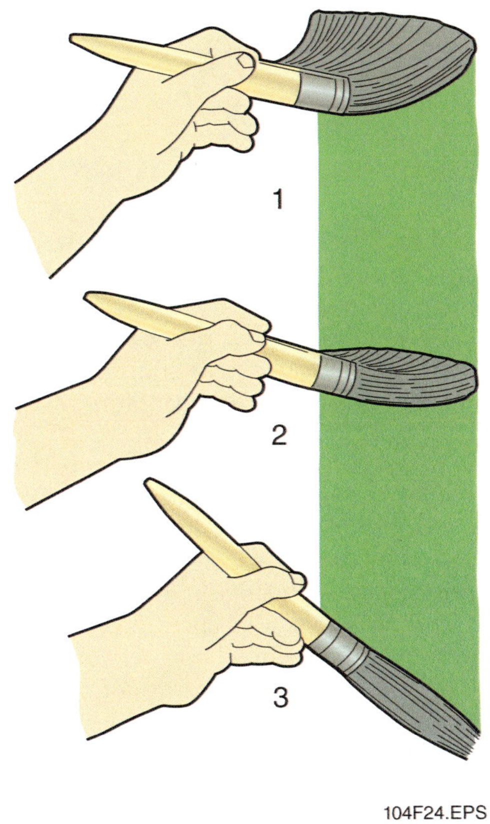

Figure 24 ◆ Feathering with a brush.

13.2.0 Rollers

Rollers can cover large flat or nearly flat areas including tank shells, roofs, floors, ceilings, and walls. Rollers are two to three times faster than brushes because they hold much more coating and each pass covers a larger area. They are used when spraying is not available, allowed, or practical. They are available in numerous types, sizes, and shapes. However, corners must be cut in by brush and the WFT will vary depending on roller pressure and the coating load used during application. Multiple coats may need to be rolled on to obtain a specified DFT.

Rollers are available in many sizes including the standard 1½ in core (3.8 cm) dip roller frames. Additionally, rollers are often used along with telescoping extension poles. The telescoping poles are available in sizes covering 1 ft (30 cm) to 2 ft (60 cm) through 8 ft (2.4 m) to 16 ft (5 m). There

Figure 25 ◆ Cutting in corners.

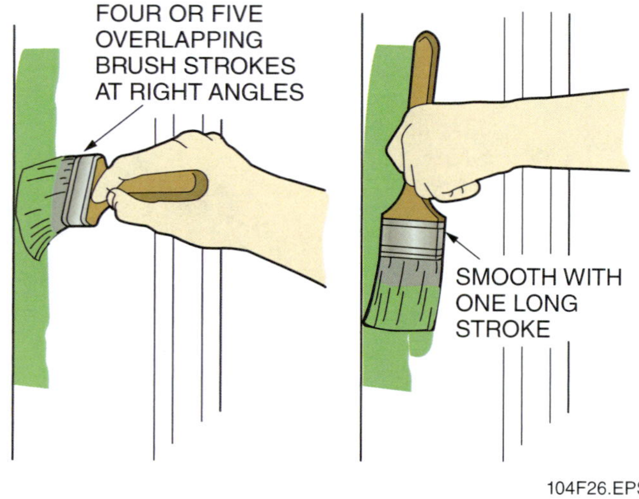

Figure 26 ◆ Cutting in before rolling.

are also some frames that will accept 18 in (45 cm) long and jumbo 2¼ in core (5.7 cm) roller covers (see *Figures 29* and *30*).

Many varieties of dip roller covers are used with the frames. They are available in lengths up to 18 in (45 cm) with core diameters of 1½ in (3.8 cm) and 2¼ in (5.7 cm). The 18 in (45 cm) long rollers are used for large areas to speed production. The 2¼ in (5.7 cm) roller covers carry a large coating load to also speed production. The most common roller covers have synthetic or synthetic blend fiber naps with thicknesses that range from ⅜ in (1 cm) to 1½ in (3.8 cm), depending on the roughness of the surface being coated. They can be used with both solvent- and water-based coatings. *Figure 31* shows roller covers used to create special finish effects.

On larger assignments, professional deep dip roller trays and dip buckets with built-in roll-off ramps and ribs are often used to remove excess coating after dipping the roller (*Figure 32*). Some buckets require a separate plastic or wire grid for roll-off purposes. The buckets are usually preferred for large projects.

A specialized type of roller is the pressure-fed roller (*Figure 33*) that is supplied with coating material from an airless spray pump. This type of roller speeds up application, and roller trays and

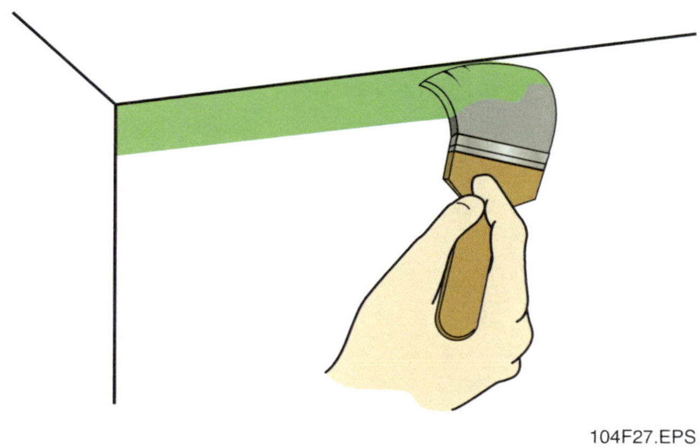

Figure 27 ◆ Cutting in corners with a bead.

Figure 28 ◆ Drawing a paint bead.

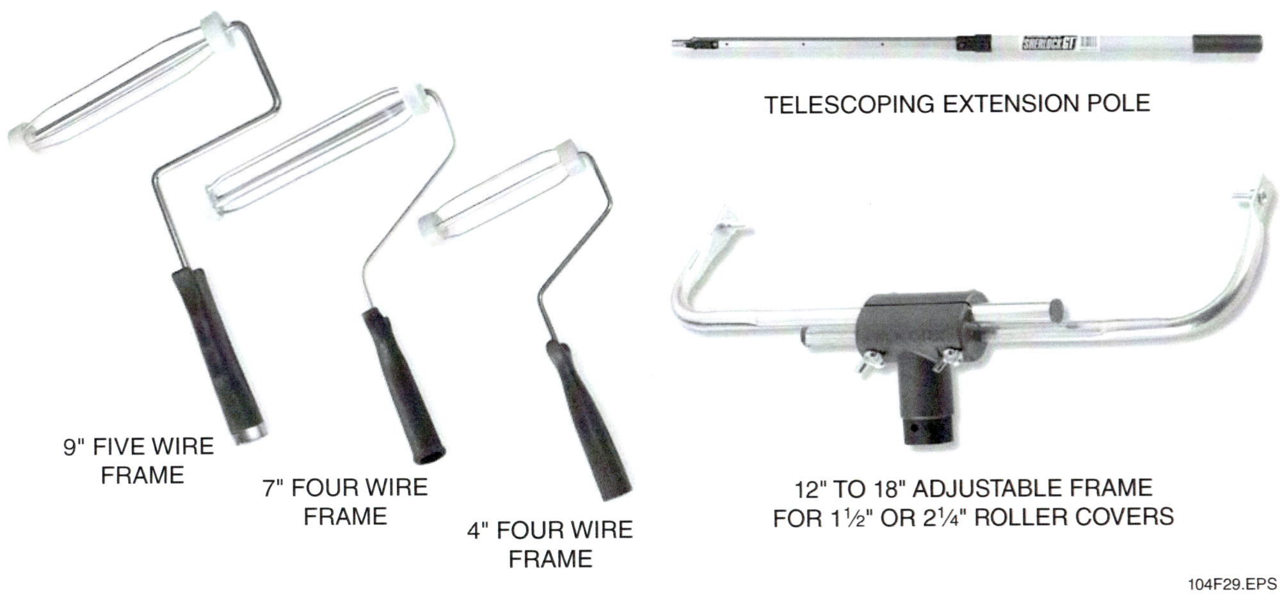

Figure 29 ◆ Dip roller frames and extension pole.

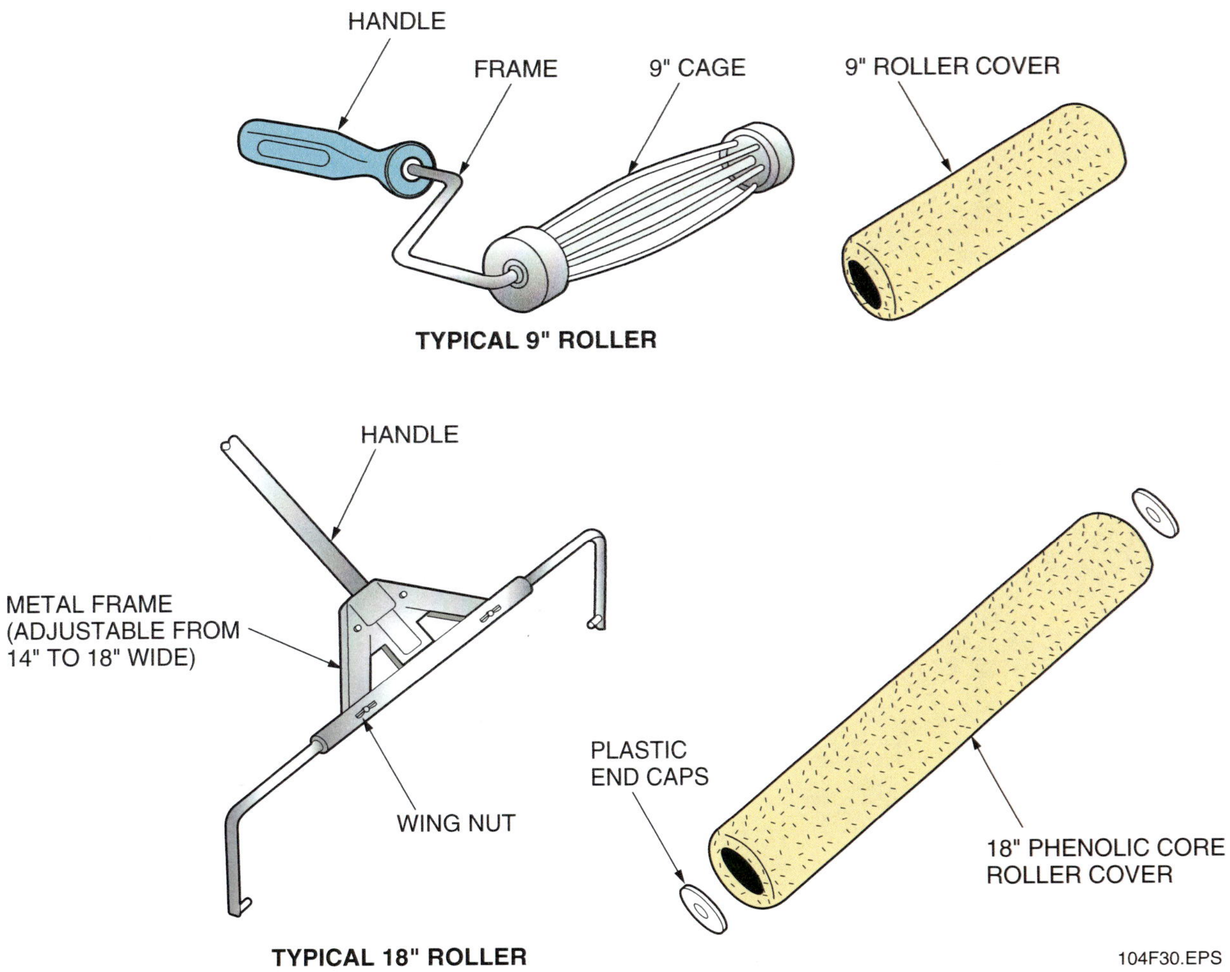

Figure 30 ◆ Parts of a dip roller.

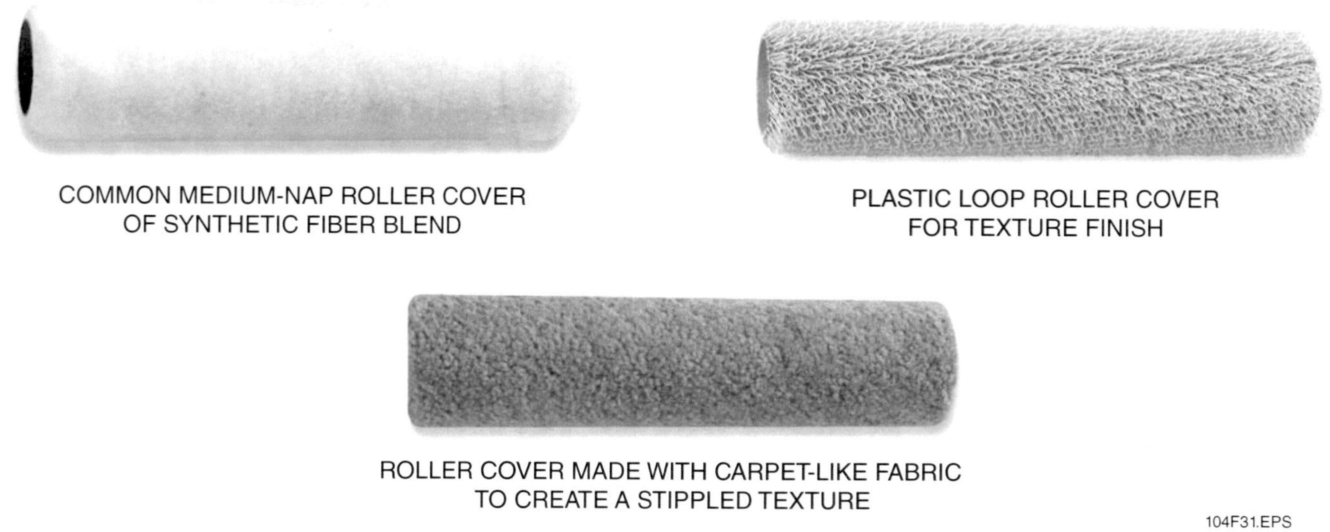

Figure 31 ◆ Dip roller covers.

Figure 32 ◆ Dip roller trays and buckets.

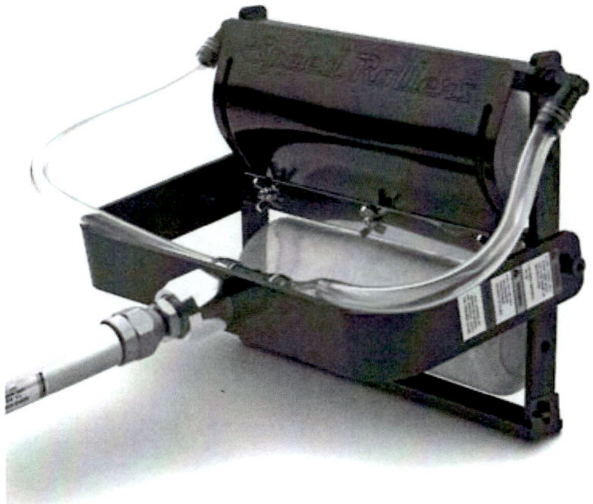

104F33.EPS

Figure 33 ♦ Pressure-fed roller.

buckets are not needed. The coating is distributed to the top roller by the operator using a variable trigger control (at the base of an extension pole) that is connected to the airless sprayer pump. The bottom roller allows spreading and backrolling at the same time. Drip and splash guards help prevent spatter and mess.

13.2.1 Limitations

The use of rollers has a few limitations. Corners must be cut in by brush. WFT varies depending on roller pressure and the coating load used during application. Also, multiple coats may have to be applied to obtain a specified DFT.

13.2.2 Roller Techniques

When applying coatings with a roller there are several things to keep in mind. Wrap the nap of new roller covers with tape; then pull the tape off to remove the loose nap. Make sure the roller has no loose nap, dirt, and/or particles that could mar the finished surface coating.

To use a dip roller, fill a clean professional tray reservoir completely, or a bucket about halfway with the coating material. Dip the roller in the coating up to the frame and roll it back and forth on the roll-off ramp or grid to spread the coating over the entire roller and remove any excess so the roller does not drip when removed from the container.

To use a pressure-fed roller (*Figure 34*), follow the manufacturer's instructions to prepare it. Then, load the roller with coating material. Break in the roller cover by first rolling it on scrap or disposable material. This works the coating into the nap and removes any air that may be trapped. Then, reload the roller.

Before you begin applying the coating, look at the area to be coated. Visually break the area up into about 3 or 4 ft (1 or 1.3 m) wide sections (*Figure 35*). Next, cut in the corners of the first section with a 2 in (5 cm) or wider brush. If using a dip roller, first load the roller, then unload the roller in an uncoated 3 or 4 ft (1 or 1.3 m) square in the section using a W or M pattern (*Figure 36*). Although it may seem logical to roll in straight lines, lap marks usually show up in the finish.

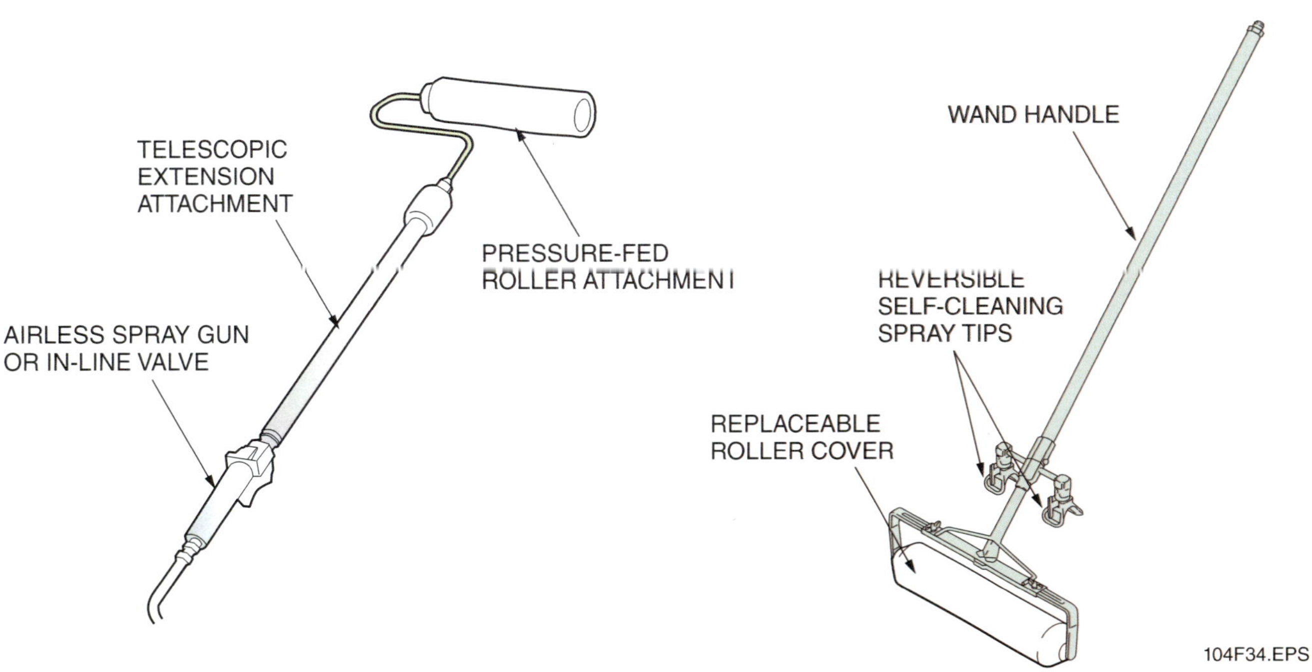

104F34.EPS

Figure 34 ♦ Pressure-fed roller and spray attachments.

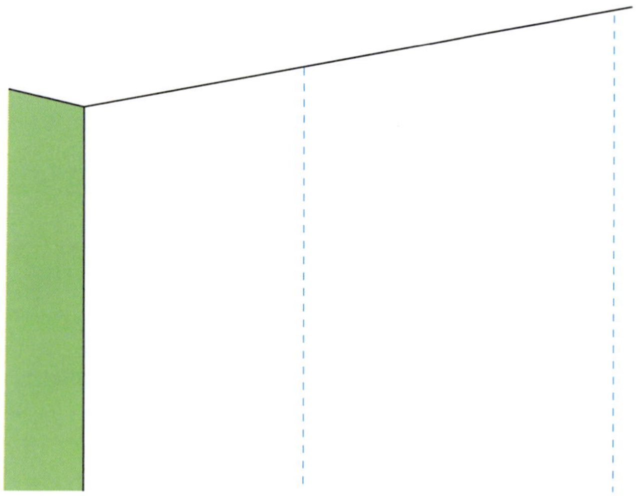

104F35.EPS

Figure 35 ◆ Breaking an area to be coated into sections.

104F36.EPS

Figure 36 ◆ Unloading a roller in a new uncoated section.

NOTE

Remember not to swivel the roller on the wet coating to change direction. Always lift the roller when changing direction and lift it before stopping the rolling motion.

If a pressure-fed roller is used, larger sections can be coated. Straight rolling can be used for initial application because the amount of coating can be closely controlled, and spreading is easier and faster. However, for final finishing, the coating should be rolled crisscross to prevent lap marks and make the coverage uniform. Again, the coating must be back-rolled into any wet coating.

Once the first 3 to 4 ft (1 to 1.3 m) square section is completed, repeat the process in an adjacent 3 to 4 ft (1 to 1.3 m) section. With each section remember to spread the coating from the uncoated area back into the wet previously coated area(s).

13.3.0 Mitts

Mitt application is used only for small or round objects (such as pipes, hand rails) or on objects where brushing is too slow and spraying or rolling is not possible. Mitts are usually synthetic lambskin with a coating-proof interior liner (*Figure 37*). New mitts, like brushes, should be combed or brushed to remove any loose fibers before use. Dip the mitt into a partially filled coating container and wipe off the excess coating on the lip of the container. Apply the coating in long strokes in and around objects, repeating as necessary for full coverage. The tip of a brush can be used to smooth the coating if desired.

104F37.EPS

Figure 37 ◆ Coating mitt.

13.3.1 Limitations

When using mitts for coating application, you may find that achieving consistent dry-film thickness is difficult. Sometimes using a brush along with a mitt can result in a smoother wet film. Several coats may be needed to obtain an acceptable-looking surface.

13.4.0 Trowels and Squeegees

Trowels and squeegees are used to spread coatings such as 100 percent solids on floors. *Figure 38* shows squeegee frames available from 16 to 48 in (40 to 120 cm) wide. A variety of replaceable neoprene or rubber blades can be used. The blades come with straight, beveled, toothed, or rounded edges for spreading coatings of different viscosities. Trowels (*Figure 39*) are also made

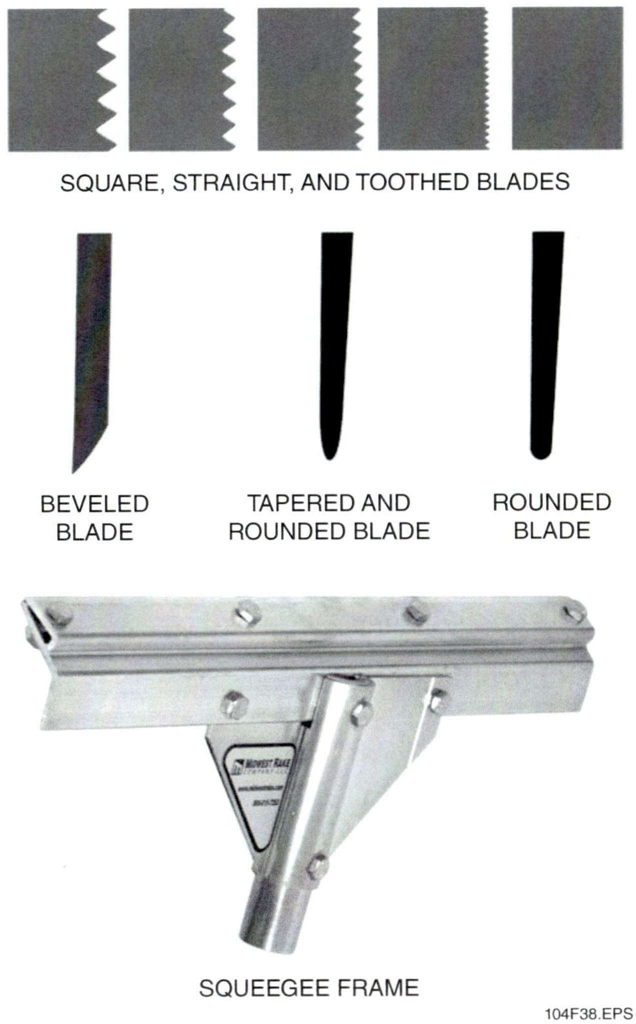

Figure 38 ◆ Squeegee with replaceable blades.

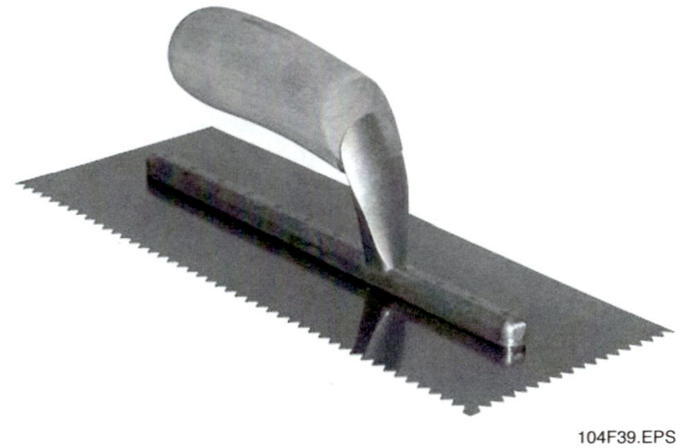

Figure 39 ◆ Toothed trowel.

in many different styles and shapes. Toothed or notched trowels and squeegees are normally used to spread floor coatings that are backrolled to achieve the desired finished thickness.

14.0.0 ◆ SPRAY APPLICATION OF COATINGS

Generally, spray application is the best method to rapidly apply coatings to a large surface area and/or complex structures, such as bridges. A highly uniform coating is possible in nearly all situations. The three major types of spray application equipment in use today are conventional, airless, and high-volume/low-pressure (HVLP), along with some variations of these types.

14.1.0 Spray Application Methods

Conventional air spray, airless spray, and HVLP spray are available with, or can be modified for, certain special uses. These include air-assisted airless spray, air-assisted HVLP spray, multi-component spray, electrostatic spray, powder coating spray, metalizing spray, and hot spray applications.

- *Conventional air spray* – This method uses an air stream to atomize the coating and propel it onto the workpiece.
- *HVLP* – This uses the same method as conventional air spray to atomize the coating, but at much lower pressure (10 psi [70 kPa] or less) with much greater air flow.
- *Airless spray* – In this technique, the coating is pressurized and forced through the tip of the gun to atomize the coating.

Some variations of these methods include:

- *Air-assisted airless spray* – This is used to obtain a finer finish.
- *Electrostatic spray* – Electrically charged particles of coating are directed toward a surface having a charge opposite that of the particles. In this way, the particles are attracted to the surface and tend to deposit with little overspray. It is mainly used on production lines, although sometimes it is used in the field.
- *Thermal spray* – In this method, powders or wires are liquefied (made molten) by heat and directed toward the surface. The material solidifies to form the coating.

14.2.0 Conventional Air Spray

Conventional air spray guns atomize the coating by a stream of compressed air; it is carried to the surface on a current of air. Both the air and the coating enter the gun through separate passages (channels). They are mixed and driven through a fluid nozzle and an air cap. Some of the air supplied to the gun is routed through an adjustable needle valve to external jets in the air cap to control the spray pattern. *Figure 40* shows a manual and a fixed-position production air spray gun. The fixed-position production gun is activated by a remote-controlled pneumatic piston at the rear of the gun.

14.2.1 Pros and Cons

Conventional air spray results in uniform high-quality finishes at a moderate production rate. It is also very versatile and easy to adjust.

Conventional air spray systems also have some disadvantages. They are less efficient than HVLP and airless spray systems. There is a lot of coating overspray, spray bounce-back, and billowing spray dust. Most coatings must be thinned to atomize them, and there are difficulties when some high-solids coatings are sprayed. The use of conventional air spray guns also requires extensive masking and/or protective covering.

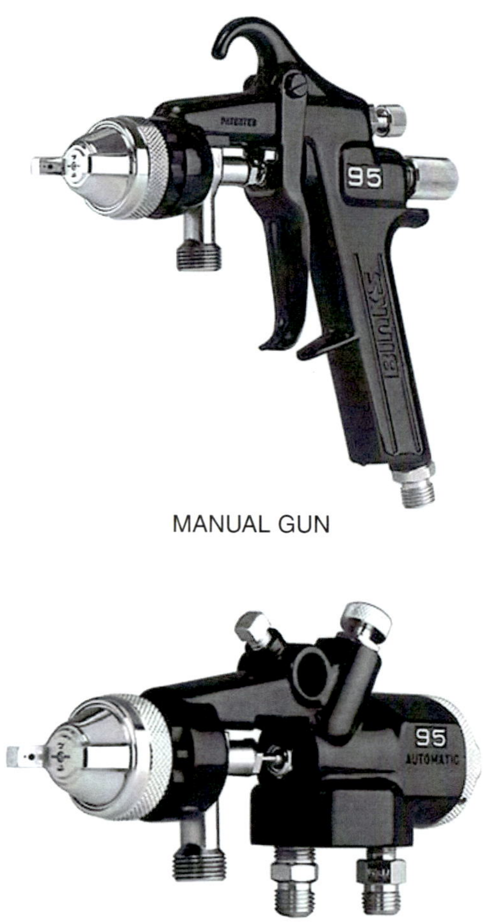

MANUAL GUN

FIXED-POSITION PRODUCTION GUN

104F40.EPS

Figure 40 ◆ Conventional air spray guns.

14.3.0 Spray Gun Components

Typical spray gun components (*Figure 41*) and hook-ups are listed below.

- For unfiltered air input, use an extractor (to remove water and oil) as well as a regulator for the gun air.
- The pressure tank air from the extractor requires a regulator at the input for the fluid pressure supply to the gun.
- For filtered air input, an extractor is normally not required and both regulators are mounted on the pressure tank. Pressure tanks are also available with air-powered agitators to constantly mix the coating, if required.

The fluid needle, fluid nozzle, and air cap selection for the gun are based on the type of coating and its viscosity. These components can be ordered with the gun or purchased separately, in accordance with charts in the manufacturer's instruction manual. Follow the warnings, cautions, and operating instructions in manufacturer's instruction manuals for the various items of equipment before attempting to apply coatings.

14.3.1 Setup for a Conventional Air or Airless Spray Gun

Step 1 With either gun, first lock the trigger. The manufacturer's instructions explain how.

Step 2 Hook up the hoses to the gun and the coating source. The conventional sprayer has two hoses, one going through the handle to supply air, the other between the trigger and the nozzle to supply the coating. The hoses simply thread onto the male thread of the gun. The different hoses are usually different sizes and different colors. The fluid hose is larger, usually ⅜ in (1 cm) NPT. The hoses may have a quick coupling, with a short length of hose remaining threaded onto the gun. *Figure 42* shows typical hook-ups using a pressure tank for the coating materials.

Step 3 Turn on the compressor after you have completely hooked up the gun to the air and fluid supply. Be sure that the hoses are tightened as specified in the manufacturer's instructions.

Step 4 Even if you know the equipment is clean, the best practice is to flush the system first with solvent. Place a small amount

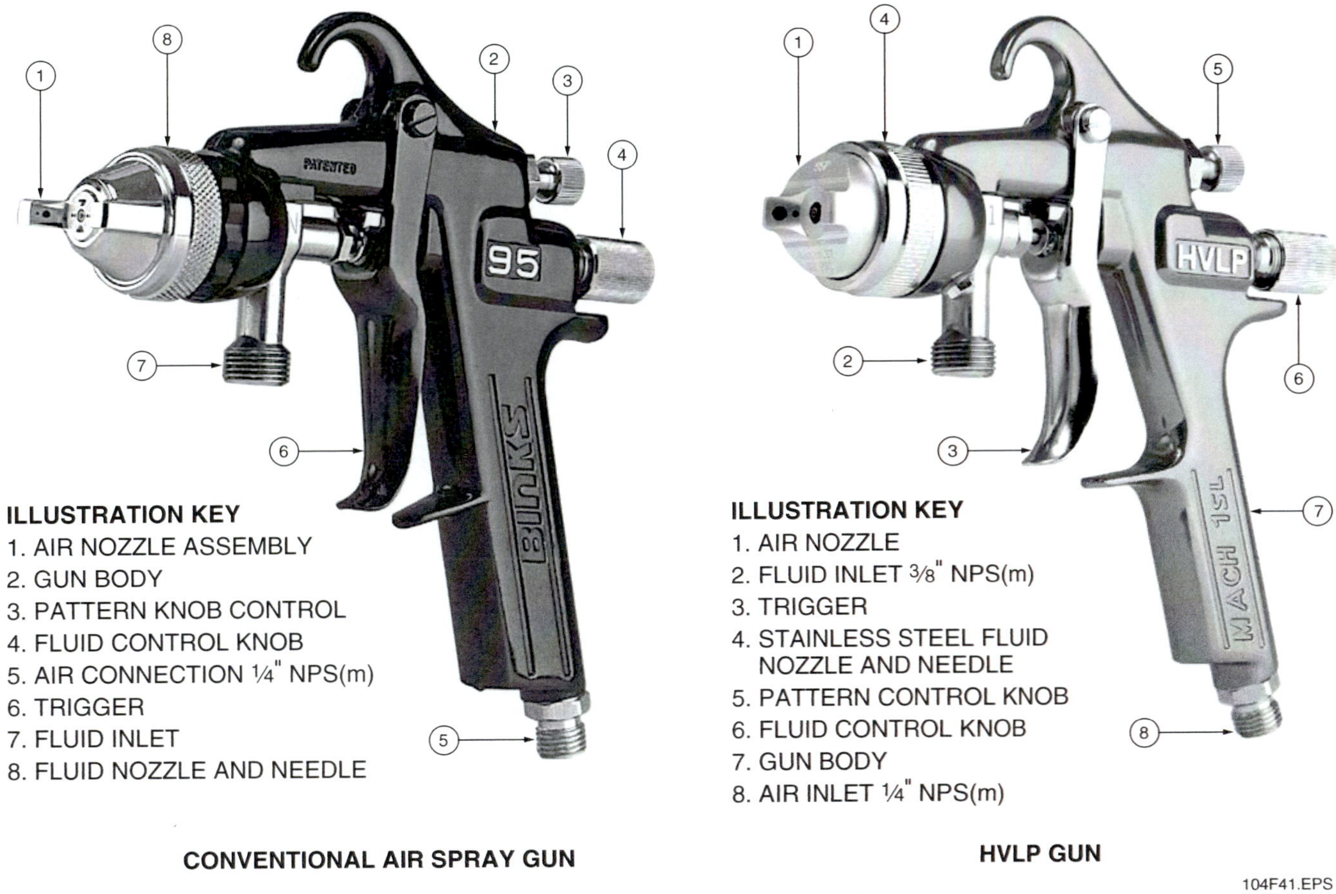

Figure 41 ◆ Typical air spray gun components.

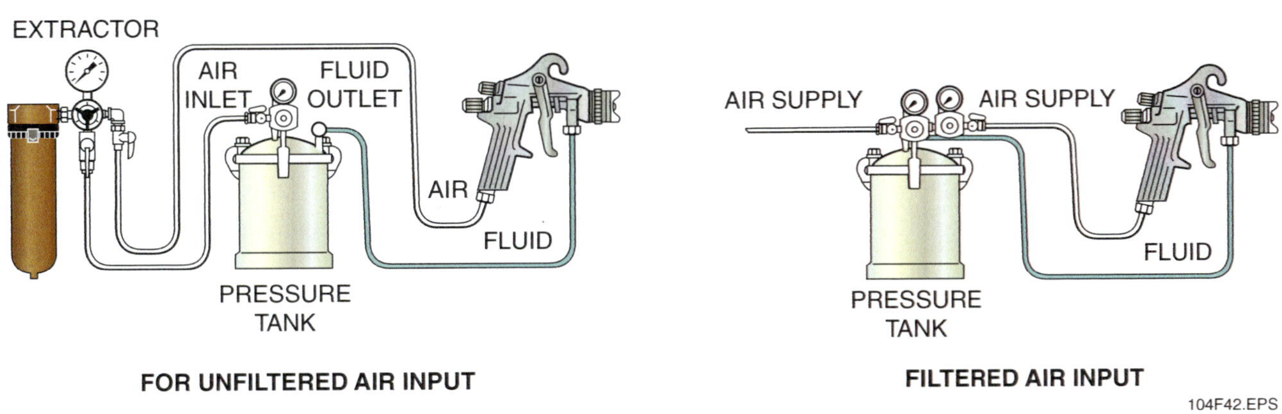

Figure 42 ◆ Typical pressure tank hookups.

of solvent in the container to be used for fluid supply. Bring the compressor online, release the trigger lock, and flush the gun and the fluid line with solvent to clean out any remaining coating material, and to make certain there are no clogs. Be certain that the solvent is not sprayed anywhere it can cause a problem.

Step 5 When the gun is running clear, turn off the compressor, release the pressure from the hoses by pulling the trigger, and set the trigger lock. Fill the fluid container, or transfer the fluid siphon hose to the coating container if you have been pulling solvent from another container.

Step 6 Turn the compressor back on, release the trigger lock, and begin the spraying procedure.

14.3.2 Conventional Spray Gun Sequence of Steps

Step 1 After the coating is mixed to the proper viscosity, pour it into the pressure tank and seal the tank.

Step 2 Next, pressurize the gun and tank to the manufacturer's recommended pressures for the fluid nozzle, needle, and air cap used.

Step 3 Close the air adjusting valve and open the fluid adjustment valve while triggering the gun. Adjust the fluid valve until a fluid stream of 16 to 24 in (40 to 60 cm) exits the gun.

Step 4 Open the air pattern knob and test spray to select the desired spray pattern (from round to full fan).

Step 5 If the spray is too fine after the spray pattern is set, reduce the air pressure; if the spray is too coarse, increase the air pressure.

Handling the gun properly is essential to obtaining a good finish. The following list describes how to handle a gun on flat surfaces, and *Figure 43* shows the technique.

- The gun should be held perpendicular (at a right angle) to the surface being coated while it is moved parallel with the surface.
- The stroke movement must be started before the trigger is pulled, and the trigger must be released before the stroke is ended.
- The distance between the gun and the surface should be 6 to 12 in (15 to 30 cm) depending on the coating material and atomizing pressure.
- The stroke (pass) speed should be high enough that coating does not sag or run after application. If the speed is too high or too slow, adjust the pressure and/or fluid flow slightly at the gun.
- Material deposited should always be even and wet. Each stroke should overlap the former stroke by about 50 percent to obtain a uniform finish, as shown in the figure.

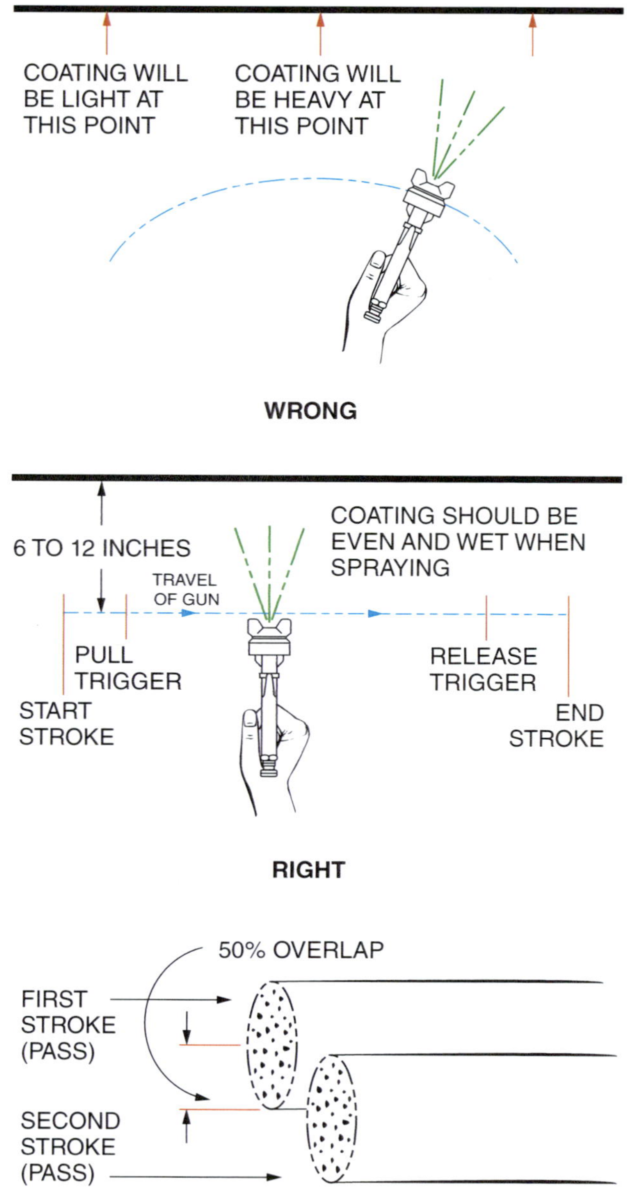

Figure 43 ◆ Basic gun handling for flat surfaces.

14.3.3 Troubleshooting Conventional Spray Guns

All equipment can have problems from time to time. Spray equipment manufacturers normally provide troubleshooting tables in their operating and service literature. *Table 1* outlines troubleshooting procedures for conventional spray guns.

> **NOTE**
> If a spray gun sputters, check for a loose fluid nozzle.

Table 1 Troubleshooting Conventional Spray Guns (1 of 3)

Problem	Cause	Solution
Top-heavy pattern	1. Horn holes partially plugged 2. Obstruction on bottom of fluid tip 3. Dirt on air cap seat or fluid tip seat	To remedy the defects, detemine whether obstruction is on air cap or fluid tip. This is done by making a solid test spray pattern, then rotating the cap one-half turn and spraying another pattern.
Bottom-heavy pattern	1. Horn holes partially plugged 2. Obstruction on bottom side of fluid tip 3. Dirt on air cap seat or fluid tip seat	If the defect is inverted, obstruction is on the air cap. Clean the air cap. If not inverted, it is on the fluid tip. Check for fine burr on the edge of the fluid tip (remove with 600 grit wet or dry sandpaper) or for dried paint just inside the opening (remove by washing).
Heavy right-side pattern	1. Right-side horn holes partially clogged 2. Dirt on right side of fluid tip	Inspect and correct.
Heavy left-side pattern	1. Right-side horn holes partially clogged 2. Dirt on right side of fluid tip	Inspect and correct.
Heavy center pattern	1. Setting of the spreader adjustment valve too low 2. Atomizing pressure too low or material too thick 3. With pressure feed, fluid pressure too high for the atomization air being used or material flow in excess of the cap's normal capacity 4. Tip too large or too small for the material used	To correct the defects readjust atomizing pressure, fluid pressure, and spray width adjustment until the desired spray is obtained.
Split spray pattern	Air and fluid not properly balanced	Reduce width of spray pattern by means of the spreader adjustment valve or increase fluid pressure. This latter adjustment increases speed and the gun must be handled much faster.
Dry spray	Sprayed coat is short of liquid material because: 1. Air pressure is too high 2. Material is not reduced or thinned correctly (suction feed only) 3. Gun is too far from work or out of adjustment	1. Decrease air pressure. 2. Reduce or thin according to directions; use proper thinner or reducer. 3. Adjust distance to work; clean and adjust gun fluid and spray pattern controls.
Starved pattern, Spotty, uneven pattern	Slow to build due to: 1. Inadequate material flow 2. Low atomization air pressure (suction feed only) 3. Gun motion too fast	1. Back fluid adjusting screw out to first thread. 2. Increase air pressure; rebalance gun. 3. Move at moderate pace, parallel to work surface.

Table 1 Troubleshooting Conventional Spray Guns (2 of 3)

Problem	Cause	Solution
Unable to get round spray	Fan adjustment stem not sealing properly	Clean or replace.
Will not spray	1. No air pressure at gun air inlet 2. Internal-mix air cap used with suction feed 3. Fluid pressure too low with internal-mix cap and pressure tank 4. Fluid tip not open enough 5. Fluid too heavy for suction feed or viscosity too great	1. Check air lines. 2. Change external air cap. 3. Increase fluid pressure at tank. 4. Open fluid-adjusting screw. 5. Change to pressure feed or reduce fluid (i.e., add thinner).
Dripping from fluid tip	1. Dry packing 2. Sluggish needle 3. Tight packing nut 4. Spray head misaligned on type-MBC guns, causing needle to bind	1. Lubricate packing. 2. Lubricate. 3. Adjust. 4. Tap all around spray head with wood and rawhide mallet and retighten locking bolt.
Excessive overspray	1. Too much atomization air pressure 2. Gun too far from surface 3. Improper stroking, i.e., arcing, moving too fast	1. Reduce air pressure. 2. Check distance. 3. Move at moderate pace, parallel to work surface.
Excessive fog	1. Too much or quick drying thinner 2. Too much atomization air pressure	1. Remix. 2. Reduce air pressure.
Fluid leakage from packing nut	1. Packing nut loose 2. Packing work or dry	1. Tighten, but not so tight as to grip needle. 2. Replace packing or lubricate.
Thin, sandy, coarse finish that dries before it flows out	1. Gun too far from surface 2. Too much air pressure 3. Quick drying thinner	1. Check distance. 2. Reduce air pressure. 3. Remix.
Tails in spray fan	1. Inadequate fluid delivery 2. Fluid not atomizing	1. a. Increase fluid pressure. b. Change to larger tip orifice size. c. Reduce fluid viscosity. 2. a. Clean gun and filter(s). b. Reduce number of guns using pump. c. Install sapphire insert.
Pattern expanding and contracting (surge)	1. Pulsating fluid delivery 2. Suction leak	1. a. Change to a smaller tip orifice size. b. Install pulsation chamber in system or drain existing one. c. Reduce number of guns using pump. d. Increase air supply to air motor. 2. a. Remove restrictions in system, clean tip screen or filter if used. b. Inspect for siphon hose leak.

Table 1 Troubleshooting Conventional Spray Guns (3 of 3)

Problem	Cause	Solution
Hour glass fan shape	Inadequate fluid delivery	1. a. Change to a smaller tip orifice size. b. Install pulsation chamber in system or drain existing one. c. Reduce number of guns using pump. d. Increase air supply to air motor. 2. Remove restrictions in system, clean tip screen or filter if used.
Fluid leakage from fluid-packing unit	A loose packing nut, work packing, or dry fluid-needle packing	Lubricate packing with a few drops of light oil. Tighten packing nut to prevent leakage, but not so tight as to grip the fluid needle. Replace the packing when it is worn. Tighten the packing nut until it grabs and holds the needle; then back off the packing nut until the needle is free to travel into the fluid tip.
Air leakage from front of gun	1. Foreign matter on valve or seat 2. Worn or damaged valve or seat 3. Broken air-valve spring 4. Sticking valve stem due to lack of lubrication 5. Bent valve stem 6. Packing nut too tight 7. Gasket damaged or omitted	Inspect and correct.
Fluid leakage from front of the gun (pressure feed)	1. Worn or damaged fluid tip or needle 2. Lumps of dirt lodged in fluid tip 3. Packing nut too tight 4. Broken fluid-needle spring 5. Wrong size needle or tip	Inspect and correct.
Jerky or fluttering spray	Applying to both suction and pressure feed: 1. Lack of sufficient material in container 2. Tipping container at excessive angle 3. Obstructed fluid passageway 4. High-viscosity coatings, requiring addition of thinner 5. Loose or cracked fluid pick-up tube in cup or tank 6. Loose fluid tip or damaged tip seat	Inspect and correct.

14.3.4 Shutdown for Conventional Air Spray

At the end of the day or shift, the breakdown is very similar for conventional, airless, and HVLP systems.

Step 1 Return any remaining fluid coating to an airtight container.

Step 2 Clean out the supply container for the gun.

Step 3 When the container is cleaned, put a sufficient quantity of solvent in to clean the hose and gun. Run the sprayer into a clean, **grounded** container until the spray is completely clear.

Step 4 Set the trigger lock on the gun.

Step 5 Wipe up any spray on the gun or hoses; turn off the compressor or pump motor.

Step 6 Relieve the pressure, and detach the hoses from the pot and gun. Coil the hoses carefully and stow the equipment correctly.

Dispose of any contaminated solvent according to the requirements shown on the MSDS for that material.

14.4.0 High-Volume/Low-Pressure (HVLP) Spray

With HVLP spray guns, the coating is atomized in the same method as conventional air guns, but at much lower pressure (10 psi [70 kPa] or less) and with much greater air flow. They are used primarily for industrial production facilities where volatile organic compound (VOC) emissions are present at high levels.

14.4.1 Pros and Cons of HVLP

Highly efficient transfer results in less coating overspray, bounce-back, and dust (*Figure 44*). It also results in a very uniform and high-quality finish, but production rates are lower than conventional air spray. *Figure 45* shows a manual and a fixed-position production HVLP gun. The production gun's high-volume low-pressure air is furnished from an external supply and regulator. The gun is remotely controlled through a pneumatic piston at the rear of the gun.

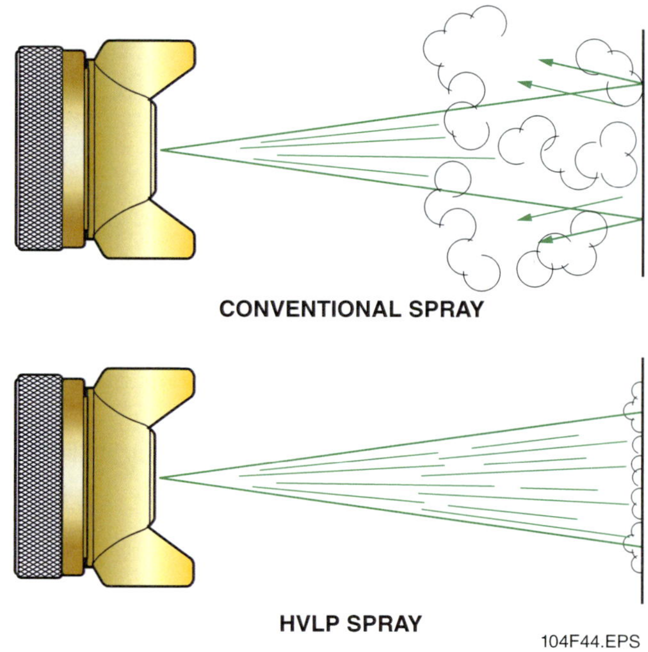

Figure 44 ◆ Conventional spray versus HVLP air spray.

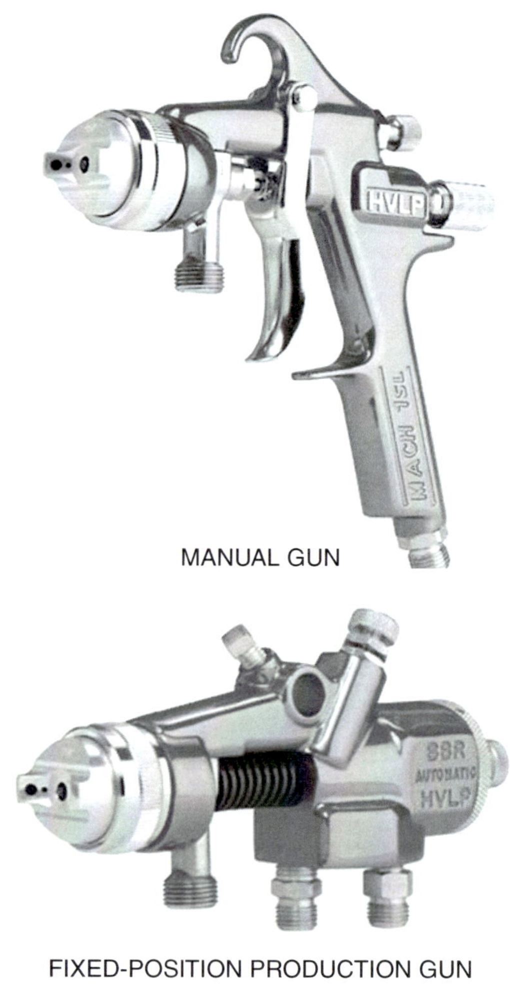

Figure 45 ◆ HVLP spray guns.

14.4.2 Components of the HVLP Spray Gun

As previously mentioned, the typical components (*Figure 46*) and hook-ups for HVLP and conventional air spray are essentially the same. For unfiltered air input, an extractor to remove water and oil is required along with a regulator for the gun air.

The pressure tank air from the extractor requires a regulator at the input for the fluid pressure supply to the gun. For filtered air input, an extractor is normally not required and both regulators are mounted on the pressure tank. Pressure tanks are also available with air-powered agitators to constantly mix the coating if required.

The fluid needle, fluid nozzle, and air cap selection for the gun are based on the type of coating and its viscosity. These components can be ordered with the gun or purchased separately, as needed. To select the correct component, follow the charts in the manufacturer's instruction manual. Applicators must follow the manufacturer's warnings, cautions, and operating instructions before attempting to use the equipment.

14.4.3 Setup and Pressure Control with HVLP Spray Guns

For manual guns, pressure reduction normally happens inside the gun. High-pressure low-volume air is directed through an internal opening. Once there it expands and becomes high-volume low-pressure air (fan air), and is mixed with the coating fluid before going through the fluid nozzle.

Some of the low-pressure air is directed to the air cap to control the pattern. Because of the high volume of air used, the openings for the fluid nozzle and air cap of these guns low-velocity spray to the surface being coated. This method has much higher **transfer efficiency** and lower volatile organic compound emissions than conventional air spray. The distance from the surface for HVLP is less than for conventional air spray because of the slower particle velocity.

Once the spray pattern is set, the fluid pressure should be increased if the spray is too fine; the fluid pressure should be decreased if the spray is too coarse. The fluid control can be used to make fine adjustments to the fluid flow.

HVLP application techniques are essentially the same as those used for conventional air spray.

14.4.4 Shutdown for HVLP

At the end of the day or shift, the breakdown is very similar for conventional, airless, and HVLP systems.

Step 1 Return any remaining fluid coating to an airtight container.

Step 2 Clean out the supply container for the gun.

Step 3 When the container is cleaned, put a sufficient quantity of solvent in to clean the hose and gun. Run the sprayer into a clean, grounded container until the spray is completely clear.

Step 4 Set the trigger lock on the gun.

Step 5 Wipe up any spray on the gun or hoses, turn off the compressor or pump motor.

Step 6 Relieve the pressure, and detach the hoses from the pot and gun. Coil the hoses carefully and stow the equipment correctly.

Dispose of any contaminated solvent according to the requirements shown on the MSDS for that material.

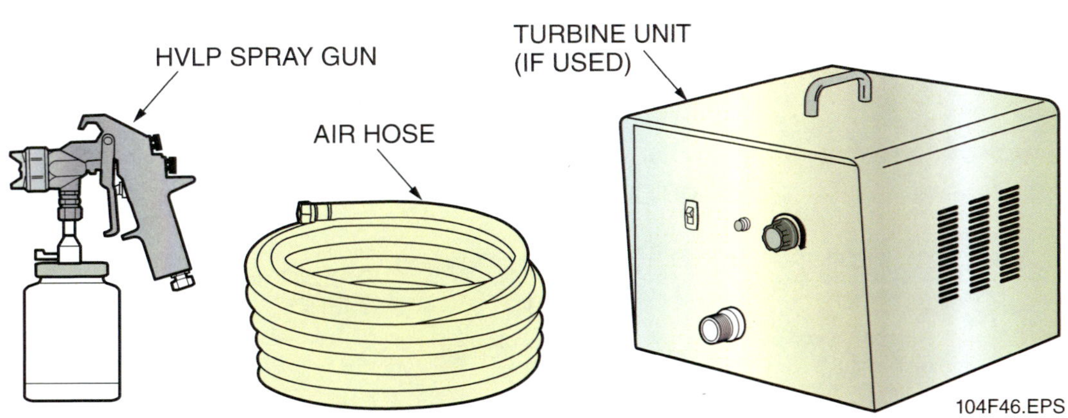

Figure 46 ◆ Components of a typical turbine HVLP spray system.

14.5.0 Airless Spray Guns

Airless spray guns are so named because the coating is not atomized by air. Instead, the fluid coating is pumped under high pressure through an **orifice** (a small specially shaped opening or hole) in the spray gun (*Figure 47*). The tip controls the flow of the coating and creates the spray pattern. As the fluid passes through the orifice at high pressure (typically 2,000 to 3,000 psi) it separates into a fine spray mist. The velocity of the spray propels it to the surface. The higher the viscosity of the coating material, the higher the pressure needed to atomize it.

14.5.1 Components of Airless Spray Units

Figure 48 shows a typical basic airless sprayer unit. *Figure 49* shows the components of typical industrial airless and air-assisted airless spray guns. The pumps on both airless and air-assisted airless sprayer units are air powered. Some versions of these units have pumps and guns designed for zinc-rich coating applications. For safety reasons, these units must be grounded to prevent static discharge. This includes mandatory metal siphon barrels or pails. The hoses that attach the guns to the units are electrically conductive so that the guns must be grounded as well. If an air-supply hose from an external air source is used for the air motor and is equipped for the air-assisted gun as well, it must also be electrically conductive.

WARNING!

Airless and air-assisted airless sprayers can operate at pressures of 7,500 psi (527 kg/cm2) or more at the gun tip. Severe injury can result if any flesh is exposed to the spray from the gun tip. Make sure the tip guard is installed. Treat the gun as a loaded weapon. Never point the gun at anyone or yourself when the unit is operating.

Always cover all exposed flesh, wear gloves, proper face protection, and any required respirator equipment when operating the equipment. Never attempt to find a leak with your hands, even when wearing gloves. The high-pressure leaks from hoses, gun fittings, or any other components of these units can be nearly invisible, but can cause severe injury.

Never attempt to do any of the following until the unit is shut down, depressurized, and the trigger safety lock-out is engaged as described in the manufacturer's instruction manual:

- Service the unit
- Change tips
- Install extension wands
- Change guns

Always follow the manufacturer's safety instructions and any cautions or warnings marked on the equipment itself. If an injection injury occurs, the person should be taken immediately to a hospital. Do this even if the injury seems minor; delays of only a few minutes could mean losing a limb or a life.

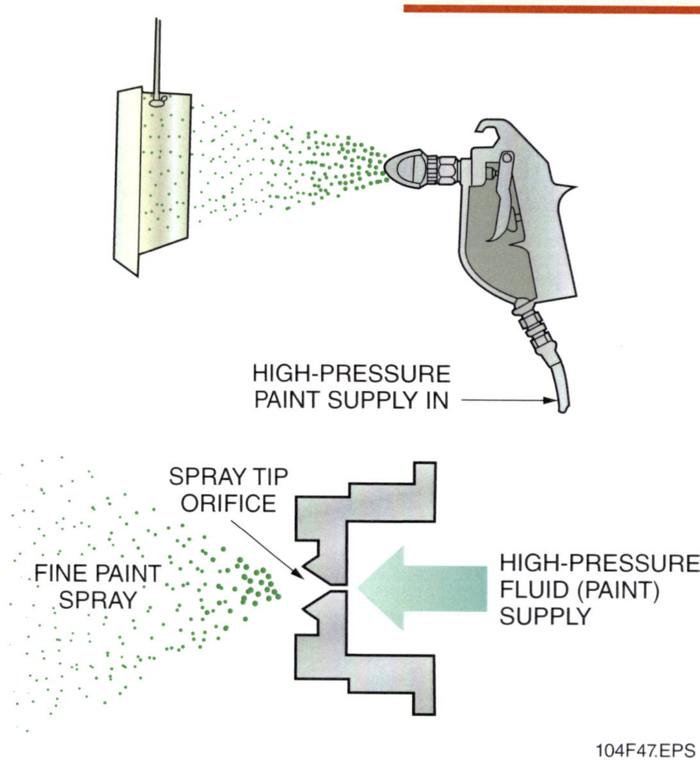

Figure 47 ♦ Airless spray gun and orifice.

There are many different models of pumps and spray guns, and a great many manufacturers of equipment to choose from. The setup, operation, maintenance, and cleaning of these units is a complicated process. Therefore, it is especially important to make sure you understand the manufacturer's manuals and follow their instructions when using the equipment. Airless spray has the highest production rate of the three systems, which reduces overspray and bounce-back.

14.5.2 Limitations of Airless Spray Guns

One of the limitations of airless spray guns is that there is little control of the quantity of coating applied. To maintain a uniform coating, gun distance and travel speed over the surface to be coated are critical. Another limitation is that the finish is of a lower quality than conventional and HVLP spray systems. Also, the spray equipment requires regular maintenance.

A major drawback of airless spray guns is their safety hazards. The high-pressure coating stream that shoots out of the spray tip can inflict serious injury if it contacts any exposed part of the body.

AIRLESS SPRAY UNIT

104F48.EPS

Figure 48 ◆ Typical basic airless spray unit.

14.5.3 Setup of Airless Spray Guns

Airless spray guns atomize the coating without using compressed air. The coating is pumped at high pressure through a spray tip, a precisely shaped and sized orifice (opening) at the front of the gun. The operating pressure and spray tip can be changed for different spray patterns and

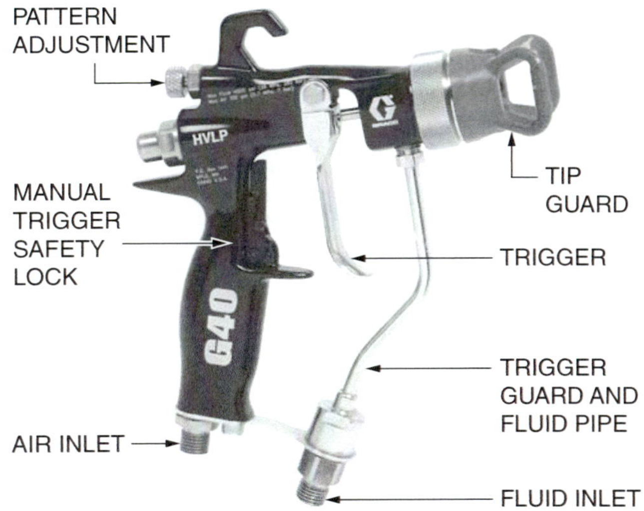

AIR-ASSISTED AIRLESS SPRAY GUN

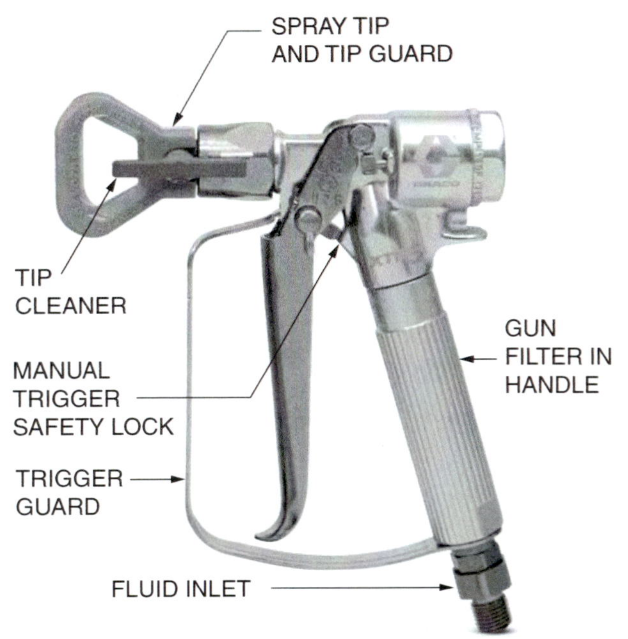

AIRLESS SPRAY GUN

104F49.EPS

Figure 49 ◆ Typical industrial airless and air-assisted spray gun components.

for different coating viscosities. To use the correct tip and operating pressure, always check the instructions that come with both the coating and the spray equipment. Some manufacturers equip the spray tip with a manual cleaning device that allows operators to clean the tip while it is attached to the gun.

The coating is atomized as it leaves the spray tip and is forced toward the surface. After priming the unit according to the manufacturer's instructions, slowly increase the air-motor pump speed by adjusting the air-regulator; this will increase the gun pressure. While you slowly increase the gun pressure, check the spray pattern frequently. Immediately stop increasing the pressure as soon as the coating is completely atomized (no splatter or droplets). This will ensure maximum coating transfer with the least possible dusting.

Follow these steps to set up an airless spray gun:

Step 1 Lock the trigger lock on the gun.

Step 2 Release the pressure from the pump. The pressure-relief valve will be somewhere on the pump, usually on the front. When the pump is turned off, the pressure must be relieved for the gun to be safe. The valve may be labeled as prime/pressure-relief valve, or some similar name.

Step 3 Fill the container to be pumped from, and attach the fluid siphon hose.

Step 4 Make sure the pressure-relief valve is closed.

Step 5 Turn on the pump (in the case of an air-driven pump, bring the compressor online) and check for leaks.

Step 6 Release the trigger lock, and begin spraying.

Always follow the manufacturer's instructions and all safety precautions when using a sprayer.

14.5.4 Airless Spray Gun Technique

Basic gun handling when spraying a flat surface is essentially the same as for conventional air spray or HVLP. Airless spray guns require a 12-inch (30 cm) distance from the surface and a faster stroke speed because when they are used for high production rates, they need a greater material flow. If the stroke speed is too fast, decrease the tip orifice size. If the stroke speed is too slow, increase the fluid pressure slightly or increase the tip orifice size. The thickness of the coating is also controlled by both the tip size and shape of the fan angle. Using a tip with a larger orifice size but with the same fan angle will increase the thickness of the paint. Using a tip with the same orifice size but a larger fan angle will decrease the thickness of the paint.

Figure 50 shows a manual and a fixed position production airless gun. OSHA requires the manual airless spray guns be equipped with a tip guard to help prevent close contact with the spray from the tip. OSHA also requires a trigger guard to help prevent accidental coating release, and either an automatic trigger lock that has to be disengaged with each use or some form of a manual trigger lock-out device. To extend the reach of applicators, pole guns and extension wands with adjustable tip mounting sockets are available for some airless units. The fixed-position production gun shown is activated by a pneumatic piston that is remotely controlled.

Airless guns operate at much higher pressures than conventional spray equipment. Follow the same flushing procedure as with the conventional sprayer.

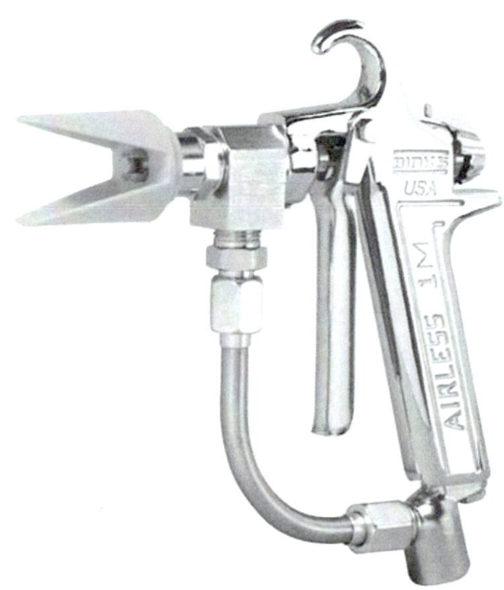

MANUAL GUN

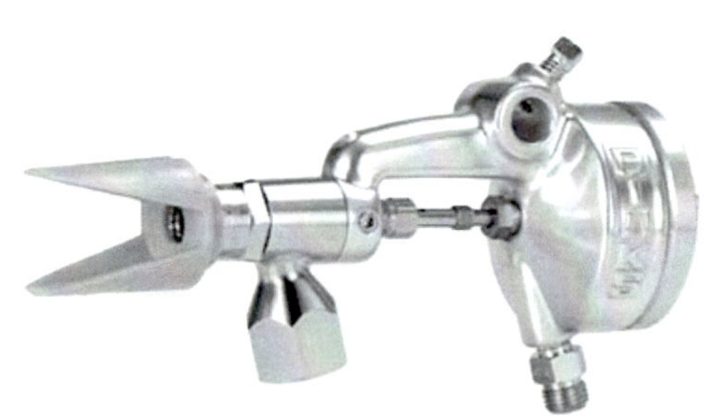

FIXED-POSITION PRODUCTION GUN

104F50.EPS

Figure 50 ◆ Airless spray guns.

14.5.5 Troubleshooting Airless Spray Guns

All equipment can have problems from time to time. Spray equipment manufacturers normally provide troubleshooting tables in their operating and service literature. *Table 2* is a troubleshooting chart for airless spray guns.

14.6.0 Air-Assisted Airless Spray Guns

The pumps on air-assisted airless sprayer units are air powered. Some versions of these units have pumps and guns designed for zinc-rich coating applications.

For safety reasons, these units must be grounded to prevent static discharge. This includes mandatory metal siphon barrels or pails. The hoses that attach the guns to the units are electrically conductive so that the guns must be grounded as well. If an air-supply hose from an external air source is used for the air motor and is equipped for the air-assisted gun as well, it must also be electrically conductive.

Because there are so many models of pumps and spray guns, and because the setup, operation, maintenance, and cleaning of each is complicated, every applicator must follow the manufacturer's instruction manuals.

Like other spray guns, the spray tips used for these units must be selected based on the type and viscosity of the coating being sprayed. Many tip sizes with various spray patterns are available. Always refer to the coating manufacturer's PDS for recommended tip and operating pressures or the manufacturer's instruction manuals to determine the tip and operating pressures. Refer back to *Figure 49* to review the components of this type of spray gun.

Table 2 Troubleshooting Airless Spray Guns

Problem	Cause	Solution
Tails, fingering	1. Inadequate fluid delivery 2. Fluid not atomizing	1. a. Increase fluid pressure. b. Change to larger tip orifice. c. Reduce fluid viscosity. 2. a. Clean gun and filter(s). b. Reduce number of guns using filter. c. Install sapphire insert.
Pattern expanding and contracting (surge)	1. Pulsating fluid delivery 2. Suction leak	1. a. Change to a smaller tip orifice. b. Install pulsation chamber in system or drain existing one. c. Reduce number of guns using pump. d. Increase air supply (volume) to air motor. e. Remove restrictions in system; clean tip screen or filter, if used. 2. Inspect for siphon hose leak.
Hourglass	Inadequate fluid delivery	1. a. Change to a smaller tip orifice. b. Install pulsation chamber in system or drain existing one. c. Reduce number of guns using pump. d. Increase air supply (volume) to air motor. e. Remove restrictions in system; clean tip screen or filter, if used. 2. Inspect for siphon hose leak.
Hourglass	Worn spray tip	Replace tip

14.6.1 *Setup for Air-Assisted Airless Spray Guns*

Step 1 Prime the unit as shown in the manufacturer instruction manual. Increase the pump air motor speed slowly by adjusting its air regulator; this will increase the gun pressure from the pump.

Step 2 In an air-assisted system, it is not the fluid pressure that completes atomization of the coating, but the addition of atomizing air. So, increase the fluid pressure until the material does not finger and only the tails on the outside of the spray pattern remain. Then add just enough atomizing air pressure to eliminate the tails.

Step 3 To modify the spray fan, turn the pattern adjustment knob.

Step 4 The thickness of coating is also controlled by both the tip size and shape of the fan angle. Use a tip with a larger opening but the same fan angle to increase the thickness of the paint. Use a tip with the same size opening but a larger fan angle to decrease the thickness of the paint.

14.6.2 *Techniques for Air-Assisted Airless Spray Guns*

The stroke speed is usually faster because of the increased material flow needed when these units are used for high production. If the stroke speed is too fast, decrease the tip orifice (opening) size. If the stroke speed is too slow, increase the fluid pressure slightly or increase the tip orifice size.

The thickness of coating is controlled by both the tip size and shape of the fan angle. Using a tip with a larger orifice size but with the same fan angle will increase the thickness of the paint. Using a tip with the same orifice size but with a larger fan angle will decrease the thickness of the paint.

14.6.3 *End-of-Day Breakdown*

The breakdown at the end of the day or shift for both airless and air-assisted spray guns is as follows:

Step 1 Return any remaining fluid coating to an airtight container.

Step 2 Clean out the supply container for the gun.

Step 3 When the container is cleaned, put a sufficient quantity of solvent in to clean the hose and gun.

Step 4 De-pressurize the pump and continue spraying clear solvent until all pressure has been released.

Step 5 Run the sprayer into a clean, grounded container, until the spray is completely clear.

Step 6 Turn off the compressor or pump motor.

Step 7 Set the trigger lock on the gun.

Step 8 Wipe up any spray on the gun or hoses.

Step 9 Detach the hoses from the pot and gun.

Step 10 Coil the hoses carefully and stow the equipment correctly.

Step 11 Dispose of any contaminated solvent as directed by the MSDS for that material.

14.7.0 Production Methods of Application

Below are some of the coating methods used in production facilities:

- *Hot dipping* – The piece is dipped into a bath of molten metal. When the material is molten zinc the process is called hot-dip galvanizing.
- *Fluidized bed method* – In this method powder coating is suspended in a tank (bed). Air is pumped into it so the particles will behave like a liquid. Surfaces to be coated are preheated then run through the bed. The hot surface melts the powder and fuses the coating. As the object cools afterward a uniform plastic sheathing is produced.
- *Powder spray* – Parts to be coated are heated to the appropriate temperature before or after a fusible coating powder is sprayed onto their surface. This production line technique can produce very uniform coatings.

Both the fluidized bed and powder spray methods can use electrostatic spray to transfer the materials to the workpieces.

15.0.0 ◆ APPLICATION OF SPECIALTY COATINGS

15.1.0 Plural-Component Spray Systems

Plural-component spraying is the automatic metering, mixing, and application of plural-component coatings such as polyesters, epoxies, and catalyzed urethanes. Two methods to do this are using special spray guns and using a manifold mixing system.

Special spray guns that mix the components internally or externally at the gun itself are available. These are used in many applications, including applications in which chopped glass fibers are added to the sprayed coating. In these situations, a chopper attachment is used. The chopper unit usually attaches to the top of the gun and is air powered. It chops a glass fiber rope into shreds of the desired length and propels the chopped glass fibers into the coating material spray pattern.

Manifold mixing systems automatically draw the various coating components from separate containers, combine them in the manifold, and then deliver the mixed coating to the spray gun or spray head. These systems can be used with conventional air spray, HVLP, air-assisted airless, or airless spray systems (*Figure 51*). Most mixing devices can vary the proportion in which the separate coating components are mixed. The range of proportions can be from 1:1 to 18:1 or greater. Manifold mixing systems are sometimes used in field structural steel coating applications, but are generally used in industrial production.

The principal advantage of these systems is that the coating materials are not mixed until just before the time when they will be applied. Coatings with pot lives of as little as five seconds can be applied in this manner.

15.2.0 Electrostatic Spray Systems

Electrostatic spray may be used with conventional air spray, HVLP, air-assisted, or airless spray equipment in manual or automatic settings (*Figure 52*). In electrostatic spray operations, an electrostatic charge of up to 75 KV for hand-held guns (up to 180 KV for some automated guns) is applied to the coating particles as they exit the spray gun. In most cases, the coating particles are given an electrical charge, and the workpiece is grounded. When the coating passes through the spray gun and into the air, the opposite charges between the workpiece and the coating particles attract each other so the fine particles of the coating are attracted to all sides, edges, corners, and some recesses of the object being coated. The opposite side of the object can be coated if it is not too far from the spray gun. Generally, almost all of the coating gets to the surface with very little loss.

Electrostatic spray's most widely used application is in-plant production finishing. It is not used as frequently in maintenance coatings or high-performance coatings for complex structures.

PLURAL-COMPONENT AIRLESS SPRAY RIG WITH AIR-MOTOR PUMPS

INTERNAL-MIX AIRLESS GUN WITH AIR PURGE

104F51.EPS

Figure 51 ◆ A plural-component spray rig and airless spray gun.

104F52.EPS

Figure 52 ◆ An electrostatic air-assisted airless or HVLP spray gun used for single or plural-component coatings.

The advantages of electrostatic spray include:

- Fairly complete coverage of odd shapes
- Fairly complete use of the liquid coating
- Reduced coating loss due to overspray
- Uniform film build

However, the disadvantages of electrostatic spray can include the following:

- The conductivity of coatings used is critical: they must be designed for electrostatic spray application.
- It can only be used on bare metal or suitable conductive surface. The electrostatic procedure cannot be used on a non-conductive surface. Usually, only one coat of the material is applied by electrostatic means because of the insulating characteristics of the coating material.
- Because of the voltage used, there can be a risk of electric shock during application.
- The equipment used for electrostatic spray may be more expensive than that of other methods.

15.3.0 Powder Coatings Spray Systems

Typical methods of powder coating include dipping (or fluidized bed), spraying (electrostatic), and cloud chamber applications.

- *Dipping* – When a finely-divided stream of air is passed through a powder, a solid-in-gas dispersion that behaves like a liquid is formed.

  A fluidized bed is a tank with a false bottom made of porous material. Air pressure is applied below this false bottom so that the powder contained above it is lifted and maintained in suspension. The volume of the powder doubles and, in this fluidized state, the powder-air mixture behaves as a liquid.

  When a heated article is dipped into the fluidized bed, the powder closest to it melts and adheres to the surface. The amount of powder that adheres is controlled by the temperature of the article at the time it is dipped, and its heat capacity. This then affects the final thickness of the coating. In other words, higher temperature and heat capacity equals more powder adhering to the surface which results in a thicker coating.

  After the coated article is removed from the fluidized bed, it is heated even further in an oven to make sure the coating is completely fused so it forms a smooth, coherent film. In the case of thermoset materials, the heating cycle must be sufficient to completely cure the coating.
- *Spraying* – This is another method to apply powder coatings. The powder is delivered to the electrostatic spray gun in fluidized form and is usually pressurized to cope with the high output this type of gun produces. The electrostatic hand gun (*Figure 53*) is an efficient method to apply powder coatings. It can be used to apply coating thicknesses from 2 mils (50 μm) to over 20 mils (500 μm).
- *Cloud chamber* – This is essentially a large box with openings for the workpiece to pass through as it goes into the chamber. A cloud of fluidized powder moves from storage hoppers to the cloud chamber, where it is dispersed onto the workpiece by one or more distribution heads.

15.4.0 Metalizing (Thermal) Spray Systems

This corrosion protection technique sprays a molten metal coating onto an object. Metalizing spray guns, also called thermal spray guns, are used. They can spray many kinds of metal as long as the metals are in either wire or powder form. Metalizing spray systems can apply the principal metals used in corrosion protection: aluminum and zinc. Lead is sometimes applied by thermal spray to the interior of vessels that are used to contain strong chemicals.

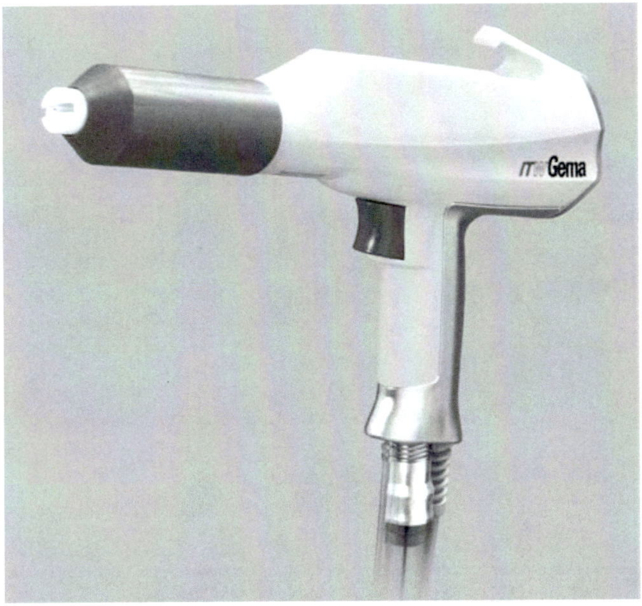

ELECTROSTATIC GUN

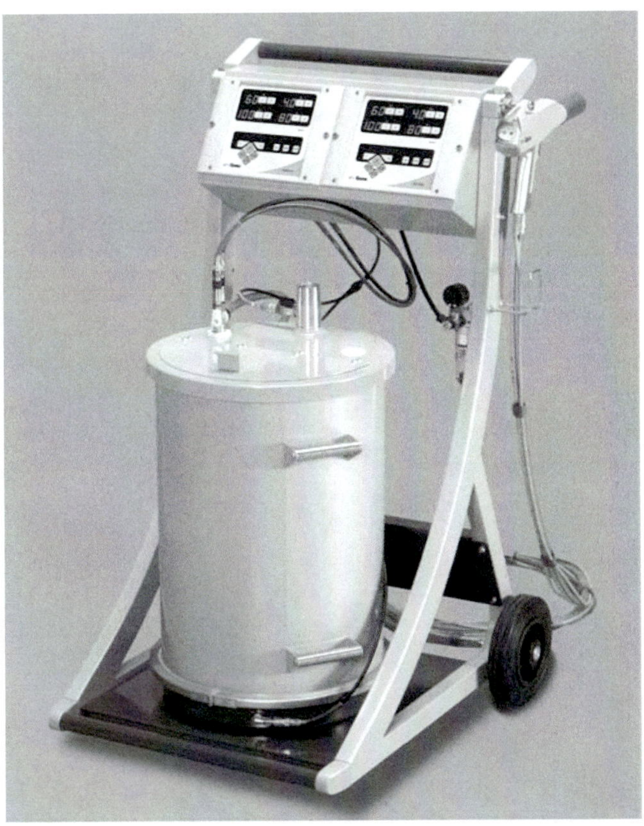

FLUIDIZER FEED

104F53.EPS

Figure 53 ◆ Fluidizer feed and electrostatic powder spray gun.

WARNING!

Lead is a toxic substance. You must wear appropriate PPE and wash thoroughly afterwards to avoid ingesting lead.

Thermally sprayed metalized coatings by manual guns are used in a wide range of settings including:

- Fresh water tanks (AWWA Standard D102)
- Industrial atmospheres
- Bridges
- Salt environments
- Ship hulls
- Tanker interiors
- Hot surfaces (where aluminum is applied by thermal spray to stacks, mufflers, steam and process piping, etc.)

The process of thermal spraying iron and steel for corrosion protection includes the following steps:

- Surface preparation
- Thermal coating application
- Seal coat application
- Inspection at appropriate intervals

Surface preparation for metalizing may require the same steps as surface preparation for other coatings. This may include abrasive blasting, or pre-cleaning to remove residues of oil, grease, dirt, or chemical salts.

Any of the methods of abrasive blast cleaning may be called for by the job specification. The abrasive blast step is very critical. Generally, the thicker the coating the greater the anchor pattern required, even up to 6 mils (203 μm) or more. Surface preparation is extremely important, because any foreign material, and even trace surface moisture, can cause a defect or premature failure of the coating. Double abrasive blasting may need to be done. After surface preparation, apply the metalized coating as quickly as possible within the time required by the specification. The impact velocity of the material sprayed by the thermal spray gun determines the hardness and density of the coating. There are a number of thermal spray guns available.

15.4.1 Flame Spray

Flame spray guns are the oldest of the thermal spray methods. They can be manual or robotically mounted and, depending on their application, can use pressurized oxygen and fuel gases (such as acetylene) to provide a very hot flame at the orifice of the gun. The flame melts a continuously fed wire and propels the melted droplets to the surface being coated. The metal wire is pulled into the gun from a remote reel by an air-powered motor within the gun. Some guns can also be fitted for powder spray. A fan pattern output can be supplied by compressed air from an air assist function within the gun. Flame spray is relatively slow and expensive due to cost of the oxygen and fuels used. It also has the lowest substrate adhesion strength.

104F54.EPS

Figure 54 ◆ Typical arc spray gun.

15.4.2 Arc Spray

Arc spray guns (*Figure 54*), which also can be manual or robotically mounted, use an electric arc (up to 350 amps) within the tip of the gun to vaporize a wire. The wire is fed by an internal air-powered motor in the gun or by a remote powered wire feed. Pressurized air fed to the gun directs and propels the vaporized material to the surface. Arc spray is much more economical and has much higher deposition rates than flame spray. Coating materials that are arc sprayed also have higher adhesion strength than flame sprayed materials.

15.4.3 Plasma, HP/HVOF, and Gas Detonation Spray

These types of guns (*Figure 55*) are used only for industrial production processes. They are used to deposit various powdered coating materials including metals to substrates such as plastic and glass with very high adhesion strengths. Since they produce very high noise levels during operation, they are usually robotically controlled in sound-proof enclosures.

Plasma spray guns use high-frequency pulsed DC power from an external source to create an arc in a gun chamber. This chamber has gas (usually argon) continuously fed into it to create a very hot plasma (up to 8,000°F or 4,427°C). Powdered coating materials (alloys, ceramics, or oxides) are injected into the hot plasma, which is then superheated as the plasma jets out of the gun tip. This very hot material is deposited on the surface to be coated. The impact velocity and heat transfer of the material is high which results in good adhesion to the substrate.

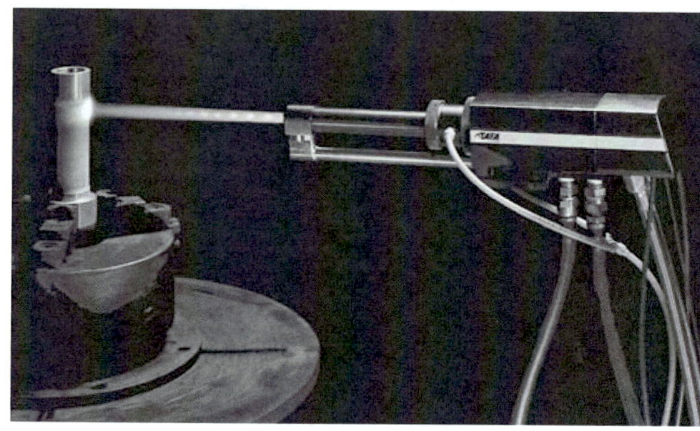

HP/HVOF SPRAY GUN

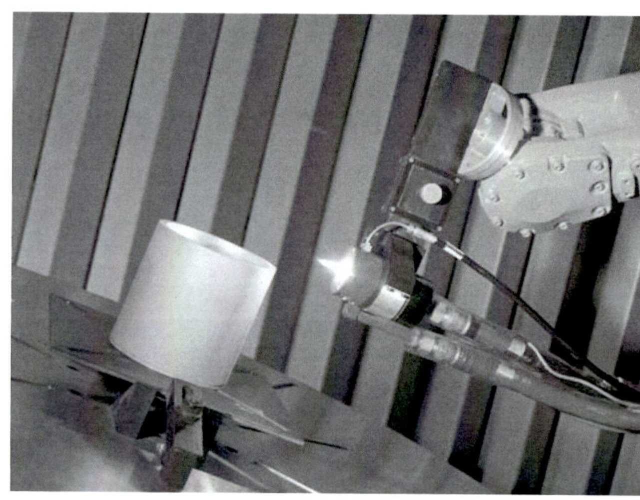

PLASMA SPRAY GUN

104F55.EPS

Figure 55 ◆ Plasma and HP/HVOF spray guns.

HP/HVOF (high pressure/high velocity oxygen fuel) spray guns use a continuously fed oxygen and gaseous fuel (propylene or hydrogen) or liquid fuel (kerosene) mix that is continuously spark-ignited in a firing chamber of the gun. This hot gas (up to 5,600°F or 3,093°C) passes through a venturi nozzle that increases its gas velocity to 7,200 ft/sec (2,195 m/sec) as it exits the tip of the gun. Powdered coating material (a carbide or super alloy) is continuously injected and superheated just after going through the venturi and is then propelled to the surface at velocities up to 3,200 ft/sec (975 m/sec) at the gun tip. The high impact velocity allows very dense layers up to ½ in (1.3 cm) thick to be deposited with very high adhesion.

Gas detonation spray guns use a method similar to an HP/HVOF gun, but the firing is accomplished in pulses instead of continuously. The powdered coating material is continuously fed to the firing chamber along with the firing gases, eliminating the need for a buffer gas, such as nitrogen. A great variety of fuel gases can be used, including natural gas. The explosion propels a heated powdered coating charge to the surface at an extremely high velocity. Surface impact velocity can exceed 1,000 ft/sec (304.8 m/sec). This results in very high density coatings with very high adhesion strength and very little heating of the substrate surface. Nearly all powdered metals, alloys, oxides, and carbides can be used and nearly all substrates can be coated including dielectric, plastic, glass and ceramic substrates.

15.5.0 Hot Spray Systems

Heating the coating liquid prior to application can be used with both air spray and airless spray applications. This is done by using either an in-line heater or a hot box, both of which heat the coating before it goes through the spray gun. The increased temperature of the liquid generally reduces viscosity, allowing the coatings to be applied more readily and at a somewhat faster speed. This method also has the advantage of cooling rapidly as the coating is propelled into the air. This increases viscosity shortly after the coating hits the surface. This helps the coating remain in place, reduces the chance of sags and curtains, and generally makes a smoother, more uniform film. The fast-evaporating solvents in a coating liquid tend to evaporate readily between the gun and the surface. This leaves a higher quantity of the slow evaporating solvents in the material. These solvents aid in leveling the coating and providing a smooth, even film.

Heating a coating (if it is formulated for hot application) means that the coating can contain higher solids, have a higher viscosity, and still be sprayed. This means more material can be applied per coat without the danger of runs or sags. Heating the coating liquid not only lowers the viscosity, but also tends to make a more uniform material for application.

When heaters are used, take care to ensure that the coating is not heated to a temperature higher than that allowed by the specification or manufacturer's recommendation.

Many suppliers carry coating heaters which come in a variety of sizes, shapes, and forms. They generally pre-heat the coating from 120°F (50°C) to possibly as high as 180°F to 200°F (82°C to 93°C). In-line heaters may be portable and attach to the applicator's belt. They are heated with electricity so they must be connected by an electric cord to an outlet. This, along with the weight of the heater itself, makes the equipment more cumbersome for the coating applicator. Another way to heat the coating is through a stationary heater positioned between the pump and the gun. Considerable heat can be lost using this method, depending on the length of the hose, although insulated and heat traced hoses are available.

Heated coatings can be applied by either airless or conventional air spray. The benefits are generally the same for both. Many shop finish-application units use airless spray and in-line heaters as standard procedure. They are also used on large, storage tanks. Some of the advantages of heated spray are as follows:

- The coating may have a more uniform viscosity.
- Even coverage without runs and sags can be achieved.
- Drying time can be sped up.
- Faster applications are possible due to the lower viscosity.
- Coatings can be sprayed readily at lower ambient temperatures.
- There may be fewer solvent fumes because there are no added solvents.
- The coating thickness per coat can be increased.
- Finer atomization, with lower application pressures, is possible.
- Improved coating adhesion with reduced viscosity can be achieved.
- Longer lengths of hoses can be used, if needed.

16.0.0 ◆ COATING APPLICATION ISSUES: HOLIDAYS/ PINHOLES

Holidays/pinholes are discontinuities in the coating layer. In other words, they are holes in the coating. These must be avoided because they cause the coating or coating system to be much less effective in protecting the surface. They can go through to the underlying coating layer or all the way to the substrate. Holidays that go through to the substrate are a particular problem. Not only are areas containing holidays prone to corrosion, but this can lead to pitting, which undermines the coating and pushes it off the surface. Examples of holidays are shown in *Figures 56* through *59*.

Holidays/pinholes make the coating more vulnerable to environmental effects and can lead to breakdown of the coating and/or corrosion of the underlying surface. Inspections made during the coating application process can detect holidays/pinholes in non-conductive cured coatings on a conductive substrate by using a variety of testing equipment. Low-voltage wet-sponge detectors and high-voltage metal brush detectors are examples of two kinds of such equipment (*Figure 60*).

Low-voltage detectors are normally used on coating systems with a DFT of 0 to 20 mils since there is no spark hazard with low-voltage detectors. Coating systems with a DFT greater than 20 mils are normally checked using high-voltage detectors. Review the coating manufacturer's manual to determine if a high-voltage meter can be used. Using an excessive amount of voltage could damage an otherwise sound coating.

104F57.EPS

Figure 57 ◆ Holiday resulting from thin coating.

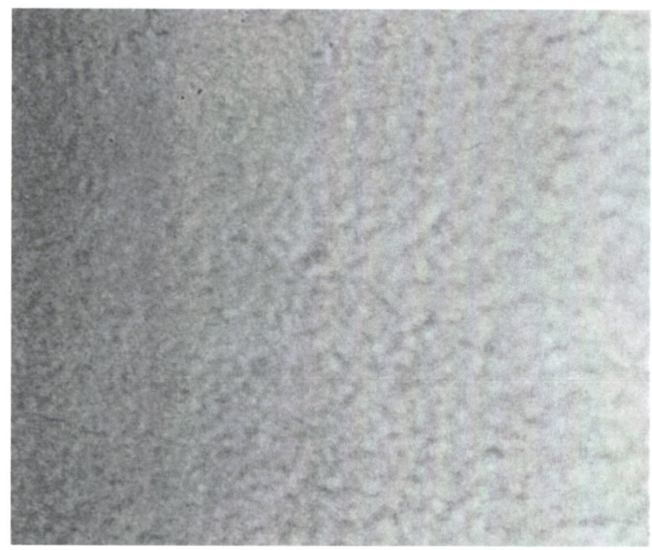

104F58.EPS

Figure 58 ◆ Holiday resulting from uneven application of coating.

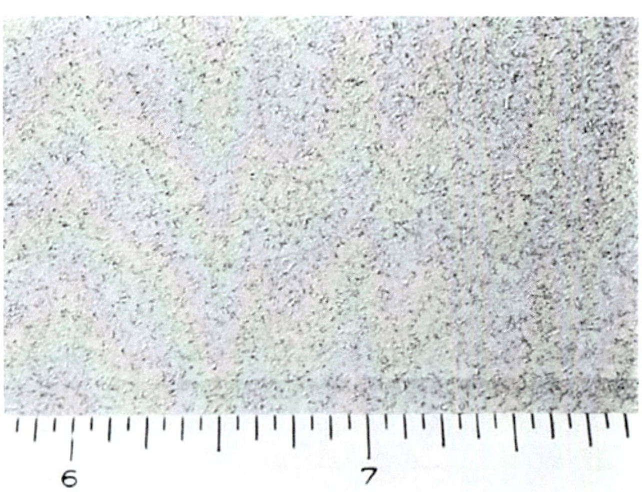

104F56.EPS

Figure 56 ◆ Holidays resulting from dry spray.

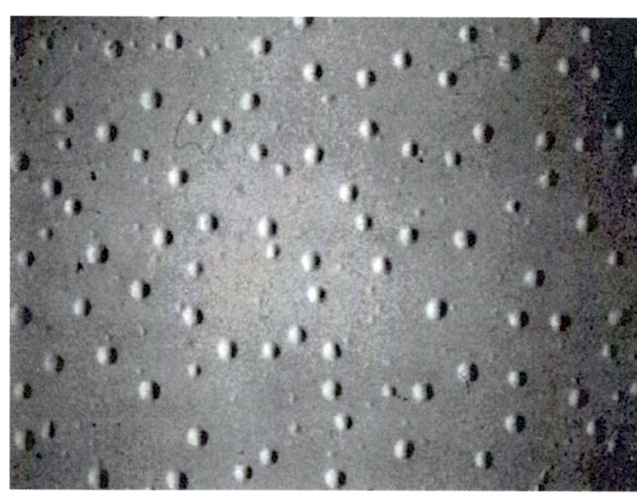

104F59.EPS

Figure 59 ◆ Holiday resulting from blistering.

WARNING!

High-voltage detectors are a shock hazard and must be used only as described in the equipment manufacturer's instructions. Avoid any contact with the metal components of these instruments during operation.

Discontinuities are usually caused by application problems. Discontinuities in spray applications can be caused by spray pressure that was too high, a gun held too close to the surface, or too much solvent used. Ideally, inspection, detection, and repair should take place before the recoating time expires. The allowable recoating time is located on the manufacturer's label or PDS. If the recoating time has expired, the surface must be allowed to cure fully. Recoating can then be done according to the manufacturer's instructions.

17.0.0 ◆ DRYING AND CURING OF COATINGS

Because coatings are selected based on their desired properties (appearance, UV resistance, chemical resistance, abrasion resistance, etc.), it is important that they are fully cured before they are exposed to their intended environment. Until the coating is fully cured, which is substantially different from being dry to the touch, it will not perform to the highest standard of its properties. Exposing the coating to the environment prior to full cure can cause premature failure of the coating.

17.1.0 Film Cure

Drying is the evaporation of fluid from a coating. Curing is the chemical reaction between coating resin and hardeners. Most coatings require both

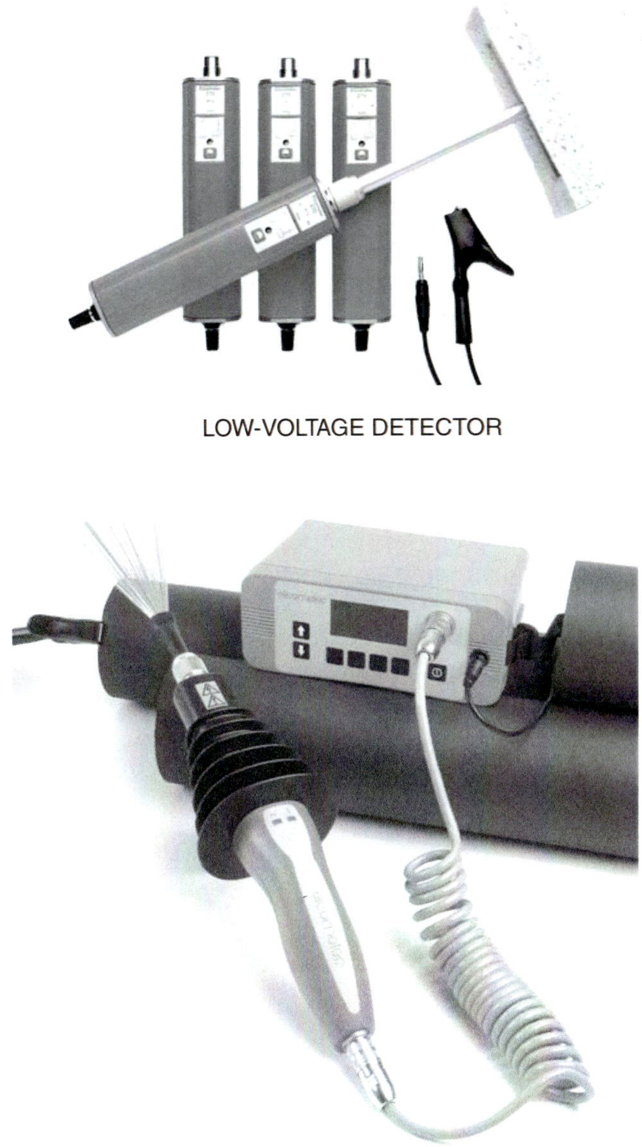

Figure 60 ◆ Holiday/pinhole detectors.

INSIDE TRACK

Coating Tests

In addition to holiday/pinhole inspection and tests, job-site inspectors will normally conduct tests on applied coatings to check various coating parameters specified in the job specification. These checks include DFT thickness tests using magnetic pull-off gauges for ferrous substrates, or eddy current instruments for nonmagnetic coatings on ferrous or nonferrous substrates. Tooke gauges may be used to test coating thickness, film adhesion, microscopic cracking, blistering, and cratering. A variety of tests can also be used to conduct pull-off tests for adhesion. Hardness tests can be done using calibrated hardness pencils or a hardness tester. The results of the inspector's tests are usually logged in daily with weekly reports that are reviewed with the contractor representative to determine if any corrective action needs to be taken.

a drying time and a curing time to achieve a solvent-free and solvent-resistant film. The time varies by the type and composition of the coating. Drying and curing times are normally dependent on humidity, temperature, air movement, and film thickness. Manufacturers usually list drying and curing times in a schedule on the label and in the PDS. The times are given for a certain humidity and film thickness at a number of ambient temperatures. The times (maximum and minimum) are related to when the coating can be touched, handled, and recoated, as well as the time needed for a total cure.

17.2.0 Cure and Recoating Intervals

If a coating is recoated before the specified recoating time, the previous coat may soften and lift, or blister. It could also develop solvent entrapment or delaminate (de-layer).

It is important to recoat within the recoat window. If too much time has occurred between coats, the surface of the previous coating may be too hard to allow adhesion so delamination or other coating failures may occur. When the recoat window has passed, allow the previous coating to cure completely. Any overcoating must be done in accordance with the manufacturer's instructions. Generally, it is a good practice to roughen up the previous coat slightly in the areas that need to be recoated to ensure adhesion. Use a spot primer next, and then feather the new coating into the existing coating.

17.3.0 Cure Issues

Sometimes a blush or white coating appears on coated surfaces shortly after application. This is caused by water in coating films that were applied while there was high humidity (over 85 percent relative humidity). If the specification allows, use a slower drying solvent or retarder to help reduce blushing. The best decision is to delay coating until the humidity drops.

Amine-cured epoxies tend to develop amine blush if the temperature is low or there is high humidity before partial curing takes place. This type of blushing is harder to detect and can reduce the adhesion of additional coats if not detected and removed. Use an amine blush check kit when environmental conditions make it likely to occur. The amine blush can be removed with a rag saturated with the solvent recommended by the specification.

18.0.0 ◆ EQUIPMENT CLEANUP

After applying coating materials, promptly clean all equipment used with an appropriate solvent. This is especially true for mixed multi-component coating materials that have a pot life.

18.1.0 Manual Equipment

For brushes, use a brush comb to help loosen and remove coatings from between bristles and from the handle. For rollers, follow the coating manufacturer's instructions regarding solvent or non-solvent cleaning.

18.1.1 Solvent Cleaners

Soak brushes and rollers in the solvent for the time recommended by the coating manufacturer and then use a brush and roller spinner (*Figures 61* through *63*) inside a bucket to remove the coating material and solvent. Repeat the soaking and spinning as necessary to remove as much of the coating material as possible.

104F61.EPS

Figure 61 ◆ Brush and roller spinner.

104F62.EPS

Figure 62 ◆ Using a brush spinner to clean a solvent from a brush.

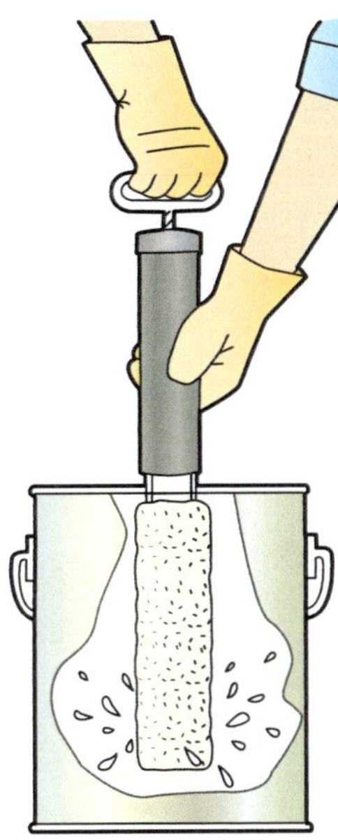

104F63.EPS

Figure 63 ◆ Using a roller spinner to clean and dry a roller.

18.1.2 Non-Solvent Cleaners

For brushes and rollers that have been cleaned with solvent other than water, wash them with soap and warm water and rinse them with clear water to remove any remaining contamination. Spin them with the spinner and hang them to dry after smoothing the brush bristles.

18.2.0 Spray Equipment

Cleanup of spray equipment is critical, especially if coating materials with a pot life have been pre-mixed before being sprayed. Because specific procedures may be required for flushing different spray equipment models from the same manufacturer or from other manufacturers, always follow the spray equipment manufacturer's instructions for flushing the equipment. Be sure to use the correct solvent recommended by the coating manufacturer.

CAUTION

Guns, hoses, paint tanks, and pumps used for spraying can all be damaged beyond repair if materials with a pot life are not immediately flushed from the equipment.

18.2.1 Breakdown at End of Day or Shift

At the end of the day or shift, the breakdown is very similar for spray systems. Any remaining fluid coating is returned to an airtight container, and the supply container for the gun is cleaned out. When the container is cleaned, put a sufficient quantity of solvent in to clean the hose and gun. Run the sprayer, preferably into a clean, grounded container, until the spray is completely clear. Wipe off any spray on the gun or hoses, relieve the pressure, and detach the hoses from the pot and gun. Coil the hoses carefully and stow the remaining equipment carefully, as well.

Because of VOC restrictions in some states, spray equipment may have to be flushed in confined areas equipped with air-exchange filtering equipment to remove the VOCs.

19.0.0 ◆ QUALITY CONTROL

The purpose of a quality control inspection is to ensure that a quality coating is being applied that meets the guidelines detailed in the specification. Proper inspection is the key to effective coating performance and should be required in the specification.

Inspection should be performed both during the application process, as well as over the life of the coating. Whether or not third-party inspection is being performed, quality control is essential for any coating project. Since these projects can involve thousands or millions of dollars, all contracts contain performance warranties. The quality of a coating depends on three very important items: coating specification, proper surface preparation and application, and proper inspection.

Common coating application failures that occur during or shortly after application, along with their remedies, are summarized in *Table 3*. Most of these failures should be detected by inspection during application or curing. Two of these failures have already been covered.

In the event of a coatings failure, the quality control records are important to determine cause and liability. Proper quality control records could potentially save the contractor thousands of dollars in rework.

20.0.0 ◆ COATINGS ON CONCRETE

Coatings adhere to concrete by penetration, so the first time concrete is coated, it will require a great deal more coating than for the same area of steel. The type of coating applied to concrete is determined by its function. Two types of concrete coatings are: bituminous coatings and chlorinated rubber coatings. Bituminous coatings are solutions of coal tar or asphalt that are used for waterproofing. Chlorinated rubber coatings are used for coating concrete water tanks and swimming pools.

Epoxy coatings are used because they resist a variety of chemicals. These can be thin-film or thick-film coatings.

Table 3 Common Coating Application Failures

Description	Cause	Remedy
SAGS Also called runs or curtains. Excess flow of paint.	Spray gun too close to work; too much thinner; too much paint; or surface too hard or glossy to hold paint.	Before cure, brush out excess paint and modify spray conditions. After cure, sand and apply another coat.
ORANGE PEEL Hills, valleys in paint resembling skin of an orange.	Paint too viscous; gun too close to surface; solvent evaporated too fast; or air pressure too low for proper atomization.	Before cure, brush out excess paint and modify spray conditions. After cure, sand and apply another coat.
OVERSPRAY Also called dry spray. Dry, flat, pebbly surface.	Particles reaching surface not wet enough to level because of too rapid solvent evaporation; gun too far from surface; or paint particles falling outside spray pattern.	Before cure, remove by dry brushing followed by solvent wiping. After cure, sand and apply another coat.
COBWEBBING Thin, stringy paint, spider web-like particles.	Solvent evaporating too rapidly. Most common with fast-evaporating lacquers, such as vinyls and chlorinated rubbers.	Use a slower evaporating solvent or apply when cooler.
CRATERING Also called pitting. Small, uniform indentations in film.	Air pockets trapped in wet film during spraying.	Sand or blast to smooth finish and apply additional coats.
FISH EYES Separation or pulling apart of wet film to expose underlying finish or substrate.	Application over oil, dirt, silicone, or incompatible coating.	Sand or blast remove, brush-apply a fresh coat plus topcoat.
BLUSHING Flat finish with milky appearance.	Moisture condensation in high humidity with fast evaporating or unbalanced thinner in spray application.	Sand or blast remove; respray with retarder added to thinner.
UNEVEN GLOSS Nonuniform sheen, shiny spots.	Nonuniform film thickness; moisture in film; temperature change during curing; or paint applied over soft or wet undercoat.	Allow to dry and apply another finish coat under acceptable conditions for moisture and humidity.
WRINKLING Rough, crinkled surface.	Surface skinning over uncured paint because of too much thickness and/or too warm weather, especially with oil-base paints.	Scrape off wrinkles and apply thinner coat; avoid intense sunlight.
BLISTERING Small to large brokenor unbroken bubbles.	Solvent entrapment; oil, moisture or salt-contaminated surfaces; or cathodic disbonding.	Put blowers in enclosed areas to accelerate solvent release; adequate cleaning of surface contamination; proper levels of cathodic protection.
PINHOLING Tiny, deep holes exposing substrate.	Insufficient paint spray atomization; coarse atomization; or settled pigment. Mist coat not applied to zone surface before final coat.	If uncured, brush out and apply additional coat. If cured, apply additional coat.

104T03.EPS

Review Questions

1. A coating method that uses pressurized fluid for atomization is _____.
   a. a pressurized roller
   b. airless spray
   c. HVLP spray
   d. thermal spray

2. A formal document used to define job-site work to be performed is the _____.
   a. MSDS
   b. PDS
   c. coating specification
   d. quality control document

3. A manufacturer's document that defines the characteristics of a coating material is a(n) _____.
   a. MSDS
   b. PDS
   c. coating specification
   d. quality control document

4. The protective covering that is available in rolls up to 1,000 ft (304.8 m) is _____.
   a. masking tape
   b. high-density thin film sheeting
   c. clear polyethylene sheeting
   d. canvas drop cloth

5. Liquid coatings with a powder component are normally mixed by adding the powder to the liquid component _____.
   a. before the liquid component is mixed
   b. after any activator is added and mixed
   c. before the activator is added and mixed
   d. during initial mixing of the liquid component

6. Large-capacity roller covers are made in 2¼ inch (5.7 cm) core diameters with a length of _____.
   a. 9 inches (22.86 cm)
   b. 12 inches (30.48 cm)
   c. 15 inches (38.1 cm)
   d. 18 inches (45.72 cm)

7. Coating mitts are used to _____.
   a. fine finish a surface
   b. coat small and odd-shaped objects
   c. cut in corners
   d. apply epoxy on floors

8. A loaded brush should first be _____.
   a. spot unloaded on an uncoated area
   b. unloaded by a long stroke movement
   c. tapped to remove excess coating
   d. held bristles up to prevent dripping

Review Questions

Questions 9 through 12 refer to *Figure 1*.

9. The pattern adjustment knob is labeled number _____.
   a. 1
   b. 3
   c. 5
   d. 6

10. The trigger lockout is labeled number _____.
    a. 2
    b. 3
    c. 4
    d. 5

11. The control to stop and start the spray is labeled number _____.
    a. 2
    b. 3
    c. 4
    d. 5

12. The air inlet is labeled number _____.
    a. 1
    b. 2
    c. 5
    d. 6

13. For conventional air spray, the atomization of the coating is adjusted by the _____.
    a. fluid control knob
    b. pattern control knob
    c. air pressure setting
    d. fluid pressure setting

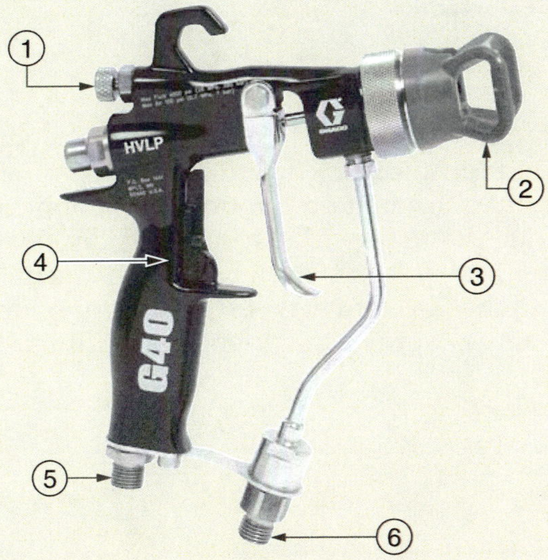

AIR-ASSISTED AIRLESS SPRAY GUN

104RQ01.EPS

Figure 1

14. WFT is measured using a(n) _____.
    a. knife edge
    b. special metal or plastic gauge
    c. viscosimeter
    d. electronic instrument

15. For manual electrostatic spray guns, the operating voltage can range up to _____ kv.
    a. 75
    b. 100
    c. 160
    d. 180

Summary

This module presented information on equipment and techniques for mixing various types of coatings. Equipment descriptions and application techniques were provided for coating application by brush, roller, mitt, and basic spray, including conventional air spray, HVLP, airless, and air-assisted airless spray equipment. In addition, this module described the environmental conditions required for applying coatings as well as quality control and inspection required during coating application. Coating application failures and remedies were discussed. Specialty spray systems and the application methods for these systems were also briefly covered.

Notes

Trade Terms Introduced in This Module

Arc: The flow of electrical current through a gas, such as air, from one pole to another.

Boxing: Intermixing the coating material if more than one container of a coating will be used.

Confined space: A work area large enough for a person to work, but arranged in such a way that an employee must physically enter the space to do work. A confined space has a limited or restricted means of entry and exit. It is not designed for continuous work. Tanks, vessels, silos, pits, vaults, and hoppers are examples of confined spaces.

Cutting in: Painting corners and the perimeter of windows, doors, and corners with a brush or trim roller before roller application of coating to walls or ceiling.

Feathering: Tapering off the edges of the coating by gradually pulling the brush bristles off the painted surface.

Flammable: Capable of igniting easily and burning rapidly; used to describe a fuel with a flash point below 100°F.

Ground: The conducting connection between electrical equipment or an electrical circuit and the earth.

Mitting: A coating application technique in which the coating is applied with a hand mitt.

Orifice: A small, specially shaped opening or hole in a spray or blast nozzle.

Overspray: Atomized coating particles that deflect from or miss the surface being sprayed. Also, spray particles that are not wet enough to fuse when they reach the surface.

Personal protective equipment (PPE): Equipment or clothing designed to prevent or reduce injuries, and/or contact with or inhalation of hazardous substances.

Pot life: The limited amount of time available for thinning (when spraying) and/or application of a coating.

Sweat-in time: The time allowed for the catalyst, converter, hardener, or activator to start a chemical reaction before thinning and/or application.

Transfer efficiency: The amount of coating that goes onto the surface compared to the total wasted due to overspray or bounceback.

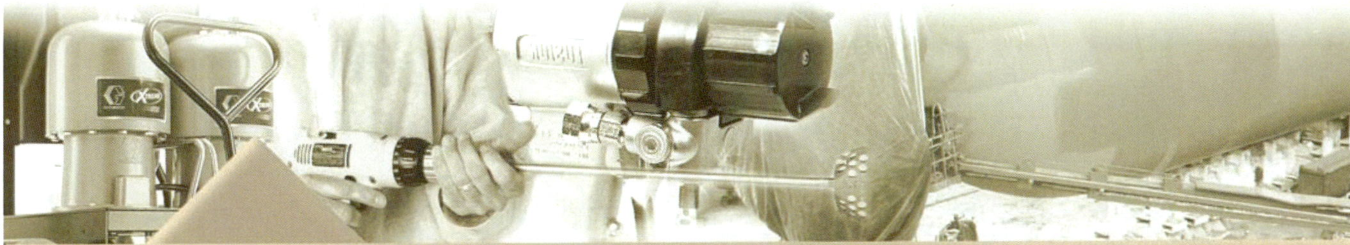

Appendix A

Sample Work Plan

Field Erected Tank Coating Specification
Sample

5.0 COATING AND FINISH

5.1 Tank Interior - All interior surfaces, including attachments, man-ways, and flange faces and bolt holes, are to be coated.

5.1.1 Detergent clean (SSPC-SP1) the interior of the tank to clean, neutralize and remove any oily residues using a low pressure water washer.

5.1.2 Preparation of welds and repair of metal defects. Detergent clean (SSPC-SP1) to remove any oily residues from mechanical repairs. Unless otherwise specified the welds shall be ground smooth and blended, as per condition "D" in NACE Standard RP0178.

5.1.3 Abrasive blast interior to remove rust and mil scale in accordance with Section 3.2.

5.1.4 Abrasive blast interior to obtain degree of cleanliness and anchor profile specified.

5.1.5 Remove spent abrasive, and clean all dust, and contaminants from substrate.

5.1.6 Application of specified lining

5.2 Tank Exterior - All exterior surfaces, including attachments (walkways and ladders) are to be coated.

5.2.1 Water wash exterior of tank to clean and neutralize.

5.2.2. Abrasive blast exterior of tank in accordance with Section 5.4.

5.2.3 Cleaning of blasted exterior

5.2.4 Application of specified coatings.

5.3 Pre-cleaning

5.3.1 Low pressure water blast entire surface with a minimum of 2,000 to 3,000 psi water to remove all foreign materials and to neutralize all surface (chemical) contaminants. Nozzle of waterblaster shall be held no more than 12" away from the substrate during washing. Oil or grease residues shall be removed by solvent cleaning (SSPC - SP 1) prior to water blasting or by utilizing detergent injection (in above water blast), followed with a fresh water rinse. An Ultraviolet Light is to be used to detect the presence of any petroleum contaminants prior to and after pre-cleaning.

5.3.2 Inspect tank interior for defects such as laminations or slivers and needed repairs. Method of repair for pits shall be approved by the Coating Manufacturer's Representative. These metal repairs shall then be performed prior to proceeding to the next step. **All metal repairs requiring welding shall strictly adhere to Company XYZ**

Specification XXXXXXXX latest revision, Sections X Welding and Y Testing. All metal surface preparation will be per NACE (RP0178), "Design, Fabrication, and Surface Finish of Metal Tanks and Vessels to be Lined for Chemical Immersion Service". All sharp edges are to be ground smooth to a minimum of condition "D". All interior welds shall be ground to a radius of l/8 inch. All weld splatter; pits or other rough surfaces are to be ground out or otherwise repaired to provide a smooth surface.

5.3.3 A pH outside of the range of 5-8 will NOT be allowed on or in the substrate. Check the pH of the surface using pH paper and distilled water. Test for pH at twelve (12) locations. Tests shall be made 90° apart around the circumference from two (2) locations each equal distance from center of the tank; this requires eight (8) tests. Make two (1) additional test in the manhole area and one (1) test in roof and two (2) on the floor. If the pH is not between 5-8, perform a second water blast with potable water. Water blast entire surface with 2,000 to 3,000 psi waterblaster to neutralize all surface (chemical) contaminants. Nozzle of waterblaster shall be moved slowly and deliberately and shall be held no more than 12" away from substrate during washing. Remove the water, wipe dry, and re-test. If no change has occurred, consult the **Coating Manufacturer** for further decontamination steps.

5.3.4 Check for soluble salts using an approved soluble salt detection kit and using the same statistical manner as described in 5.3.3 above. See Section 09910 paragraph 3.15 of for allowable quantity of salts.

5.4 Abrasive blasting

5.4.1 A test blast for the interior shall be performed on the exterior of the tank to determine the acceptability of the **Sub Contractor's** blast procedure prior to starting interior blasting.

5.4.2 Abrasive blasting shall not be conducted on surfaces that are wet or less than 5 degrees F above the dew point, or when relative humidity is above 85% or steel Temperatures are less than 50 degrees F. If surfaces should start to turn during blast cleaning operation, blast cleaning shall be stopped until conditions are favorable to resume cleaning operations. Dehumidification equipment shall be used if required to maintain schedule.

5.4.3 Abrasive blast interior of tank to a degree specified.

5.4.4 Remove all spent abrasives and foreign debris.

5.4.5 Abrasive Blasting shall be done in accordance with the Company XYZ's Technical Specification. Abrasives shall be clean and dry and shall be sized and graded so as to provide the specified surface profile as measured by TesTex replica tape or profile meter. Air pressure control is essential to achieve the desired anchor profile.

5.4.6 All traces of dust shall be removed from the blasted surface by vacuum cleaning and brushing with a stiff bristle brush. Care should be taken to avoid contaminating the surface with fingerprints or from detrimental material on the workers' clothes or atmospheric contamination. Once the vessel has been cleaned, care shall be taken by workmen to maintain a clean surface throughout all phases of the work.

5.4.7 Exterior of tank shall have surface prepared as in Items 5.3.1, 5.3.2, 5.3.3, and 5.3.4 of this section, along with abrasive blasting to the degree specified. External surface preparation shall be performed after completion and acceptance of the internal coatings work. All nozzles, man-ways etc. Shall be covered with 1/2" plywood, gaskets and shall be bolted so as not to allow damage to interior coatings or flange faces.

5.5 APPLICATION AND CURING

5.5.1 All coatings shall be applied in accordance with the **Coating Data Sheets**, the manufacturer's data sheet and this specification, as applicable. Manufacturers' recommended re-coat and cure schedule will be adhered to.

5.5.2 Prior to spraying the first coat, all interior welds and attachments are to be brush coated, using mixed material thinned 25% with the appropriate manufacturers' thinner. Manufacturer's recoat time shall be complied with before the application of the first coat.

5.5.3 Spray equipment and application procedures shall be per manufacturer's data sheet.

5.5.4 Continuous air flow during drying and curing periods shall be required to ensure adequate ventilation for safety and curing.

5.5.5 Proper surface preparation shall be completed and inspected prior to application of coating. Coating shall commence no later than four hours after surface has been approved by **Contractor** or before any "turning" of the substrate blast cleanliness grade has occurred on the area to be coated.

5.5.6 All coatings shall be thoroughly mixed prior to and during application. Thinners shall be used only as recommended by manufacturers' technical data sheets. Manufacturer's batch numbers for coatings and thinners used shall be recorded and submitted.

5.5.7 **Sub Contractor** shall see that paint is kept free of dried paint scale and other foreign matter by straining as necessary. Paint not used continuously shall be kept sealed to avoid undue evaporation and entrance of contaminants.

5.5.8 The method of paint application shall be per the application data sheet. Use of 360° spray lances shall be required to coat the internal diameter of all nozzle penetrations unless waived by **Contractor**.

5.5.9 Coatings shall be applied as a continuous film of uniform thickness, free of pores, runs, holidays and sags.

5.5.10 Special care shall be taken at flanges, penetrations, welds, laps and splices to insure an adequate paint cover. All flange faces and bolt holes are to be coated.

5.5.11 After application of each coat of material, all surface defects shall be repaired as per coating manufacturer recommendations. If dry spray occurs, it shall be removed by lightly hand sanding with 300+ grit sand paper. Then vacuum and solvent wipe. Any contamination between coats will be unacceptable. Any coating that is contaminated between coats must be stripped off and begun anew.

5.6 INSPECTION

5.6.1 Pre-acceptance inspection by **Contractor** will commence as work is performed. **Sub Contractor** must be prepared to maintain a sufficient number of personnel on site to correct any coating deficiencies.

5.6.2 **Contractor**, at his option, may inspect any and all phases of the work. He shall approve all surface preparation and metal repair prior to application of any coating.

5.6.3 Any painted surfaces found to be damaged or improperly applied shall be repaired as per manufacturer's recommendations at **Sub Contractor's** expense.

5.6.4 All phases of the work shall be available to inspection by a representative of the coating manufacturer as well as by **Contractor**.

5.6.5 The presence of **Contractor** on the job shall not relieve **Sub Contractor** of his responsibilities under this specification.

5.6.6 **Contractor** shall inspect the lining after each coat and shall approve the surface prior to each succeeding coat.

5.6.7 The air supply for abrasive blasting and spray painting shall be clean, dry and free of visible moisture and oil. A blotter test shall be required prior to the start of the work each shift to ensure clean, dry air.

5.6.8 Dry film thickness measurements shall be taken in accordance with SSPC-PA-2 with the exception that the "5 spot readings" shall be taken over an area of 25 square feet in lieu of the stated 100 square feet.

5. 6.9 Replica tape, such as TesTex or profile meter shall be used to evaluate the surface profile.

5.6.10 Coating integrity shall be checked by using a low voltage, wet sponge type holiday detector (e.g. Elcometer 270 or equivalent). A surfactant (such as Kodak Photo-Flo) shall be used during the holiday test. If repairs are required, the surfactant shall be completely removed. The surface shall be completely dry prior to making any repairs to the coating.

5.7 FIELD TECHNICAL ADVISOR

5.7.1 The manufacturer of the coating will, upon request by **Contractor**, make available an experienced Field Technical Advisor, to advise and consult with **Contractor** or **Sub Contractor**, regarding the proper installation of the coatings system.

5.7.2 The Advisor will be furnished at no cost to **Contractor**. The Advisor shall spend time as required in the field with the cleaning and application crew to assure that all recommendations are understood and that good, correct cleaning and coating practices are observed.

5.7.3 Any request for a Field Technical Advisor shall be made by **Contractor** to **Sub Contractor**.

5.8 QUALITY ASSURANCE AND INSPECTION

5.8.1 **Sub Contractor** shall be responsible for ensuring that all work is done in accordance with the manufacturer's data sheets and these technical specifications. This will include all installation, environmental and acceptance testing as required. See Section 8 of the Technical Specifications for testing and reporting requirements.

5.8.2 **Contractor** reserves the right to witness and conduct parallel testing / inspection during all phases of the work.

5.9 SUBMITTALS

5.9.1 **Sub Contractor** shall submit a list of all materials, e.g. blasting abrasive, solvents, thinners, paint, etc. for approval prior to their use.

5.10 COMPLETION

5.10.1 Upon completion of the job, **Sub Contractor** shall notify **Contractor** who will inspect the work. Any defects found will be corrected according to this specification and the manufacturer's recommendations.

5.11 WARRANTY

5.11.1 **Sub Contractor** agrees to repair or replace any coating failures as a result of improper metal preparation, surface preparation or application of coatings for a period of two (2) years from acceptance by **Contractor**.

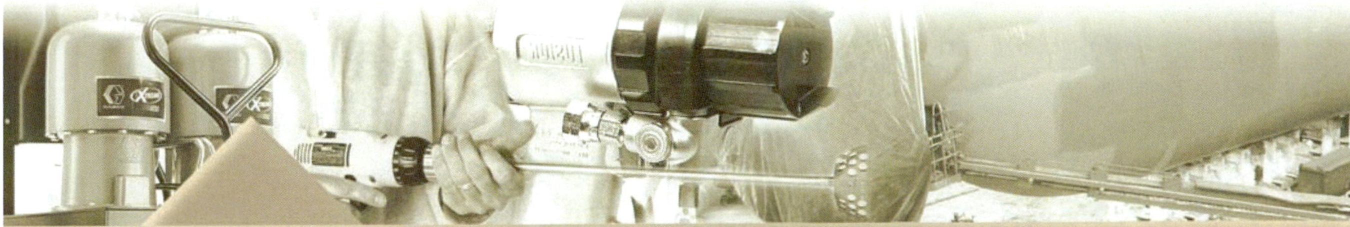

Appendix B

Product Data Sheets

Industrial & Marine Coatings

SHERWIN WILLIAMS

5.23

POLY-LON® 1900

POLYESTER POLYURETHANE

PART A	B65-500	SERIES
PART B	B65V500	HARDENER

PRODUCT INFORMATION

Revised 6/05

PRODUCT DESCRIPTION

POLY-LON 1900 POLYESTER POLYURETHANE is a heavy duty, high performance, two component, exterior/interior, VOC compliant, high solids, polyester-aliphatic urethane. When properly cured, this dries to a super tough, "wet look", high gloss, flexible finish with maximum gloss retention, color retention, and chalk resistance. Designed to withstand aggressive industrial environments and provide excellent durability against severe weather conditions, prolonged exterior exposure, abrasion, impact, and general chemical attack.

- Retains its exterior appearance over a wide range of chemical, weather, and mechanical conditions.
- Superior exterior color and gloss retention.

RECOMMENDED USES

For use over prepared metal and masonry surfaces in industrial environments such as:

- Tank exteriors
- Pipelines
- Structural steel
- Bridges
- Marine vessels
- Rolling stock
- Conveyors
- Refineries
- Walls
- Floors

Conforms to AWWA D102-03 OCS #5 & #6.

- Suitable for use in USDA inspected facilities.

PRODUCT CHARACTERISTICS

Finish: High Gloss

Color: Wide range of colors available

Volume Solids: 65% ± 2%, mixed, may vary by color

Weight Solids: 76% ± 2%, mixed, may vary by color

VOC (EPA Method 24): mixed

Unreduced:	<340 g/L; 2.80 lb/gal	
Reduced 10%:	<388 g/L; 3.23 lb/gal	

May vary by Color

Mix Ratio: 3:1 by volume, 4 gallon mix

Recommended Spreading Rate per coat:

Wet mils:	3.0 - 4.5
Dry mils:	2.0 - 3.0
Coverage:	360 - 545 sq ft/gal approximate

NOTE: Brush or roll application may require multiple coats to achieve maximum film thickness and uniformity of appearance.

Drying Schedule @ 3.0 mils wet @ 50% RH:

	@50°F	@ 77°F	@100°F
To touch:	16 hours	2 hours	30 minutes
To handle:	24 hours	10 hours	2 hours
To recoat:			
minimum:	24 hours	12 hours	2 hours
maximum:	3 days	48 hours	24 hours
To cure:	7 days	7 days	5 days

If maximum recoat time is exceeded, abrade surface before recoating. Drying time is temperature, humidity, and film thickness dependent.

Pot Life:	5 hours	4 hours	45 minutes

Sweat-in-Time: None required

Shelf Life: 12 months, unopened
Store indoors at 40°F to 100°F.

Flash Point: 102°F TCC, mixed

Reducer/Clean Up: Reducer #132, R7K132

PERFORMANCE CHARACTERISTICS

System Tested: (unless otherwise indicated)
Substrate: Steel
Surface Preparation: SSPC-SP10
2 cts. Epolon II Primer @ 2.5 mils dft/ct
1 ct. Poly-Lon 1900 @ 2.0 mils dft

Abrasion Resistance:
Method: ASTM D4060, CS17 wheel, 1000 cycles, 1 kg load
Result: 76 mg loss (average of 5 trials)

Adhesion:
Method: ASTM D3359 Method B
Result: 5B, 100% Retention
Method: ASTM D4541
Result 1200 psi

Direct Impact Resistance:
Method: ASTM D2794
Result: 100 in. lb.

Dry Heat Resistance:
Method: ASTM D2485
Result: 200°F, 250°F intermittent

Exterior Durability:
Method: 2 years at 45° South
Result: Excellent, 87% gloss retention

Flexibility: (urethane only)
Method: ASTM D522, 180° bend, 1/4" mandrel
Result: Passes

Humidity Resistance:
Method: ASTM D4585, 100°F, 2000 hours
Result: No blistering, cracking, softening or delamination

Pencil Hardness:
Method: ASTM D3363
Result: 6H

Salt Fog Resistance:
Method: ASTM B117, 1000 hours
Result: Rating 10 per ASTM D610 for rusting, less than 1/16" creepage at scribe. No blistering, cracking, softening, or delamination of the film.

Meets the requirements of SSPC Paint No. 36, Levels 2 & 3

Polyurethane 5.23

continued on back

104A02A.EPS

Industrial & Marine Coatings

SHERWIN WILLIAMS.

POLY-LON® 1900

POLYESTER POLYURETHANE

PART A	B65-500	SERIES
PART B	B65V500	HARDENER

PRODUCT INFORMATION

RECOMMENDED SYSTEMS

Steel:
1-2 cts. Epolon II Epoxy Primer @ 2.0 - 4.0 mils dft/ct
1-2 cts. Poly-Lon 1900 Polyester Polyurethane @ 2.0 - 3.0 mils dft/ct

Steel:
1 ct. Zinc Clad II Plus @ 3.0 - 5.0 mils dft
1 ct. Epolon II Epoxy Primer @ 2.0 - 4.0 mils dft
1-2 cts. Poly-Lon 1900 Polyester Polyurethane @ 2.0 - 3.0 mils dft/ct

Steel:
1 ct. Epoxy Mastic Aluminum II @ 6.0 mils dft
1-2 cts. Poly-Lon 1900 Polyester Polyurethane @ 2.0 - 3.0 mils dft/ct

Galvanizing:
1-2 cts. Epolon II Epoxy Primer @ 2.0 - 4.0 mils dft/ct
1-2 cts. Poly-Lon 1900 Polyester Polyurethane @ 2.0 - 3.0 mils dft/ct

Concrete/Masonry:
1 ct. Kem Cati-Coat HS Epoxy Filler/Sealer @ 10.0 - 20.0 mils dft
1-2 cts. Poly-Lon 1900 Polyester Polyurethane @ 2.0 - 3.0 mils dft/ct

The systems listed above are representative of the product's use. Other systems may be appropriate.

SURFACE PREPARATION

Surface must be clean, dry, and in sound condition. Remove all oil, dust, grease, dirt, loose rust, and other foreign material to ensure adequate adhesion.

Refer to product Application Bulletin for detailed surface preparation information.

Minimum recommended surface preparation:

* Iron & Steel:	SSPC-SP6/NACE 3, 1-2 mils profile
* Galvanizing:	SSPC-SP1
* Concrete & Masonry:	SSPC-SP13/NACE 6, or ICRI 03732, CSP 1-3

* Primer required

TINTING

Tint Part A with 844 Colorant at 200% tint strength. Five minutes minimum mixing on a mechanical shaker is required for complete mixing of color.

APPLICATION CONDITIONS

Temperature:	40°F minimum, 100°F maximum (air, surface, and material) At least 5°F above dew point
Relative humidity:	75% maximum

Refer to product Application Bulletin for detailed application information.

ORDERING INFORMATION

Packaging:	1 gallon mix:	4 gallon mix:
Part A:	.75 gallons	3 gallons
Part B:	1 quart	1 gallon
(premeasured components)		

Weight per gallon: 11.4 ± 0.2 lb mixed, may vary with color

SAFETY PRECAUTIONS

Refer to the MSDS sheet before use.

Published technical data and instructions are subject to change without notice. Contact your Sherwin-Williams representative for additional technical data and instructions.

DISCLAIMER

The information and recommendations set forth in this Product Data Sheet are based upon tests conducted by or on behalf of The Sherwin-Williams Company. Such information and recommendations set forth herein are subject to change and pertain to the product offered at the time of publication. Consult your Sherwin-Williams representative to obtain the most recent Product Data Information and Application Bulletin.

WARRANTY

The Sherwin-Williams Company warrants our products to be free of manufacturing defects in accord with applicable Sherwin-Williams quality control procedures. Liability for products proven defective, if any, is limited to replacement of the defective product or the refund of the purchase price paid for the defective product as determined by Sherwin-Williams. NO OTHER WARRANTY OR GUARANTEE OF ANY KIND IS MADE BY SHERWIN-WILLIAMS, EXPRESSED OR IMPLIED, STATUTORY, BY OPERATION OF LAW OR OTHERWISE, INCLUDING MERCHANTABILITY AND FITNESS FOR A PARTICULAR PURPOSE.

104A02B.EPS

Industrial & Marine Coatings

5.23A

POLY-LON® 1900

POLYESTER POLYURETHANE

PART A	B65-500	SERIES
PART B	B65V500	HARDENER

APPLICATION BULLETIN

Revised 6/05

SURFACE PREPARATION

Surface must be clean, dry, and in sound condition. Remove all oil, dust, grease, dirt, loose rust, and other foreign material to ensure adequate adhesion.

Iron & Steel

Remove all oil and grease from surface by Solvent Cleaning per SSPC-SP1. Minimum surface preparation is Commercial Blast Cleaning per SSPC-SP6/NACE 3. For better performance, use Near White Metal Blast Cleaning per SSPC-SP10/NACE 2. Blast clean all surfaces using a sharp, angular abrasive for optimum surface profile (1-2 mils). Prime any bare steel the same day as it is cleaned or before flash rusting occurs.

Galvanized Steel

Allow to weather a minimum of six months prior to coating. Solvent Clean per SSPC-SP1. When weathering is not possible, or the surface has been treated with chromates or silicates, first Solvent Clean per SSPC-SP1 and apply a test patch. Allow paint to dry at least one week before testing adhesion. If adhesion is poor, brush blasting per SSPC-SP7 is necessary to remove these treatments. Rusty galvanizing requires a minimum of Hand Tool Cleaning per SSPC-SP2, prime the area the same day as cleaned or before flash rusting occurs.

Poured Concrete

New

For surface preparation, refer to SSPC-SP13/NACE 6, or ICRI 03732, CSP 1-3. Surfaces must be clean, dry, sound and offer sufficient profile to achieve adequate adhesion. Minimum substrate cure is 28 days at 75°F. Remove all form release agents, curing compounds, salts, efflorescence, laitance, and other foreign matter by sandblasting, shotblasting, mechanical scarification, or suitable chemical means. Refer to ASTM D4260. Rinse thoroughly to achieve a final pH between 8.0 and 10.0. Allow to dry thoroughly prior to coating.

Old

Surface preparation is done in much the same manner as new concrete, however, if the concrete is contaminated with oils, grease, chemicals, etc., they must be removed by cleaning with a strong detergent. Refer to ASTM D4258. Form release agents, hardeners, etc. must be removed by sandblasting, shotblasting, mechanical scarification, or suitable chemical means. If surface deterioration presents an unacceptably rough surface, Kem Cati-Coat HS Epoxy Filler/Sealer is recommended to patch and resurface damaged concrete. Fill all cracks, voids and bugholes with ArmorSeal Crack Filler.

Always follow the standard methods listed below:

ASTM D4258 Standard Practice for Cleaning Concrete.
ASTM D4259 Standard Practice for Abrading Concrete.
ASTM D4260 Standard Practice for Etching Concrete.
ASTM F1869 Standard Test Method for Measuring Moisture Vapor Emission Rate of Concrete.
SSPC-SP 13/Nace 6 Surface Preparation of Concrete.
ICRI 03732, Concrete Surface Preparation

APPLICATION CONDITIONS

Temperature:	40°F minimum, 100°F maximum (air, surface, and material) At least 5°F above dew point
Relative humidity:	75% maximum

APPLICATION EQUIPMENT

The following is a guide. Changes in pressures and tip sizes may be needed for proper spray characteristics. Always purge spray equipment before use with listed reducer. Any reduction must be compatible with the existing environmental and application conditions.

Reducer/Clean Up .... Reducer #132, R7K132

Airless Spray

- Pressure .... 2400 - 3000 psi
- Hose .... 3/8" ID
- Tip .... .013" - .017"
- Filter .... 60 mesh
- Reduction .... As needed up to 10% by volume

Conventional Spray

- Gun .... Binks 95
- Cap 63P
- Tip 66
- Atomization Pressure ... 50 - 60 psi
- Fluid Pressure .... 20 - 30 psi
- Reduction .... As needed up to 10% by volume

Brush

- Brush .... Natural Bristle
- Reduction .... Not recommended

Roller

- Cover .... 1/4" woven with phenolic core
- Reduction .... Not recommended

If specific application equipment is not listed above, equivalent equipment may be substituted.

Polyurethane 5.23A

continued on back

104A02C.EPS

Industrial & Marine Coatings

5.23A

POLY-LON® 1900

POLYESTER POLYURETHANE

PART A	B65-500	SERIES
PART B	B65V500	HARDENER

APPLICATION BULLETIN

APPLICATION PROCEDURES

Surface preparation must be completed as indicated.

Mix contents of each component thoroughly with power agitation. Make certain no pigment remains on the bottom of the can. Then combine three parts by volume of Part A with one part by volume of Part B. Thoroughly agitate the mixture with power agitation. Allow the material to sweat-in as indicated. Re-stir before using.

If reducer is used, add only after both components have been thoroughly mixed, after sweat-in.

Apply paint at the recommended film thickness and spreading rate as indicated below:

Recommended Spreading Rate per coat:

Wet mils:	3.0 - 4.5
Dry mils:	2.0 - 3.0
Coverage:	360 - 545 sq ft/gal approximate

NOTE: Brush or roll application may require multiple coats to achieve maximum film thickness and uniformity of appearance.

Drying Schedule @ 3.0 mils wet @ 50% RH:

	@50°F	@ 77°F	@100°F
To touch:	16 hours	2 hours	30 minutes
To handle:	24 hours	10 hours	2 hours
To recoat:			
minimum:	24 hours	12 hours	2 hours
maximum:	3 days	48 hours	24 hours
To cure:	7 days	7 days	5 days

If maximum recoat time is exceeded, abrade surface before recoating. Drying time is temperature, humidity, and film thickness dependent.

Pot Life:	5 hours	4 hours	45 minutes

Sweat-in-Time: none required

Application of coating above maximum or below minimum recommended spreading rate may adversely affect coating performance.

PERFORMANCE TIPS

Stripe coat all crevices, welds, and sharp angles to prevent early failure in these areas.

When using spray application, use a 50% overlap with each pass of the gun to avoid holidays, bare areas, and pinholes. If necessary, cross spray at a right angle.

Spreading rates are calculated on volume solids and do not include an application loss factor due to surface profile, roughness or porosity of the surface, skill and technique of the applicator, method of application, various surface irregularities, material lost during mixing, spillage, overthinning, climatic conditions, and excessive film build.

Excessive reduction of material can affect film build, appearance, and adhesion.

Do not apply the material beyond recommended pot life.

Do not mix previously catalyzed material with new.

In order to avoid blockage of spray equipment, clean equipment before use or before periods of extended downtime with Reducer #132, R7K132.

Mixed coating is sensitive to water. Use water traps in all air lines. Moisture contact can reduce pot life and affect gloss and color.

E-Z Roll Urethane Defoamer is acceptable for use. See data page 5.99 for details.

Quik-Thane Urethane Accelerator is acceptable for use. See data page 5.97 for details.

Refer to Product Information sheet for additional performance characteristics and properties.

CLEAN UP INSTRUCTIONS

Clean spills and spatters immediately with Reducer #132, R7K132. Clean tools immediately after use with Reducer #132, R7K132. Follow manufacturer's safety recommendations when using any solvent.

SAFETY PRECAUTIONS

Refer to the MSDS sheet before use.

Published technical data and instructions are subject to change without notice. Contact your Sherwin-Williams representative for additional technical data and instructions.

DISCLAIMER

The information and recommendations set forth in this Product Data Sheet are based upon tests conducted by or on behalf of The Sherwin-Williams Company. Such information and recommendations set forth herein are subject to change and pertain to the product offered at the time of publication. Consult your Sherwin-Williams representative to obtain the most recent Product Data Information and Application Bulletin.

WARRANTY

The Sherwin-Williams Company warrants our products to be free of manufacturing defects in accord with applicable Sherwin-Williams quality control procedures. Liability for products proven defective, if any, is limited to replacement of the defective product or the refund of the purchase price paid for the defective product as determined by Sherwin-Williams. NO OTHER WARRANTY OR GUARANTEE OF ANY KIND IS MADE BY SHERWIN-WILLIAMS, EXPRESSED OR IMPLIED, STATUTORY, BY OPERATION OF LAW OR OTHERWISE, INCLUDING MERCHANTABILITY AND FITNESS FOR A PARTICULAR PURPOSE.

104A02D.EPS

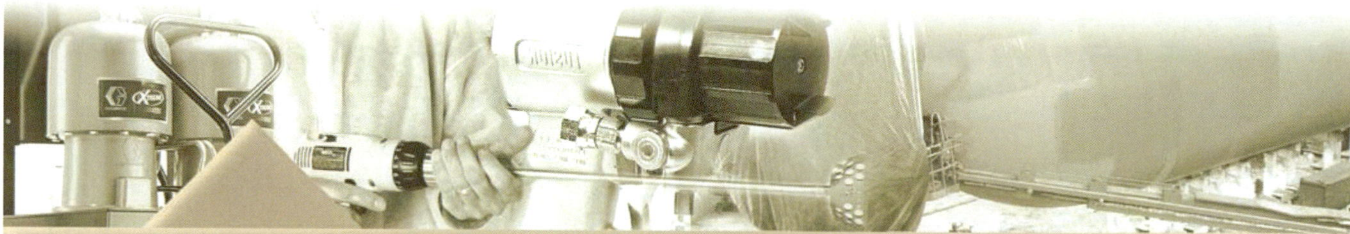

Profiles in Success

Richard McGinty

Project Manager
Pro-Tect Services Inc.
Houston, Texas

How did you choose a career in the industrial coatings field?
My stepfather was a painter. Good money kept me in the business. I came up through the field, started as a helper, painter, foreman, superintendent, general superintendent, and quality control. In the early 1990s I started working in safety and worked in corporate quality assurance and safety for many years. Now I am a project manager with a containment company.

What types of training have you been through?
Formal apprenticeship training in Houston with TOF Enterprises Inc.; NACE CIP and OSHA 500; taught at Brazosport Community College/Cooperative Education.

What kinds of work have you done in your career?
Foreman; Superintendent; Resident Engineer for CT DOT Bridge repair; Regional Quality Control and Safety; Corporate Quality Control and Safety; Expert Witness work; Overseas Applicator Trainer for Bechtel (Carboline); and Project Manager.

Tell us about your present job and what you like about it.
I need to have a sense of accomplishment. This job gives me an ongoing challenge! I am always dealing with different customers with interesting projects. I listen to my customers' problems and concerns. It gives me real pride to really understand their issues and come up with a good solution.

What factors have contributed most to your success?
Hard work and drive; the desire to move up and aspire. I take pride in my work and am not satisfied until I know that I have given each job my best effort. I am always looking forward. I keep focused on where I want to go. Each day I move closer to my goals. I regularly set new goals to keep motivated.

What advice would you give to those new to the field?
Work hard. Be safe. Don't be afraid to try new things and don't be afraid of mistakes. The key is to learn from your mistakes and not repeat them. Face the problems on the job head on and look at different solutions until you find the answer.

Always be safety conscious, and look out for your fellow workers. If you build a good reputation for safety, your company will look to you when challenging assignments are available. By taking care of those working around you, you show leadership and will be promoted.

Tell us about some of the accomplishments you have had in your career.
I was working for SIPCO Services and Marine Inc. and we were vacuum recovering grit out of supertankers. The process used PVC pipe which caused static electricity. The cost of injuries was enormous per year. I figured out a way to stop it. First, I tried grounding equipment using trial and error, but this wasn't solving the problem. Then I found non-conductive hoses that eliminated static build-up. But management was resistant to changing the process because it would increase costs. So I invited some professors from Texas A&M to visit and analyze the situation. They told management, "Your Richard McGinty has the answer."

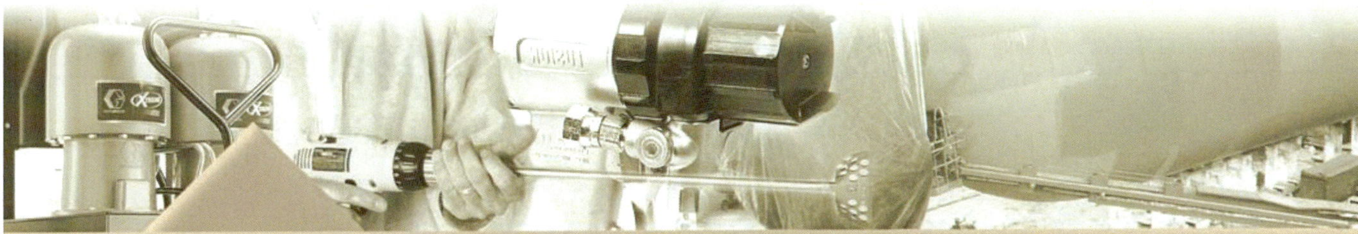

Additional Resources and References

Additional Resources and References

This module is intended to be a thorough resource for task training. The following reference works are suggested for further study. These are optional materials for continued education rather than for task training.

NACE International, www.nace.org

The Protective Coating User's Handbook, Dr. Louis D. Vincent. Houston, Texas: NACE International, 2004.

Figure Credits

Topaz Publications, Inc., 104F01, 104F19, 104F20

Painting and Decorating Contractors of America, 104F02, 104F04–104F06, 104F24, 104F26, 104F28, 104F30, 104F34, 104F44, 104F46, 104F47, 104F62, 104F63

EXTECH INSTRUMENTS, 104F03

Rogers & Associates Consulting Services, 104F07

Red Devil Inc., 104F08

Miracle Paint Rejuvenator Company, Inc., 104F09 (top)

Greg Londrigan, Hougan Mfg., 104F09 (bottom)

Graco, Inc., 104F10, 104F48, 104F49, 104F51, 104F52

Campbell Hausefeld Powered Products, 104F11

Photo courtesy of Paul N. Gardner Company, Inc., 104F11, 104F60

Norcross Corporation, 104F12, 104F13, 104F13 www.viscosity.com, 617-969-7020

Altex Coatings Limited, 104F14, 104F15

NACE International, 104F16, 104F56–104F59, 104T01, 104T02, 104T03

Berry Plastics Holding Corporation, 104F17

Northern Safety Co., Inc., 104F18

The Wooster Brush Company, 104F21, 104F29, 104F31, 104F32, 104F33, 104F37

© 2007 Bestt Liebco, used under permission, 104F22

Midwest Rake Company, LLC, 104F38, 104F39

ITW Finishing Innovations, 104F40, 104F41, 104F42, 104F43, 104F45, 104F50, 104F55

Praxair Surface Technologies, 104F54, 104F55

ITW Gema, 104F53

Shur-Line, 104F61

Elcometer, Inc., 104F60, 800-521-0635, www.elcometer.com

Sherwin-Williams Co., 104SA01, 104SA02

NCCER CURRICULA — USER UPDATE

NCCER makes every effort to keep its textbooks up-to-date and free of technical errors. We appreciate your help in this process. If you find an error, a typographical mistake, or an inaccuracy in NCCER's curricula, please fill out this form (or a photocopy), or complete the online form at **www.nccer.org/olf**. Be sure to include the exact module ID number, page number, a detailed description, and your recommended correction. Your input will be brought to the attention of the Authoring Team. Thank you for your assistance.

Instructors – If you have an idea for improving this textbook, or have found that additional materials were necessary to teach this module effectively, please let us know so that we may present your suggestions to the Authoring Team.

NCCER Product Development and Revision

13614 Progress Blvd., Alachua, FL 32615

Email: curriculum@nccer.org
Online: www.nccer.org/olf

❑ Trainee Guide ❑ Lesson Plans ❑ Exam ❑ PowerPoints Other ______________________

Craft / Level: Copyright Date:

Module ID Number / Title:

Section Number(s):

Description:

Recommended Correction:

Your Name:

Address:

Email: Phone:

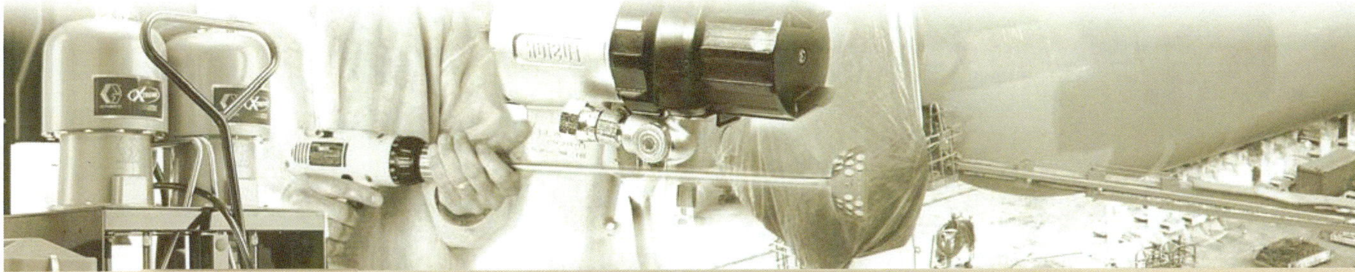

Industrial Coating and Lining Application Specialist Level One

69105-09

Health and Safety, Debris Management, Containment, and Ventilation

69105-09

Health and Safety, Debris Management, Containment, and Ventilation

Overview

Surface preparation, like abrasive blasting, produces a significant amount of dust and debris. This may involve large quantities of liquids, as is the case with waterjetting. In many cases, this debris contains hazardous materials. The control of dust, debris, and overspray is an important part of the job. Containment holds the dust and debris to a limited area where it can be collected. Once collected, it can be recycled or placed in containers for disposal. However, a containment system concentrates airborne particles in the work area. Ventilation circulates and filters the air to minimize the dust hazards and provide better visibility. Air monitoring programs verify containment and ventilation effectiveness. Containment and ventilation systems help keep unwanted materials from soiling surrounding areas. This makes cleanup easier and protects the environment.

This module addresses health and safety concerns, describes some common containment methods, and explains ventilation systems. It also describes equipment used to monitor the air for environmental protection and worker safety. These systems are required to comply with federal and state laws controlling work safety standards and environmental emissions.

Table of Contents

Figures

Tables

Objectives

When you have completed this module, you will be able to do the following:

1. Describe the required personal protective equipment (PPE).
2. List the reasons and methods for containment, ventilation, and dehumidification.
3. Name the industry standards for containment and ventilation.
4. Name containment classes.
5. List methods to verify containment effectiveness.
6. List consequences of containment failure.
7. List and demonstrate safety procedures associated with disturbing coatings containing toxic materials.
8. Discuss methods to stay safe while working in a containment area.

Trade Terms

Airlock
Permissible exposure limit (PEL)
Permit-required confined space

Required Trainee Materials

1. Pencil and paper
2. Appropriate personal protective equipment

Prerequisites

Before you begin this module, it is recommended that you successfully complete *Core Curriculum*, Modules 00101-04 and 00106-04; and *Industrial Coating and Lining Application Specialist Level One*, Modules 69101-09 through 69104-09.

This course map shows all of the modules in the first level of the Industrial Coating and Lining Application Specialist curriculum. The suggested training order begins at the bottom and proceeds up. Skill levels increase as you advance on the course map. The local Training Program Sponsor may adjust the training order.

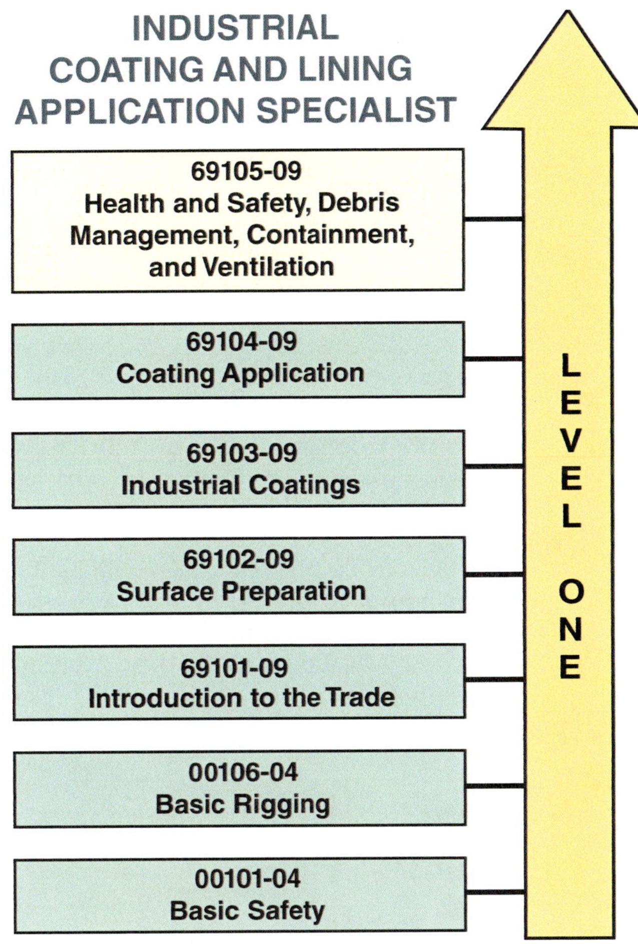

105CMAP.EPS

1.0.0 ◆ INTRODUCTION

Proper surface preparation is critical to successful coating applications. Some surface preparation methods, like abrasive blasting, produce dust and debris. This debris may contain hazardous chemicals. It is important to control debris, wastewater, dust, and overspray. If not controlled, these materials can pollute the environment, soil surrounding surfaces, and harm people. The use of personal protective equipment is vital to the coating professional. Further, a good containment system prevents the release of debris to the environment and collects wastes so they can be recycled or disposed of properly. In addition to good work practices, containment is required in many instances.

Containment confines the debris, dust, and chemical vapors to a limited area (*Figure 1*). The air within a containment zone must be treated to remove the dust and chemicals. A ventilation system circulates and filters the air. This reduces the potential for flammable or explosive conditions within the containment area and reduces dust so personnel can work more effectively. An air monitoring program checks the air quality inside the containment area to ensure that safe levels are maintained. In addition, air quality is checked outside of the containment area before surface preparation begins, to establish a baseline, and then during surface preparation operations to ensure that the containment system is effectively preventing the release of debris.

Many coatings contain toxic chemicals. Several laws prohibit the release of toxic chemicals to the environment on any job site. Fines and penalties for illegal release of toxic chemicals can be severe. If the job is located in an existing facility, it is important that coating operations do not affect other areas where people are working.

2.0.0 ◆ HEALTH AND SAFETY, AND PERSONAL PROTECTIVE EQUIPMENT

2.1.0 Health and Safety

The most common danger for coating professionals is physical accidents. Flying debris can cause injury, especially to the eyes and skin. In addition to vision loss, an eye injury can lead to other accidents, like falling from scaffolding or dropping equipment. A high concentration of dust can also impair your vision, increasing the chance of accidents. Appropriate personal protective equipment can protect your skin and eyes from harm.

105F01.EPS

Figure 1 ◆ Containment tent for water bridge.

Most solvents used during surface preparation operations are flammable, and many industrial coatings contain solvents. Many solvents contain volatile organic compounds (VOCs) that easily evaporate into the air and cause pollution. In addition, some of the resins and binders are also flammable. When a coating is in the liquid state, flames or sparks can set it on fire. You must prevent any possible source of flames or sparks from contacting solvents or liquid coatings. Containment is one way to prevent sparks or flames from getting into the area. Tools that can generate sparks and flames should never be used near solvents.

Solvents in coatings evaporate easily into the air. When some vapors reach a certain concentration, known as the lower explosive limit (LEL), a flame or spark can cause an explosion. Explosions can cause severe injury or death.

Working in a containment area or other enclosed space poses significant danger from solvent evaporation. Without air circulation, the concentration of solvent vapors in the air can increase rapidly. Ventilation and monitoring are key factors in preventing explosions from solvent vapors.

The primary hazard from solvents and other chemicals in industrial coatings is that they are toxic. Many solvents can harm the skin on contact. Xylene and organic acid in esters can cause skin irritation. Prolonged contact with turpentine can cause blisters. Organic alkalis or mineral acids can destroy the skin, leaving open sores.

Certain individuals are sensitive to some chemicals, such as those in epoxy coating and curing agents, and can develop a skin problem. This condition, known as dermatitis, results in scaling areas and blisters on the body, and may incapacitate a worker for weeks. Always wear protective clothing and gloves when handling chemicals.

Some people can develop allergies to solvents and coatings. Redness at the site where a solvent or coating has touched the skin, and sneezing, are signs of a mild allergic reaction. Breathing difficulty and swelling of the throat are signs of a severe life-threatening allergic reaction. Allergies usually develop after repeated exposure to the substance, so don't ignore any of these symptoms, even if you have worked with a coating or solvent before. Leave the work area and seek medical assistance immediately should you develop any symptoms of exposure.

In addition to skin contact, solvents enter the body through the lungs when vapors are mixed in the air. Exposure can lead to nausea, headaches, unconsciousness, convulsions, and even death. In addition to the immediate effects, toxic chemicals can build up in the body over time. Some chemicals that enter the bloodstream through the lungs are stored in the tissue before they can be eliminated. These chemicals injure the central nervous system, lungs, and kidneys.

Many hazards will have an immediate effect on worker health, such as exposure to excessive levels of carbon monoxide. Carbon monoxide can be deadly in a matter of minutes. Other hazards may not have any noticeable effect on the body until years after the exposure. For example, repeated exposure to noise has a cumulative effect rather than an immediate effect.

Part of worker responsibility for safety is to follow good safety practices while working. Report any unsafe conditions to your supervisor. Workers have an obligation to themselves, their families, and their co-workers to report any unsafe conditions.

To be a safe worker, make sure you're alert while working. This can be difficult when performing repeated tasks. The repetition can be monotonous and lull the worker into a trance-like state. To combat this, it is important that workers do the following:

- Get adequate rest. Caffeine is no substitute for sleep.
- Eat on a regular schedule so that your body has the fuel it needs to work and think.
- Avoid any prescription or over-the-counter medications that can cause drowsiness.
- Avoid any recreational drug use or excessive alcohol use when you are off duty.
- Never use drugs or alcohol on the job.

2.2.0 Protective Clothing and Personal Protective Equipment

When working around chemical or coating removal debris, workers need to wear protective clothing. Protective clothing not only protects work clothing, but also protects the skin from chemical splashes and debris contamination. The type of clothing required will depend on the operation. Be sure to check the material safety data sheet (MSDS) for specific information.

Additionally, take steps to protect your family's health. Dust particles that cling to clothing can contain materials that are hazardous to the health of people, especially children. Cleaning is not only important for the worker, but there are documented cases where workers brought contamination home, which caused health problems for family members. Protect your family by observing the following practices:

- Shower immediately after work. If workplace showers are available, take advantage of them.
- Change into clean clothing before socializing with the family.
- If possible, immediately launder work clothing after removing it. If this is not possible, keep work clothing separate from all other clothing and launder it separately.
- After laundering work clothing, run the washing machine through an empty cycle to rinse it of any residual debris.

PPE protects the worker from exposure to toxic materials, *i.e.*, lead or other heavy metals. After work is completed, the worker must be clean before eating, drinking, smoking, using the rest room, or applying cosmetics. Occupational Safety and Health Administration (OSHA) standards require the worker to at least remove dust and debris from exposed skin. Generally, a simple rinse with clean water will be effective. The wash water must be collected and may be considered hazardous waste.

During surface preparation or coating work, many different types of tools and chemicals are used. When working with chemicals, be certain to check the MSDS for specific information about the types of PPE necessary to wear. Remember that the primary way to prevent injury is to think first. It is important to stop and consider what type of accidents could happen before performing any job. Using common sense and knowing how to use PPE will greatly reduce the chances of injury.

The best PPE is of no use to workers unless these four rules are followed:

- Regularly inspect it.
- Properly care for it.
- Use it properly when it is needed.
- Never alter or modify it in any way.

PPE commonly used in industrial coating includes these items:

- Hard hats
- Eye protection
- Safety harnesses
- Gloves
- Safety shoes
- Hearing protection
- Respiratory protection

2.2.1 Hard Hats

The outer shell of a hard hat can protect the head from a hard blow. The webbing inside the hat maintains a space between the shell and the head. Adjust the headband so that the webbing fits and there is at least 1 inch (2.54 cm) of space between the head and the shell (*Figure 2*). Do not alter a hard hat in any other way.

2.2.2 Safety Glasses, Goggles, and Face Shields

Wear eye protection whenever there is even the slightest chance of an eye injury. Regular safety glasses will protect the worker from objects such as large chips, particles, sand, or dust flying at the worker from the front. Side shields are available for protection from the sides. In some cases, a face shield is necessary. Safety goggles give best protection from all directions (*Figure 3*).

2.2.3 Safety Harnesses

During surface preparation and coating work, workers will often need to work above ground and will need to wear a safety harness (*Figure 4*). Safety harnesses are heavy-duty harnesses that buckle around the body. They have leg, shoulder, chest, and pelvic straps. Safety harnesses have a D-ring attached to one end of a short section of rope, called a lanyard. The other end of the lanyard must be attached to a strong anchor point located above the work area. A qualified person will select the anchor point. The lanyard must be long enough to let the worker perform the job, but short enough to prevent a fall of more than 6 feet (1.8 m).

Use a safety harness and lanyard when you are working in the following situations:

- More than 6 feet (1.8 m) above ground or according to company policy
- Near a large opening in a floor
- Near a deep hole
- When there is a danger of falling on protruding rebar

A properly used safety harness can save a worker's life. Carefully inspect the harness prior to each use. Check that the buckles and D-ring are not bent or deeply scratched. Check the harness for any cuts or rough spots. If any damage is discovered, turn in the harness for testing or replacement.

2.2.4 Gloves

On many industrial coating jobs, workers must wear heavy-duty gloves to protect their hands (*Figure 5*). Work gloves are usually made of cloth, canvas, or leather. Never wear cloth gloves around rotating or moving equipment.

Gloves help prevent cuts and scrapes when you handle sharp or rough materials. Heat-resistant gloves are sometimes used for handling hot materials. Latex or other nonporous gloves protect worker hands from chemicals, paints, or other coatings. Choose the right gloves for the job using the MSDS.

Inspect work gloves before putting them on. Replace gloves when they become worn, torn, or soaked with oil or chemicals. When you're using latex or other nonporous gloves, you may want to wear two pairs of gloves. In doing this, when the outer pair becomes soiled, you can remove them without getting chemicals or debris on your skin.

2.2.5 Safety Shoes

The best shoes to wear on an industrial coating site are steel-toed, puncture-resistant safety shoes. Steel-toed boots protect toes from falling objects. Some boots also have a steel plate that keeps nails and other sharp objects from puncturing the foot. The next best footwear material is heavy leather. Never wear canvas shoes or sandals on a job site. Canvas shoes can catch fire easily and they do not protect the worker from sharp objects. Always replace boots or shoes when the sole tread becomes worn or the shoes have holes, even if the holes are on top.

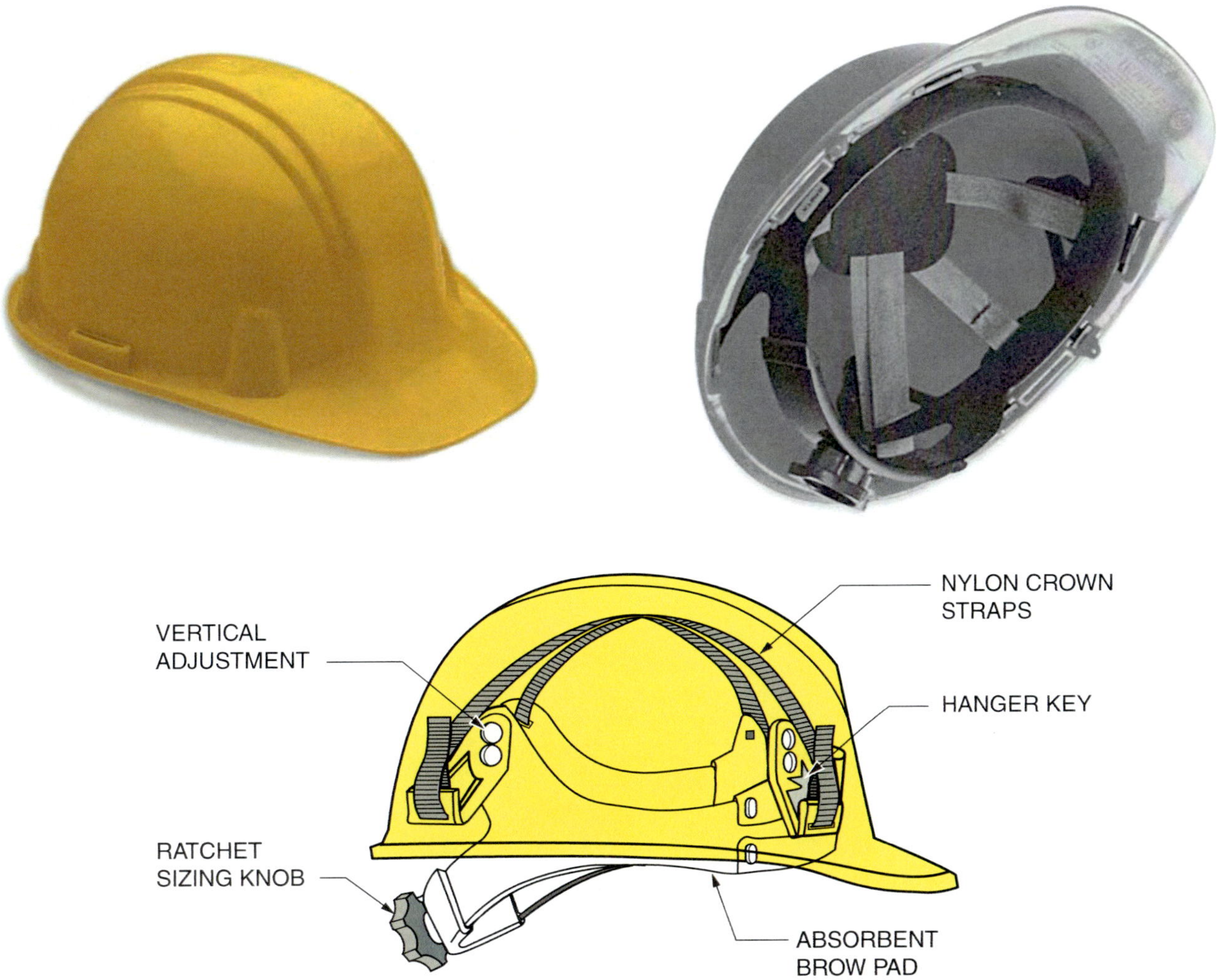

105F02.EPS

Figure 2 ◆ Typical hard hat.

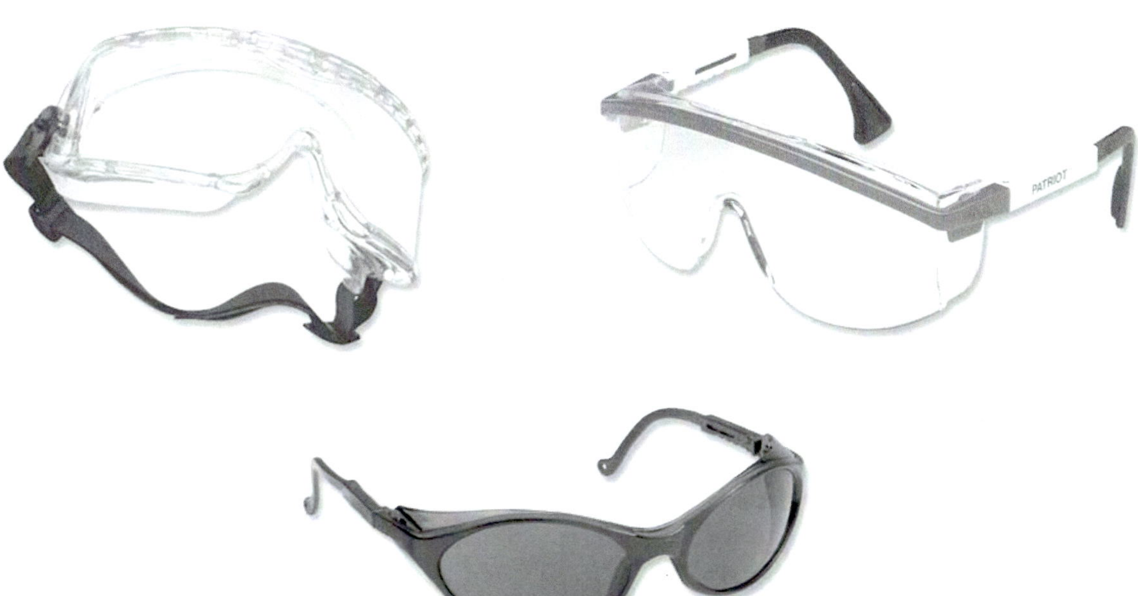

105F03.EPS

Figure 3 ◆ Typical safety goggles and glasses.

2.2.6 Hearing Protection

Exposure to loud noise over a long period of time can cause hearing loss, even if the noise is not loud enough to cause pain.

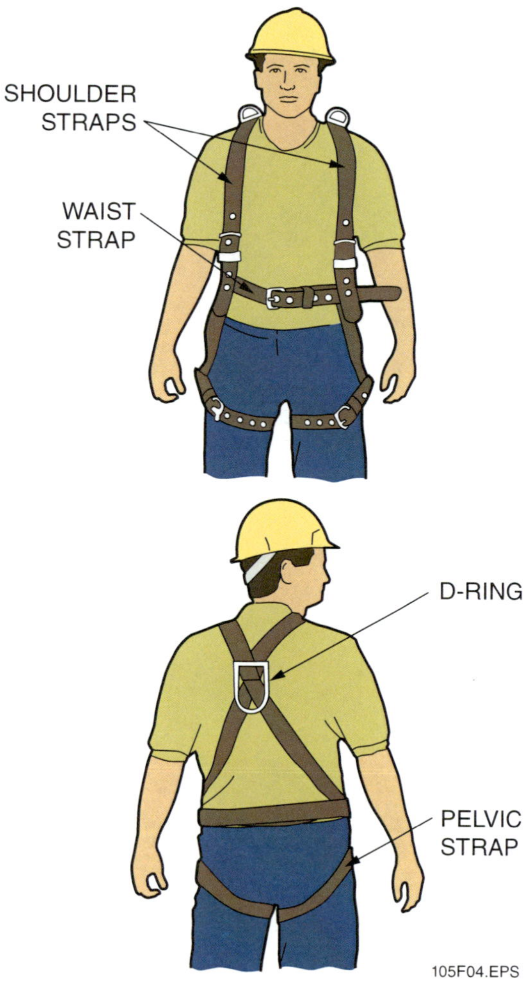

Figure 4 ♦ Typical safety harness.

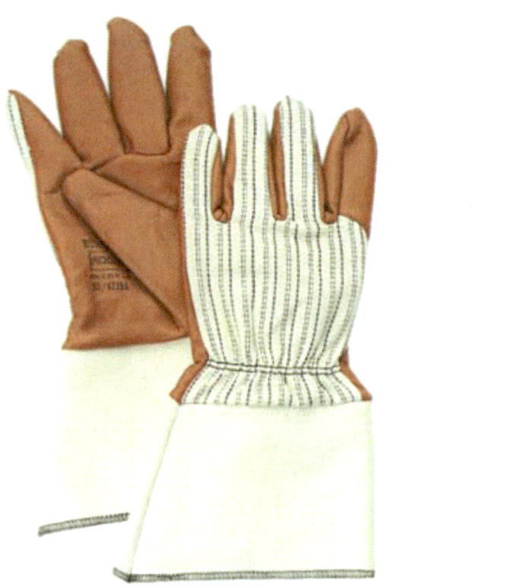

Figure 5 ♦ Work gloves.

Most industrial companies follow OSHA rules in deciding when hearing protection must be used. Specially designed earplugs that fit into your ears and filter out noise are one type of hearing protection (*Figure 6*). You need to clean earplugs regularly with soap and water to prevent ear infections.

Earmuffs are another type of hearing protection. Earmuffs are large, padded covers for the entire ear. You must adjust the headband on earmuffs for a snug fit. If the noise level is very high, you may need to wear both earplugs and earmuffs.

Hearing loss caused by noise can be prevented by limiting exposure and using appropriate PPE. Note that even at noise levels of 90 decibels, repeated unprotected exposure will result in hearing loss over the long term. Short, very loud noises, like those from impact tools, can do the most harm in the short term. Use this rule: if loud voices need to be used to speak to someone three feet (.9 m) away to hear you, the site is too noisy and hearing protection should be worn.

2.2.7 Respiratory Program and Protection

The proper respirator must be chosen for whatever contaminant is present and its concentration level. The respirator must fit properly and be used according to the manufacturer's instructions. It must be worn during all exposure times.

An employer must have a respiratory protection program with the following components:

- Standard operating procedures for selection and use
- Employee training
- Regular cleaning and disinfecting
- Sanitary storage
- Regular inspection
- Annual fit testing
- Pulmonary function testing

Coating professionals are responsible for wearing respiratory protection when needed. In certain concentrations, vapors or fumes can be eliminated by using air-purifying devices as long as

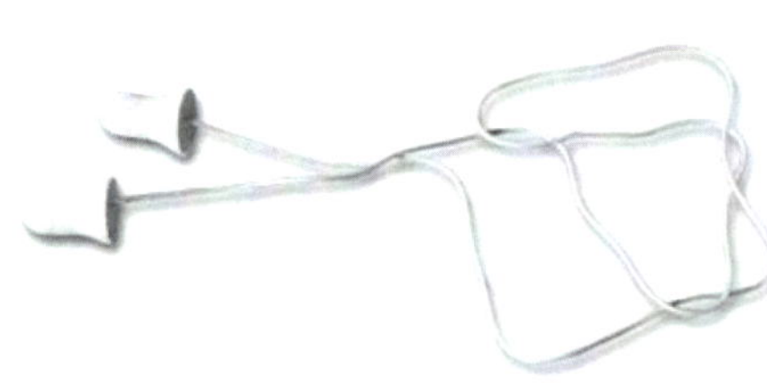

Figure 6 ♦ Ear plugs for hearing protection.

the oxygen levels are acceptable. Always check the cartridge on the respirator to ensure it is the correct type for the air conditions and contaminants found on the job site.

When selecting a respirator, coating professionals must first determine what hazardous ingredients are contained in the material and what exposure level is dangerous, and then choose the proper respirator to protect at those levels. Always read the product's MSDS. It identifies the hazardous ingredients and should list the type of respirator and cartridge recommended for use with the product.

All half-mask (air-purifying) respirators have limitations which include the following:

- They do not completely eliminate exposure to contaminants, but they will reduce the exposure to an acceptable level.
- They do not supply oxygen and must not be used in areas where the oxygen level is below 19.5 percent.
- They must not be used in areas where chemicals are not easily detected (poor warning signs, such as no taste or odor).

Leave the area immediately, go to an area where the air is fresh, and seek any necessary assistance if the following conditions arise:

- Difficulty breathing
- Dizziness or nausea
- Development of a smell or taste the of the chemical
- Other noticeable effects of exposure

Positive and Negative Fit Checks

Respirators are useless unless properly fit-tested to each individual. To obtain the best protection from a respirator, coating professionals must perform positive and negative fit checks each time they use a respirator. These fit checks must be repeated until a good face seal is achieved. To perform a positive fit check, follow these steps:

Step 1 Adjust the face piece for the best fit, then adjust the head and neck straps to ensure good fit and comfort.

Step 2 Block the exhalation valve with your hand or other material.

Step 3 Breathe out into the mask.

Step 4 Check for air leaks around the edges of the face piece.

Step 5 If the face piece puffs out slightly for a few seconds, a good face seal has been obtained.

To perform a negative fit check, do the following:

Step 1 Block the inhalation valve with your hand or other material.

Step 2 Breathe in.

Step 3 Check for air leaks around the edges of the face piece.

Step 4 If the face piece caves in slightly for a few seconds, a good face seal has been obtained.

Respirator Inspection

To achieve proper protection, respirators must be inspected and complete with all parts. Respirators must be cleaned every day. Failure to do so will limit their effectiveness and offer little or no protection. Remember, only a clean and complete respirator will provide the worker with the necessary protection. Follow these general guidelines:

- Inspect the condition of the respirator before and after each use.
- Do not wear a respirator if the face piece is bent, or if it is worn and cracked, because a proper face shield cannot be achieved.
- Do not wear respirators if any part is missing.
- Replace worn straps or missing parts before use.
- Do not expose respirators to excessive heat or cold, chemicals, or sunlight.
- Clean and wash the respirator each day. Remove the cartridge and filter, hand wash the respirator using mild soap and a soft brush, and let it air dry overnight.
- Sanitize the respirator each week. Remove the cartridge and filter, and then soak the respirator in a sanitizing solution for at least two minutes. Thoroughly rinse with warm water and let it air dry overnight. Store the clean and sanitized respirator in its resealable plastic bag. Do not store the respirator face down. This will bend the face piece.

Types of Respirators

Federal law specifies which type of respirator to use for various breathing hazards. Respirators are grouped into three main types based on how they work to provide protection from contaminants. They are air-purifying respirators, supplied-air respirators (SAR), and self-contained breathing apparatus (SCBA).

Air-Purifying Respirators

Air-purifying respirators provide the lowest level of protection. They are made for use only in atmospheres that have enough oxygen to sustain life (at least 19.5 percent). Air-purifying respirators use special filters and cartridges to remove specific gases, vapors, and particles from the air. The respirator cartridges contain charcoal, which absorbs certain toxic vapors and gases. When you detect any taste or smell, it indicates the charcoal's absorption capacity has been reached. This means the cartridge can no longer remove the contaminants. The respirator filters remove particles such as dust, mists, and fumes by trapping them within the filter material. Filters should be changed at the recommended intervals. Depending on the contaminants, cartridges can be used alone or in combination with a filter/pre-filter and filter cover. Air-purifying respirators should be used for protection against only the types of contaminants listed on the filters and cartridges. Also, refer to the National Institute for Occupational Safety and Health (NIOSH) approval label affixed to each respirator carton and replacement filter/cartridge carton. Respirator manufacturers typically classify air-purifying respirators into four groups: no-maintenance, low-maintenance, reusable, and powered air-purifying respirators (PAPRs).

No-maintenance and low-maintenance respirators are typically used for residential or light commercial work that does not call for constant and heavy respirator use. No-maintenance respirators are typically half-mask respirators with permanently attached cartridges or filters. The entire respirator is discarded when the cartridges or filters are full. Low-maintenance respirators are generally half-mask respirators that use replaceable cartridges and filters. However, they are not designed for constant use.

Reusable respirators are made in half-mask and full-face piece styles (*Figure 7*). These respirators require the replacement of cartridges, filters, and respirator parts. They also require a complete respirator maintenance program.

Powered air-purifying respirators (PAPRs) are made in half-mask, full-face piece, and hood styles. They use battery-operated blowers to pull outside air through the cartridges and filters attached to the respirator. The blower motors can be either mask or belt-mounted. Depending on the cartridges used, they can filter particles, dust, fumes, and mists along with certain gases and vapors. Many models also have an audible and visual alarm that is activated when the airflow falls below the required minimum level. This feature gives an immediate alert when a filter is full or the battery charge is low.

105F07.EPS

Figure 7 ◆ Reusable air-purifying respirators.

Supplied-Air Respirators

Supplied-air respirators supply air for extended periods of time through a high-pressure hose connected to an external source of air, such as a compressor, compressed-air cylinder, or pump. They provide a higher level of protection in atmospheres where air-purifying respirators are not adequate. Supplied-air respirators are typically used in toxic atmospheres. Some can be used in atmospheres that are immediately dangerous to life and health (IDLH) as long as they are equipped with an air cylinder for emergency escape. An atmosphere is considered IDLH if it poses an immediate hazard to life or produces immediate, irreversible, and debilitating effects on health. There are two types of supplied-air respirators: continuous-flow and pressure-demand.

A continuous-flow supplied-air respirator provides air in a constant stream. One or two hoses are used to deliver the air from the air source to the face piece. Unless the compressor or pump is specially designed to filter the air, or a portable air-filtering system is used, the unit must be located where there is breathable air. Continuous-flow respirators are made with tight-fitting half-masks or full-face pieces. They are also made with hoods. The air flow can be adjusted either at the air source (fixed flow) or on the unit's regulator (adjustable flow).

The pressure-demand supplied-air respirator is similar to the continuous flow type, except that it supplies air to the face piece by a pressure-demand valve that responds when fresh air is needed. It typically has a two-position exhalation valve that allows the wearer to switch between the pressure-demand and negative-pressure modes to make it easier to go in and out of a work area and move around within it.

Self-Contained Breathing Apparatus (SCBA)

SCBAs provide the highest level of respiratory protection. They can be used in oxygen-deficient atmospheres (below 19.5 percent oxygen), in poorly ventilated or confined spaces, and in IDLH atmospheres. These respirators provide about 30 to 60 minutes of air from a compressed-air cylinder worn on the back. An emergency escape breathing apparatus (EEBA) is a smaller version of a SCBA cylinder. EEBAs are used to escape from hazardous environments and generally provide a five- to ten-minute supply of air (*Figure 8*).

2.3.0 Additional Safety Requirements

WARNING!
Use a respirator when there is a danger of a respiration hazard or suffocation.

2.3.1 Water Cleaning Safety

In addition to the standard equipment, if operators work in the area of a water cleaning operation they may need to wear waterproof suits, rubber boots, head protection, and protective goggles or a full-face shield. Some operators must wear protective foot and shin guards, gloves, and a full-face shield. The waterproof clothing worn for this work is not sufficient to stop a direct swipe with a high-pressure water jet.

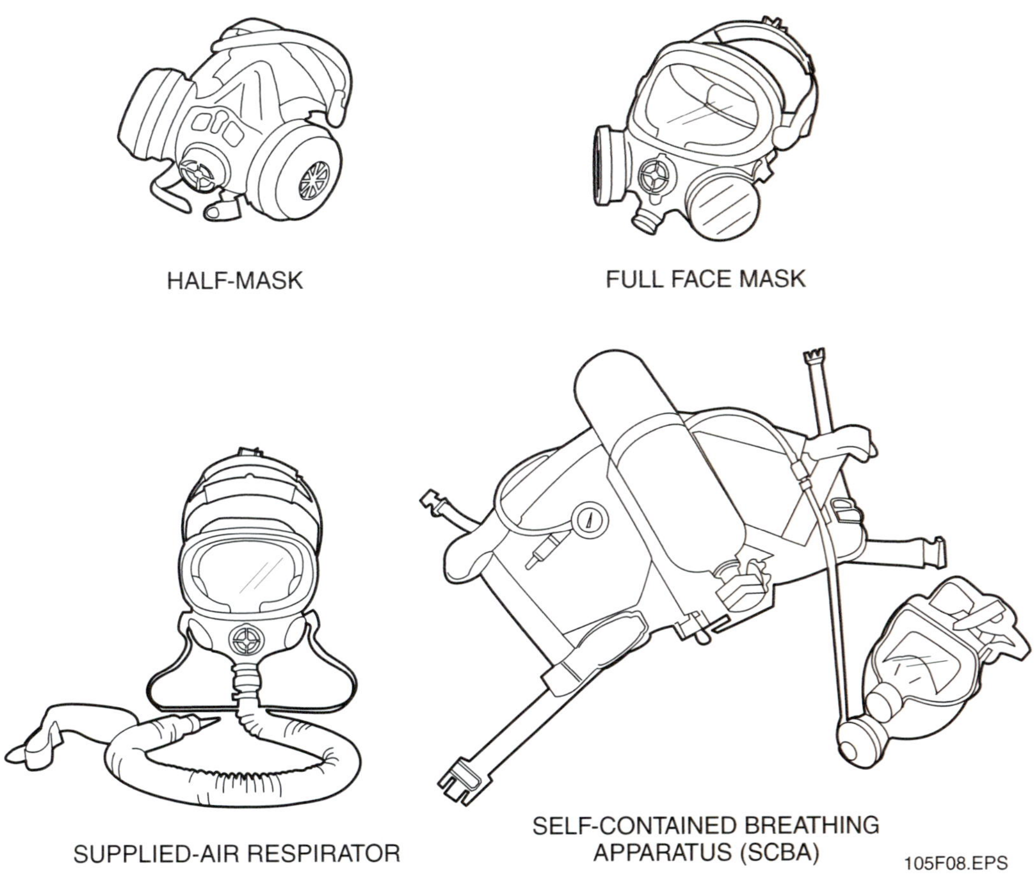

Figure 8 ◆ Respirators.

The sound level of a waterjetting operation is frequently much louder than 90 decibels, so hearing protectors or earplugs are required. In addition, if working where toxic substances are being removed from surfaces, breathing equipment or masks may be necessary.

2.3.2 Confined Spaces

Containment structures are considered confined spaces, which mean spaces that are large enough work in but have limited means of entry or exit. Confined spaces present unique hazards to workers. Not only are they often difficult to access but they may have limited lighting and ventilation, making it dangerous for any work to be performed. In addition, industrial coating work can expose workers to dust and chemical vapors.

When a work area is both a confined space and a place where workers are exposed to health or safety hazards, the work area is called a **permit-required confined space**. Workers must have a written authorization to enter a confined space. In addition, workers must be informed of the potential hazards involved with working in the space, the precautions to be taken, and what to do in the case of an emergency.

Workers must never enter a permit-required confined space until trained to do so. As mentioned previously, any confined spaces require special ventilation systems that remove dust and chemical vapors from the containment atmosphere. Because it is impossible to fully remove all harmful atmospheric contaminants from the work area, workers will most likely need to wear some type of respirator and protective clothing (*Figure 9*). Confined spaces are often monitored for things that are harmful to workers or the coating operation, such as oxygen and carbon monoxide levels, temperature, humidity, volatile organic compounds (VOC), etc.

Always know worker roles when working in any confined space. Don't take anything for granted. Questions should be asked and answered before entering a confined space. Know the location of exits and any monitoring devices.

2.3.3 Decontamination

Personal protective equipment protects the worker from exposure to toxic materials while working. After the work is completed, workers must decontaminate themselves before eating, drinking, smoking, using the rest room, or applying cosmetics. OSHA standards require workers to at least remove dust and debris from any exposed skin. Generally, a simple rinse with clean water will be effective. The wash water must be collected and may be considered hazardous waste.

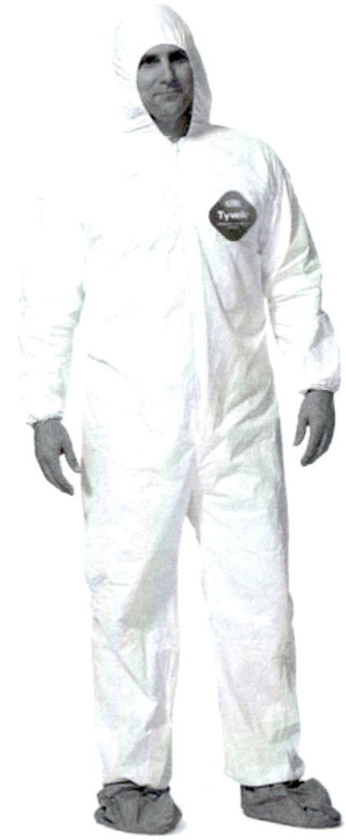

105F09.EPS

Figure 9 ◆ Protective clothing.

There are various types of portable facilities designed to decontaminate workers and equipment. Portable hand wash stations are available as individual units or four-person stations. Individual showers are designed as walk-through units (*Figure 10*). One end is placed at the exit of the containment area. The worker walks through, cleans off, and walks out into the clean zone. This confines the dust and debris to a limited area.

2.3.4 Hazardous Communication

An industrial coating professional will be exposed to a wide variety of chemicals that can be hazardous. Workers need to take steps to protect themselves from these hazards. The Occupational Safety and Health Administration (OSHA) requires that all employers educate their employees about hazardous chemicals at the job site as part of a Hazardous Communication (HazCom) program.

This program requires that all hazardous chemicals at a workplace be clearly identified. It also requires that suppliers of hazardous chemicals provide an MSDS for each hazardous substance in a shipment. This information must be available to all workers at the work site (*Figure 11*).

The employer is required to supply workers with any safety equipment that is needed in order to work around hazardous chemicals. But it is the worker's responsibility to check the MSDS and evaluate the work environment to determine the necessity of protective equipment.

All MSDSs do not look alike, so it will take some practice to quickly find pertinent information. OSHA requires that each MSDS provide the following information:

- Product identification
- Physical and chemical characteristics
- Physical hazards
- Health hazards
- Routes of entry or exposure
- Exposure limits
- Cancer-causing potential
- Precautions for handling and use
- Emergency and first aid procedures
- Date the MSDS was prepared or changed
- Preparer name and where additional information can be found

The MSDS details the product's hazards, and will clearly list the types of recommended PPE and the conditions under which it is needed.

2.4.0 Federal Laws and Industry Standards

There are several environmental and safety regulations that apply to surface coating operations. The use of toxic chemicals is regulated by state and federal laws. Moreover, the release of chemicals into the environment through spills or waste disposal is strictly controlled. Federal laws determine the legality or illegality of discharges to the air, water, and land. Certain wastes, especially those containing toxic chemicals like lead, cadmium, or chromium, must be taken to approved facilities for disposal. These wastes must be tracked using a hazardous waste manifest. Workers are protected by both OSHA and Right-to-Know laws. State and local codes mirror federal law and also control disposal of waste to the sewer system. All of these laws restrict chemical releases, and while they do not specifically mandate containment, the penalties for failure to contain toxins are severe.

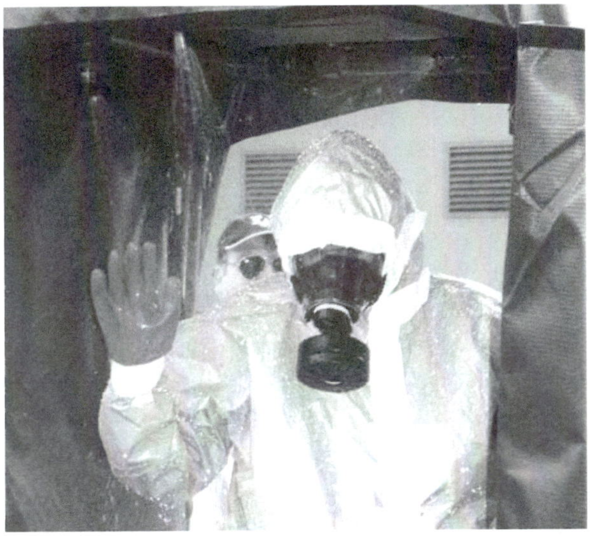

105F10.EPS

Figure 10 ◆ Portable shower.

105F11.EPS

Figure 11 ◆ MSDS compliance center.

Coatings that contain toxic or regulated chemicals are more strictly controlled than non-toxic coatings. However, even if non-toxic coatings are being applied, coatings removed during surface preparation may contain toxins. State laws reflect federal laws, and while they may be more stringent, they can never reverse federal law. Highlights of the federal laws that may affect surface preparation and coating operations include the following:

- *Clean Air Act* – Regulates the emissions of potentially hazardous air pollutants including lead, volatile organic compounds, and particulates, and requires the use of best management practices for the prevention of air pollution.
- *Clean Water Act* – Regulates most discharges into bodies of water to maintain water quality standards.
- *Comprehensive Environmental Response Compensation and Liability Act* – Defines a spill as a release into the environment of more than the reportable quantity of certain chemicals, including lead, benzene, and toluene.

- *Toxic Substances Control Act* – Covers asbestos, polychlorinated biphenyls (PCBs), and lead-based paints.
- *Resource Conservation and Recovery Act* – Establishes a cradle-to-grave control program for waste containing certain chemicals including lead, chromium, and solvents.
- *Emergency Planning and Community Right-to-Know Act* – Requires businesses to report on toxic chemical emissions, report releases of extremely hazardous chemicals, and disclose the quantities and types of toxic chemicals stored on site.
- *Occupational Safety and Health Act* – Establishes worker safety standards on chemical exposures, including lead and solvents, physical hazards, personal protective equipment, and other safety regulations.

2.4.1 Industry Standards

Most of the time, the requirements to install containment and ventilation are mandated by OSHA, the U.S. Environmental Protection Agency (EPA), state and local agencies, and business owners.

Standards are guidelines published by government and non-government organizations that describe procedures and methods for performing certain tasks. When a government agency such as OSHA or the EPA sets guidelines and standards, they are often based on law and must be strictly followed.

Professional organizations like NACE International, the Society for Protective Coatings (SSPC), and the American National Standards Institute (ANSI) also issue standards that make recommendations about all aspects of coating operations. Some standards are jointly issued by two or more organizations. Although these standards are not enforceable law, they become the accepted procedures in the industry. Contracts may specify best management practices or refer to specific industry standards.

This module focuses on the containment and ventilation requirements detailed in *SSPC Technology Guide No. 6, Guide for Containing Surface Preparation Debris Generated During Paint Removal Operations.* ANSI, the American Industrial Hygiene Association (AIHA), and the National Fire Protection Association (NFPA) also produce standards on many different topics. These organizations address the need for containment and ventilation in both surface preparation and spray application activities. Standards that address containment and ventilation in surface coating applications include the following:

- *ANSI/AIHA Z9.4 1997*, Abrasive Blasting Operations – Ventilation and Safe Practice
- *ANSI/AIHA Z9.6 1999*, American National Standard – Exhaust Systems for Grinding, Polishing, and Buffing
- *ANSI/AIHA Z9.2-2001*, American National Standard for Fundamentals Governing the Design and Operation of Local Exhaust Ventilation Systems
- *ANSI/AIHA Z9.3-1994*, Standard for Spray Finishing Operations – Safety Code for Design, Construction, and Ventilation
- *NFPA 33 1995*, Standard for Spray Application Using Flammable or Combustible Materials

Companies will need to determine the applicable standards and laws that apply to their operations. The greatest need for containment and ventilation tends to be during surface preparation activities, especially abrasive blasting.

2.4.2 Worker Protection Standards

Containment and ventilation systems serve to protect workers and the environment. A containment and ventilation system must minimize worker exposure to airborne contaminants and provide sufficient visibility within the enclosure so workers can safety perform their jobs. A containment and ventilation system must also adequately contain dust and debris created during surface preparation operations to prevent its release into the environment.

OSHA sets enforceable standards to protect employee health in many different areas, including exposure to toxic chemicals. The standards limit exposure to hazardous conditions and toxic chemicals. Overall there is a general duty for employers to provide a safe workplace. Additionally, OSHA limits a worker's exposure to airborne toxins by setting a **permissible exposure limit (PEL)**. Ventilation and respiratory protection are required in work areas with a chemical concentration above the PEL. OSHA has set limits on worker exposure to solvents, lead, cadmium, chromium, and other toxic chemicals. Regulations regarding lead exposure can be found in *OSHA Standard 1910.1025 App A* at www.osha.gov.

One of the primary concerns in coating operations is the surface preparation of structures previously coated with lead or other toxic metals. Containment confines dust, debris, and airborne toxins to the work area. In some circumstances, worker safety standards require ventilation and dust handling controls, respirator use, and air monitoring.

There are specific safety measures for the removal of lead-based paints. The *OSHA Technical Manual, Chapter 3,* details the requirements for controlling lead exposure in the construction industry. It is available online and provides guidance on engineering and work practice controls to limit exposure to lead. The engineering controls include isolation and ventilation. Isolation methods contain lead contamination and protect nearby work areas, bystanders, and the environment. However, erection of a sealed containment structure increases the exposure of the workers within the structure. Ventilation is required to reduce exposure. Ventilation methods include exhaust ventilation systems with filtration of exhaust air, dust collection shrouds on power tools, and vacuum blast nozzles.

Work practice controls guide the way the task is performed. These include good housekeeping, personal hygiene, and administration. Good housekeeping practices include regular cleaning to remove accumulated lead dust. Personal hygiene practices prevent workers from bringing contaminants out of the isolation area. This includes separate decontamination areas, washing facilities, eating facilities, showers, and changing areas. Administrative controls include training, supervision, and scheduling to limit exposure.

2.5.0 Surface Preparation Methods

There are four types of surface preparation methods:

- Abrasive blast cleaning
- Water blasting or waterjetting
- Chemical stripping
- Power tool and hand tool cleaning (with and without shrouds)

During surface preparation operations, workers will use two of the methods described above. Each method creates different types debris and represents different types of hazards to workers and the environment.

2.5.1 Abrasive Blast Cleaning

The most common method of surface preparation is dry abrasive blasting. This method is typically used on metals, but can be used on other surfaces. Abrasive blasting applies to a stream of small abrasive particles under pressure to a surface, removing rust, mill scale or other contaminants. This is commonly referred to as sand blasting, but sand is seldom used as an abrasive today because it contains a substance called free silica that is released during blasting. Free silica is dangerous to human health.

Abrasive blasting creates debris by removing old coatings, which can contain hazardous materials, such as lead. In addition, this type of surface preparation operation can generate waste from the abrasive particles that are used. Sometimes the abrasives are collected, cleaned, and reused so the containment system must support this operation. If the abrasive is not reused, the containment system must collect the abrasive.

Abrasive blasting creates a lot of dust, making it difficult for workers to breathe and see. Abrasive blasters are required to wear protective clothing and hoods with respirators. Containment and ventilation systems must keep the air clean.

2.5.2 Water Blasting or Waterjetting

Waterjetting is the use of a high-pressure water stream, with or without chemicals, to clean surfaces. It is also used to strip existing coatings. This method may produce debris from old coatings and will generate a huge wastewater stream. Wastewater contaminated with coating debris and/or chemicals must be captured and treated.

This method has advantages over dry abrasive blasting, especially with respect to worker air quality. With waterjetting, respiratory requirements may be more flexible than with other methods of surface preparation. However, waterjetting has its own hazards. Even water under low pressure is powerful enough to scrape skin. Water under high pressure can be lethal.

2.5.3 Chemical Stripping

Solvent cleaning removes all visible oil, grease, soil, dust, and other contaminants from surfaces. This type of cleaning uses detergents, solvents, acids, water, and other agents to clean surfaces. This method can generate sludge from old coatings, as well as solvent waste. Some cleaning agents generate hazardous fumes during the chemical stripping process, so ventilation is necessary. Whether a containment and ventilation system is used or not, always consult the MSDS for any products used. It will list hazards, precautions, and the types of PPE workers need to use when applying the product.

2.5.4 Hand and Power Tool Cleaning

Hand and power tool cleaning uses hand tools and electric or air powered tools to prepare surfaces. This method of cleaning creates a dangerous environment because coating chips and dust fly through the air. The containment and ventilation system used during this method of cleaning must control and contain flying debris, dust, and coating chips, as well as maintain visibility.

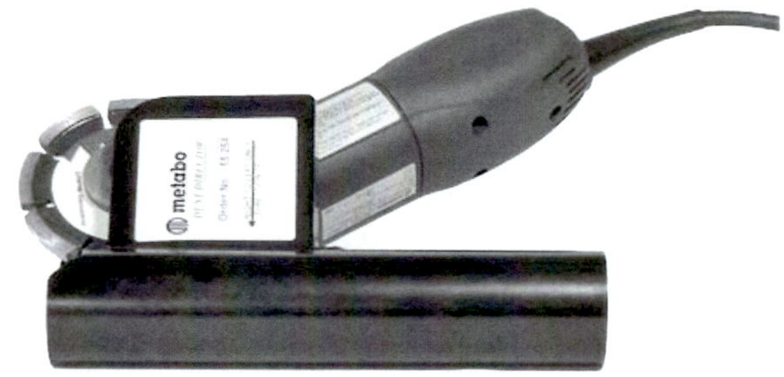

105F12.EPS

Figure 12 ◆ Grinder with a vacuum shroud.

3.0.0 ◆ DEBRIS MANAGEMENT, CONTAINMENT, AND VENTILATION

3.1.0 Debris Collection Methods

Surface preparation operations create a lot of debris that needs to be collected for recycling or disposal. There are two types of debris collection methods: localized and general.

3.1.1 Localized Collection at Point of Cleaning

For localized collection at the point of cleaning, coating removal and debris collection are simultaneous. This collection method has the advantage of significantly decreasing airborne dust and debris, and works well on large open structures and horizontal flat structures, such as decks.

In localized collection, the coating removal tools have a shroud around the working head (*Figure 12*), so debris is contained at the surface where it is generated. Masks, brushes, and rubber seals are used to seal the shroud to the surface being cleaned. While working with hand and power tools, the vacuum system will most likely be a stand-alone unit. During dry abrasive blasting, the vacuum system may be attached to the blast gun or to the abrasive recovery system.

3.1.2 General Area Collection

For general area cleaning, debris is removed from the containment area rather than at the point of cleaning. Two primary methods for general area cleaning are channeling and bulk collection. In a channeling system, debris is continuously removed from the containment area. Bulk collection is performed periodically.

Channeling systems use gravity and a series of tarps, funnels, panels, and other materials. The debris falls into the containment system where it is channeled to the collection area. If the blast medium is to be reused, the materials are channeled to a recovery or recycling unit. The channeling system may also be independent of the containment structure.

The channeling system is made of materials appropriate to the debris being collected. For example, systems for the collection of wastewater from waterjetting must be impermeable to water. Channeling systems for chemical stripping must be made of chemically resistant materials.

Bulk collection requires regular gathering of debris from the containment area. This can be done manually by sweeping and shoveling the debris into suitable containers. It may also be done with equipment, such as a vacuum, vacuum trucks, or conveyors. If the blast media is to be reused, the collected material is treated in a recovery or recycling unit.

3.2.0 Containment and Ventilation

A good containment and ventilation system can prevent contamination of the local area and the environment, human exposure to coatings or waste products, and long-term health problems.

The primary reasons for using containment systems are:

- Prevention of debris from entering the environment
- Prevention of air, soil, and water contamination
- Collection of rust and old paint produced from surface preparation activities
- Compliance with strict federal, state and local laws

Surface preparation methods like abrasive blasting produce a significant amount of dust and debris. This may involve substantial quantities of liquids, as is the case with waterjetting. In many cases, this debris contains hazardous materials and chemical vapors. Controlling the dust, debris, overspray, and chemical vapors is an important part of the job. Containment confines the hazards to a limited area where it can be collected. Once collected, it can be recycled or placed in containers for disposal.

However, a containment system concentrates airborne particles or chemical vapors in the work area. For this reason, ventilation is used to circulate and filter the air to minimize the dust hazards and provide better visibility, as well as to reduce the chance of flammable or explosive conditions. Air monitoring programs are used to verify containment of the hazards, check the ventilation's effectiveness, and ensure that safe levels are maintained. Containment and ventilation systems also help keep unwanted materials from soiling surrounding areas. Air quality outside of the containment area should be checked before work begins to establish a baseline. This makes cleanup easier and protects the environment.

Not only is containment a good work practice, it is required in many instances since many coatings contain toxic chemicals. Several laws prohibit the release of toxic chemicals to the environment on any job site. Fines and penalties for illegal release of toxic chemicals can be severe. If the job is located inside a facility where other people are working, it is important that coating operations do not affect those areas.

The following section describes some common containment methods and ventilation systems. It also describes equipment used to monitor the air for environmental protection and worker safety. These systems are required to comply with federal and state laws controlling work safety standards and environmental emissions.

3.2.1 Containment System Classes

Containment systems are classified depending on the degree of emissions control required. Class 1 provides for the greatest control, and Class 4 the lowest. Within each Class, the standard specifies containment and ventilation requirements for different surface preparation methods.

Surface preparation methods are designated by letter as follows:

- A – Abrasive blast cleaning
- C – Chemical stripping
- W – Water blasting or waterjetting (with and without abrasives or vacuum shrouds)
- P – Power tool and hand tool cleaning (with and without shrouds)

For example, Class 2W containment requirements are for a high level of emissions control for a waterjetting operation.

Ventilation systems have several purposes. One is to help prevent the release of emissions into the atmosphere. Another is to provide a workable environment for personnel with or without additional respiratory equipment. Ventilation systems also balance the air pressure between the inside of the containment structure and the outside environment to prevent the containment structure from collapsing.

Class 1 Containment

Class 1 provides the highest level of emissions control and ventilation. Class 1 containment systems for various surface preparation techniques are shown in *Table 1*. Class 1 ventilation requirements are shown in *Table 2*.

Class 2 Containment

Class 2 provides a high level of emissions control. Class 2 containment systems for various surface preparation techniques are shown in *Table 3*. Class 2 ventilation requirements are shown in *Table 4*.

Class 3 Containment

Class 3 provides a moderate level of emissions control. Class 3 containment systems for various surface preparation techniques are shown in *Table 5*. Class 3 ventilation requirements are shown in *Table 6*.

Class 4 Containment

Class 4 provides the lowest level of emissions control. Class 4 containment systems for various surface preparation techniques are shown in *Table 7*. Class 4 ventilation requirements are shown in *Table 8*. Note that Class 4 containment is not applicable for chemical stripping and hand and power tool cleaning.

Table 1 Class 1 Containment

Surface Preparation Method	Containment Material	Penetrability	Support Structure	Joints	Entryway
Abrasive	Rigid or flexible	Air impenetrable	Rigid or flexible	Full seal	Airlook or resealable
Water	Rigid or flexible	Air impenetrable and water impermeable	Rigid or flexible	Full seal	Resealable
Power/hand tools*	Rigid or flexible	Air impenetrable	Rigid or flexible	Full seal	Resealable
Chemical	Rigid or flexible	Air impenetrable and water impermeable and chemical resistant	Rigid or flexible	Full seal	Overlap

*Ground covers or hanging tarps may be used when using hand tools or vacuum shrouded power tools.

105T01.EPS

Table 2 Class 1 Ventilation Requirements

Surface Preparation Method	Air Make-up	Input Air Flow	Air Pressure	Air Movement	Exhaust Dust Filteration
Abrasive	Controlled	Forced or natural	Instrument verification	Minimum specified	Air filteration
Water	Open	Forced or natural	Visual verification	Minimum specified	Air filteration
Power/hand tools*	Controlled	Forced or natural	Visual verification	Minimum specified	Air filteration
Chemical	Open	Forced or natural	Not required	Not specified	Air filteration

*Ground covers or hanging tarps may be used when using hand tools or vacuum shrouded power tools.

105T02.EPS

Table 3 Class 2 Containment

Surface Preparation Method	Containment Material	Penetrability	Support Structure	Joints	Entryway
Abrasive	Rigid or flexible	Air impenetrable	Rigid or flexible	Full seal	Resealable or overlap
Water	Rigid or flexible	Water impermeable	Rigid or flexible	Full seal	Overlap
Power/hand tools*	Rigid or flexible	Air impenetrable or air penetrable	Rigid or flexible	Full seal	Overlap or open seam
Chemical	Rigid or flexible	Water impermeable and chemical resistant	Rigid or flexible	Full seal	Overlap

*Ground covers or hanging tarps may be used when using hand tools or vacuum shrouded power tools.

105T03.EPS

Table 4 Class 2 Ventilation Requirements

Surface Preparation Method	Air Make-up	Input Air Flow	Air Pressure	Air Movement	Exhaust Dust Filteration
Abrasive	Controlled or open	Forced or natural	Visual verification	Minimum specified	Air filteration
Water	Open	Natural	Not required	Not specified	Not required
Power/hand tools*	Open	Natural	Not required	Not specified	Not required
Chemical	Open	Natural	Not required	Not specified	Not required

*Ground covers or hanging tarps may be used when using hand tools or vacuum shrouded power tools.

105T04.EPS

Table 5 Class 3 Containment

Surface Preparation Method	Containment Material	Penetrability	Support Structure	Joints	Entryway
Abrasive	Rigid or flexible	Air impenetrable or air penetrable	Rigid or flexible	Full seal or partial seal	Overlap or open seam
Water	Rigid or flexible	Water impermeable	Minimal	Partial seal	Open seam
Power/hand tools*	Rigid or flexible	Air penetrable	Minimal	Partial seal	Overlap or open seam
Chemical	Rigid or flexible	Water impermeable and chemical resistant	Rigid or flexible	Full seal	Open seam

*Ground covers or hanging tarps may be used when using hand tools or vacuum shrouded power tools.

105T05.EPS

Table 6 Class 3 Ventilation Requirements

Surface Preparation Method	Air Make-up	Input Air Flow	Air Pressure	Air Movement	Exhaust Dust Filteration
Abrasive	Controlled or open	Forced or natural	Not required	Not specified	Air filteration
Water	Open	Natural	Not required	Not specified	Not required
Power/hand tools*	Open	Natural	Not required	Not specified	Not required
Chemical	Open	Natural	Not required	Not specified	Not required

*Ground covers or hanging tarps may be used when using hand tools or vacuum shrouded power tools.

105T06.EPS

Table 7 Class 4 Containment

Surface Preparation Method	Containment Material	Penetrability	Support Structure	Joints	Entryway
Abrasive	Rigid or flexible	Air penetrable	Minimal	Partial seal	Open seam
Water	Rigid or flexible	Water permeable	Flexible	Partial seal	Open seam

105T07.EPS

Table 8 Class 4 Ventilation Requirements

Surface Preparation Method	Air Make-up	Input Air Flow	Air Pressure	Air Movement	Exhaust Dust Filteration
Abrasive	Open	Natural	Not required	Not specified	No control on exhaust
Water	Open	Natural	Not required	Not specified	Not required

105T08.EPS

3.2.2 Containment Enclosure Components

There are several aspects to consider when constructing a containment enclosure. These include the containment material, material penetrability, support structures, joints, and entryways. *SSPC Technology Guide No. 6, Guide for Containing Surface Preparation Debris Generated During Paint Removal Operations*, refers to these components by type. This section briefly describes each aspect of a containment system. A containment system requires structural stability and strength, so a registered professional engineer must be consulted during the design and before the erection of this type of system.

Containment Material (Type A) and Material Penetrability (Type B)

Containment materials make up the walls, ceilings, and floors of a containment system. They are classified as rigid (Type Al), such as plywood, aluminum, rigid metal, plastic, or similar materials, or flexible (Type A2), such as screens, tarps, drapes, plastic sheeting, or similar materials. They are further classified by air penetrability (Types B1, B2a and B2b), water permeability (Types B3a and B3b), and chemical resistance (Type B4). Flexible materials have the advantage of easily containing oddly shaped and protruding areas.

Support Structures (Type C)

The support structures are the framework upon which the containment material is mounted. These structures are classified as rigid (Type C1), flexible (Type C2), and minimal (Type C3).

Rigid structures permit no movement and may consist of scaffolding or framework to which the containment material is attached. This type of structure needs to stay in place until the surface preparation, and possibly application jobs, are finished. Type C support structures need to be able to withstand the typical weather conditions at the job site.

Flexible structures allow a small amount of movement and may consist of cables, chains, and similar materials. The cables or chains are usually affixed to the ground and to a point above the area being cleaned. This allows the containment structure to be raised and lowered as necessary for weather conditions.

Minimal structures provide only enough structure to attach the containment material and can consist of cables and ropes.

Joints (Type D)

Effective containment requires sealing the joints between the containment materials, as well as the joints between the containment material and the supporting structure. Joints are classified as fully sealed (Type D1) or partially sealed (Type D2). Fully sealed joints require all seams to be overlapped and completely sealed with stitching, caulking, tape, heat, or other devices. Partially sealed joints require the seams to be overlapped and that all seams be attached so that they will not easily come apart, but they don't need to be completely sealed.

Entryways (Type E)

Entryways provide a means by which workers may enter and leave a containment and ventilation system. All entryways must be accessible from both directions when in use—that is, not locked so that the door can be opened from only one side. Entryways are classified as through an **airlock** with resealable doors (Type E1), through a resealable door (Type E2), through overlapping door tarps (Type E3), and through open seams (Type E4).

Entryways through an airlock with resealable doors (Type El) provide the most containment. An airlock is a sealed space with two doors. One door accesses the outside environment, while the other accesses the containment area. The doors must be far enough apart that one person cannot open both at the same time. The doors must be capable of being repeatedly used while providing a good seal.

Entryways through a resealable door (Type E2) have a door that must be capable of being repeatedly used while providing a good seal, but without an airlock.

Entryways through overlapping door tarps (Type E3) do not have doors. Instead, tarps are hung over the entry, but the tarps are huge with multiple folds and layers to reduce the chances of dust escaping the containment area. Entryways through open seams (Type E4) provide the least amount of containment. Entry is gained through unsealed seams in the containment material.

3.2.3 Ventilation

The primary reasons for using ventilation include:

- Creation of air movement in the containment area or confined space

- Improved visibility for the worker and provides clean air for breathing
- Aid in keeping debris from escaping

Dehumidification is removing moisture vapor from the air to lower than its dew point. This helps control the work environment, and helps prevent steel corrosion and flash rusting.

Environmental (ambient) conditions, such as humidity and temperature, can either benefit or harm surface preparation and coating operations. For example, the appropriate conditions can help a coating cure better and more quickly, which gets the equipment or area back into service more quickly. These ambient conditions affect even the long-term performance of coatings.

As mentioned previously, containment is a method to minimize or prevent materials from soiling areas surrounding the work area. A good containment system prevents materials from entering the environment and aids in the collection of debris for disposal. A containment system includes tarps and screens held in place by scaffolds and other supports, as well as ground covers, water booms, and equipment shrouds.

The ventilation systems used to create air movement through the work area can include both mechanical and natural ventilation. Fans, hoods, filters, dust collectors and ducts are parts of a mechanical ventilation system.

3.2.4 Containment Ventilation System Components

Ventilation system components include ducts, baffles, louvers, flap seals, valves, filters, and fans. These components are used to maintain airflow through the containment system and to control the pressure difference between the interior and exterior at the containment area. The ventilation system controls the air makeup, input airflow, air pressure internal to the containment system, and air movement through the containment area. In addition, it is used to control exhaust dust filtration.

Adequate ventilation is critical to the safety of workers in the containment structure. The area may be monitored for temperature, humidity, VOC and carbon monoxide levels. Ventilation may also be important to the surface preparation or coating operation. Many metals will rust almost immediately when their bare surfaces are exposed to high levels of humidity. All coatings have optimum temperature and humidity conditions for application. Just as with the containment components, *SSPC Technology Guide No. 6, Guide for Containing Surface Preparation Debris Generated During Paint Removal Operations*, refers to ventilation components by type.

Air Makeup (Type F)

Air makeup refers to how the air supply is input into the containment system. Controlled (Type Fl) uses baffles, louvers, filters, and similar devices to prevent debris from escaping the containment system. Open (Type F2) uses entry points that are not protected by any devices.

Input Airflow (Type G)

Air input flow refers to how the air supply flows into the containment area. Air input flow can be forced (Type Gl) or natural (Type G2). Forced airflow uses variable speed fans and blowers to force air into the containment area. When air is forced in, it increases air pressure in the structure and this can force contaminants out of the confinement area. With forced input airflow, it is important to maintain a negative air pressure in the confinement area by balancing the input airflow with the exhausted air. Natural airflow does not use any devices to input air into the containment area.

Air Pressure (Type H)

Air pressure refers to the air pressure on the inside of the structure relative to the air pressure on the outside of the structure. A slightly negative air pressure inside the structure is necessary to help prevent contaminants from escaping the confinement area. Too much negative pressure inside flexible material and flexible support structures can collapse the containment area.

Air pressure may be verified with instruments (Type H1) or visually (H2). To visually verify negative pressure, first note the wind conditions outside the structure, and then look at the material on the walls, floors, and ceilings of the containment structure. When they are all concave (curved in) there is negative pressure in the structure. When some are concave but others are convex (pushed out), the effect is most likely due to wind.

When there is no requirement to verify negative pressure, air pressure is designated as Type H3.

Air Movement (Type I)

Air movement inside the containment area is needed for worker safety and to prevent debris from being released into the environment. When air movement is specified (Type 11), it is usually given as the minimum velocity of air in a downdraft, or a minimum volume of air per minute for a specified volume of containment space. Velocity is specified in feet per minute (fpm); volume is specified in cubic feet per minute (cfm). When velocity is specified, it refers to the velocity of airflow near workers. When there is no special requirement for air movement, none will be specified (Type 12).

Exhaust and Dust Filtration (Type J)

Exhaust and dust filtration refers to the outflow of air and debris from the containment area. When air is exhausted with either fans or blowers (Type J1), or naturally (Type J2), it is difficult to control debris without filters. When fans and blowers are used, debris can be directed into wet or dry debris collectors.

3.2.5 Containment Effectiveness

Protrusions

Effective containment requires sealing the joints between the containment materials, as well as the joints between the containment material and the supporting structure. Joints are classified as fully sealed (Type D1) or partially sealed (Type D2). Fully sealed joints require all seams to be overlapped and completely sealed with stitching, caulking, tape, heat, or other devices. Partially sealed joints require the seams to be overlapped and attached so that they will not easily come apart. However, they do not need to be completely sealed.

Emissions Control

Emissions control is critical to protecting workers, the environment, and other people living and working in the area of coating operations. It is important to assess the degree to which containment measures are effective. To do this, the air is monitored during the surface preparation operation. The water and soil are tested before the job begins and after the job is finished. *SSPC Technology Guide No. 6, Guide for Containing Surface Preparation Debris Generated During Paint Removal Operations*, refers to emissions testing and monitoring by various methods.

Visible Emissions and Opacity Scale (Method A)

In Method A of emissions assessments, a trained certified observer will periodically check the containment area for visible emissions. The observer will rate the opacity (cloudiness) or concentration of any emissions on a scale from 0 to 100 percent. The contract or project specification will state the allowable opacity of emissions in terms of time. In addition, visible emissions will be rated by the frequency with which they occur. The frequency is rated by level:

- Level 0 – No visible emissions
- Level 1 – Random emissions occurring for not more than 5 minutes of an 8-hour workday
- Level 2 – Random emissions occurring for not more than 24 minutes of an 8-hour workday
- Level 3 – Random emissions occurring for not more than 48 minutes of an 8-hour workday
- Level 4 – No restriction on frequency of time of emissions

Ambient Air Monitoring for PM-10 (Method B)

In Method A of emissions assessments, the emissions are monitored for particles not greater than 10 micrometers in size. To accurately measure particulate emissions during surface preparation, air monitoring must be performed before work begins to establish ambient particulate matter for the area.

Occupational Monitoring for Regulated Area Emissions for Lead (Method C)

When performing surface preparation operations on surfaces that have been previously coated with lead-based material, the air outside of the containment area must be monitored. When levels reach 30 micrograms of particulate matter per cubic meter of air, OSHA requires that the area be isolated and posted to keep unsuspecting workers out of the area. Hazardous waste disposal will be discussed in more detail later in this module.

EPA Ambient Air Monitoring for Toxic Metals (Method D)

When performing surface preparation operations on surfaces that have been previously coated with toxic metal-based material, the air near the operation must be monitored for the presence of the metals. The EPA establishes safe exposure levels for each metal.

Soil Analysis for Toxic Metals (Method E)

Method E is used to determine whether a containment system was effective after the job has been completed. It involves analyzing soils for the presence of toxic metals. Soil samples must be taken at various areas in and around the work area both before and after the surface preparation operation.

Water and Sediment Analysis for Toxic Metals (Method F)

Method F is used to determine whether a containment system was effective after the job has been completed. It involves analyzing water and sediment for the presence of toxic metals. Samples must be taken at various areas in and around the work area both before and after the surface preparation operation.

Visual Assessment of the Work Site (Method G)

Method G assessment will be done most often—at least once daily. This is a visual inspection of the work site to evaluate its cleanliness. The employer is responsible for leaving a work site in at least as good condition as it was found. Before work begins, the area is inspected to locate any existing debris. Existing debris should be removed from the site before work begins. If it is not, its location should be noted to avoid disagreements over responsibility when the work is complete. While work is being performed, the site is inspected daily for debris that could indicate leaks in the containment system. Once the job is complete, the area is inspected again to ensure that there is no new debris on the site.

4.0.0 ◆ CONTAINMENT FAILURE AND HAZARDOUS DEBRIS

Containment and ventilation systems can fail on the inside of the containment area (failing to remove dust from the air), or on the outside (leaking abrasives onto the ground). When containment and ventilation systems fail, the results can be immediately obvious (abrasive dust and coating debris flying through the air), or they can be unnoticeable (lead-based coating debris settling onto nearby soil).

When containment systems fail, it can result in the dumping of unsightly but non-hazardous debris on the ground and dust in the air. This type of failure, while costly to clean up, is preferable to a failure involving hazardous wastes. Often, containment and ventilation systems are used when surface preparation work involves removing coatings that contain hazardous material, such as paints containing heavy metals such as lead, cadmium, and chromium. Coatings containing these metals and others like them must not be released into the environment because they are hazardous to human health. Heavy metals are not eliminated from the body, so they tend to accumulate in tissue. Children are especially vulnerable.

When hazardous debris is released into the environment, it can contaminate soil and water. Dust settling on nearby structures can affect the health of anyone who comes in contact with it. Therefore, steps must be taken to prevent the accidental release of toxic materials. When people are exposed to low levels of lead on a regular basis, they are at risk of developing chronic lead poisoning.

4.1.0 Lead Poisoning

Chronic overexposure to lead may have no symptoms at all. When symptoms do appear, severe irreversible organ damage may have already occurred. Lead damages blood-forming, nervous, and reproductive systems. It can cause severe kidney and brain damage. Common symptoms of chronic lead poisoning are loss of appetite, a metallic taste in the mouth, anxiety, constipation, nausea, pallor, overwhelming tiredness, weakness, insomnia, headaches, muscle and joint pain or soreness, muscle tremors, numbness, dizziness, hyperactivity, and severe abdominal pain.

Industrial coating professionals are at risk for both chronic and acute lead poisoning when working with lead-based paint debris—even in a containment area and while wearing a respirator. Common symptoms of acute lead poisoning are similar to chronic lead poisoning and many other illnesses. They include loss of appetite, nausea, vomiting, severe stomach cramps, constipation, difficulty in sleeping, fatigue, moodiness, headaches, joint or muscle aches, anemia, and decreased sexual drive. If you experience these symptoms after you have been exposed to toxic paint debris, see a doctor. The doctor will perform a blood test that measures the amount of the substance in the blood. Untreated lead poisoning in an adult damages the nervous system and can lead to death. In children and developing fetuses, it can lead to mental retardation, sterility, and death.

Containment structures reduce environmental pollution by capturing debris and used blasting materials. Unfortunately, they also increase worker exposure. Ventilation should be provided to reduce the airborne concentration of lead and increase visibility. Outside the structure, air must be monitored to ensure that the emissions do not contain an excessive amount of hazardous debris.

Workers can reduce their exposure to toxic substances by wearing disposable or washable protective clothing, changing from work clothing before leaving the work site, and showering and shampooing at the work site. It is also beneficial to launder work clothing separately from other clothing. Vehicles should be parked away from the work area.

To reduce the exposure to lead, wear protective clothing and a respirator when in the containment area. Wash your hands and face before eating, drinking, or smoking. Never eat, drink, or use tobacco products in the containment area. To alert others of lead-contaminated work areas, post warning signs (*Figure 13*).

Figure 13 ♦ Lead work area warning sign.

Review Questions

1. Flames or sparks can cause an explosion when certain vapors reach a concentration known as the _____.
   a. permissible exposure limit (PEL)
   b. lower explosive limit (LEL)
   c. upper explosive limit (UEL)
   d. lower exposure limit (LEL)

2. Fire and explosions are a danger during industrial coating operations for all of the following reasons *except* _____.
   a. solvents are flammable
   b. coatings contain solvents
   c. solvents evaporate
   d. solvents are toxic

3. Allergic reactions to a substance usually occur _____.
   a. on the first exposure
   b. after repeated exposures
   c. to a mild degree
   d. to a severe degree

4. The most common danger of surface preparation work in a contained area is/are _____.
   a. fire
   b. accidents
   c. explosions
   d. health hazards

5. When leaving a containment area, what is the minimum that a worker must wash?
   a. the face
   b. all exposed skin
   c. the face and hands
   d. the hands

6. All coatings contain toxic chemicals.
   a. True
   b. False

7. When working with chemicals, be sure to check the _____.
   a. PDS
   b. MSDS
   c. NIOSH
   d. OSHA

8. OSHA requires that _____.
   a. all hazardous chemicals stay locked up
   b. hazardous chemicals not be used
   c. all hazardous chemicals at a workplace be clearly identified
   d. hazardous chemicals only be used in containment

9. OSHA limits a worker's exposure to airborne toxins by setting a _____.
   a. permissible exposure limit
   b. upper inhaled threshold
   c. permissible ingested limit
   d. upper exposure threshold

10. All of the following are surface preparation methods *except* _____.
    a. asphalting
    b. chemical stripping
    c. hand tool cleaning
    d. water blasting

11. The most common method of industrial surface preparation is _____.
    a. chemical stripping
    b. dry abrasive blasting
    c. power tool cleaning
    d. asphalting

12. When is it acceptable for a worker to enter a confined space?
    a. When there is no odor from vapors.
    b. When wearing appropriate PPE.
    c. When there is written authorization.
    d. When working around non-toxic substances.

Review Questions

13. During chemical stripping operations, consult the MSDS for the product being used for _____.
    a. directions about using the product
    b. the type of respirator filter needed
    c. the product expiration date
    d. the product lot number

14. The containment class that provides for the greatest emissions control is Class _____.
    a. 1
    b. 4
    c. A
    d. M

15. All of these are classifications for support structures *except* for _____.
    a. rigid
    b. minimal
    c. maximum
    d. flexible

Summary

Surface preparation techniques, such as abrasive blasting, produce a large amount of dust and debris, while waterjetting produces a considerable quantity of wastewater. In many cases, the debris contains hazardous materials that could endanger public health if released into the environment. The control of dust and debris is an important part of the coating professional's job. Containment systems confine the dust and debris to a limited area where they can be collected. Once collected, they can be recycled or placed in containers for disposal. However, a containment system concentrates airborne particles in the work area, placing workers at risk. Ventilation circulates and filters the air to minimize the dust hazards and provide better visibility. Air monitoring programs verify containment and ventilation effectiveness. These systems help to protect workers and the environment.

Notes

Trade Terms Introduced in This Module

Airlock: An airtight compartment between the entry to a containment area and the entry to the outside that is designed to prevent the escape of dust and debris.

Permissible exposure limit (PEL): Regulatory limits on the amount or concentration of a substance in the air. They may also contain a skin designation.

Permit-required confined space: A confined space that has real or potential hazards including atmospheric, physical, electrical, or mechanical, defined by OSHA.

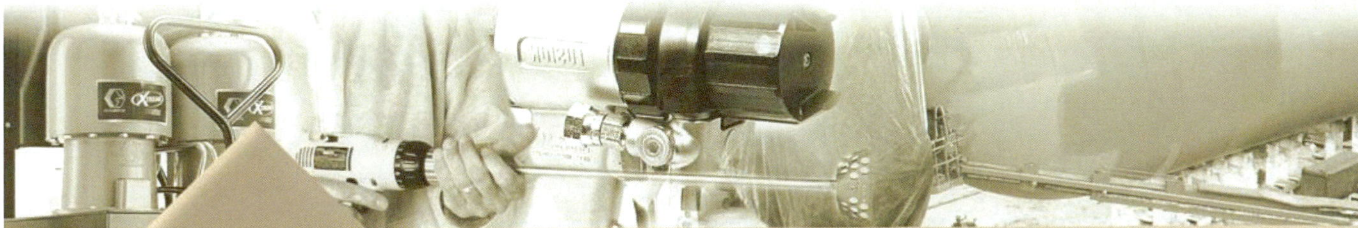

Additional Resources and References

Additional Resources

This module is intended to be a thorough resource for task training. The following reference works are suggested for further study. These are optional materials for continued education rather than for task training.

Preventing Lead Poisoning in Construction Workers, NIOSH ALERT: April 1992 DHHS (NIOSH) Publication No. 91-116a.

The Paint and Coatings Resource Center (PCRC), www.paintcenter.org

Coating Inspector Program Level 1 © NACE International, 2003, January 2007.

United States Environmental Protection Agency (7404), EPA Fact Sheet EPA-747-F-98-006, December 1998.

Figure Credits

Picture Courtesy of Detroit Tarp, Inc.,105F01

North Safety Products, 105F07

Painting and Decorating Contractors of America, 105F08

Dupont™ Building Innovations™, 105F05, 105F06, 105F09

Army Medicine, 105F03, 105F10

Metabo Corp., 105F02, 105F12

Datum Filing Systems, Inc., 105F11

Centers for Disease Control and Prevention, 105F13

NCCER CURRICULA — USER UPDATE

NCCER makes every effort to keep its textbooks up-to-date and free of technical errors. We appreciate your help in this process. If you find an error, a typographical mistake, or an inaccuracy in NCCER's curricula, please fill out this form (or a photocopy), or complete the online form at **www.nccer.org/olf**. Be sure to include the exact module ID number, page number, a detailed description, and your recommended correction. Your input will be brought to the attention of the Authoring Team. Thank you for your assistance.

Instructors – If you have an idea for improving this textbook, or have found that additional materials were necessary to teach this module effectively, please let us know so that we may present your suggestions to the Authoring Team.

NCCER Product Development and Revision

13614 Progress Blvd., Alachua, FL 32615

Email: curriculum@nccer.org
Online: www.nccer.org/olf

❑ Trainee Guide ❑ Lesson Plans ❑ Exam ❑ PowerPoints Other ________________

Craft / Level: Copyright Date:

Module ID Number / Title:

Section Number(s):

Description:

Recommended Correction:

Your Name:

Address:

Email: Phone:

Glossary of Trade Terms

Aggregate: Crushed rock or gravel that is screened to size for use in concrete.

Airlock: An airtight compartment between the entry to a containment area and the entry to the outside that is designed to prevent the escape of dust and debris.

Alloy: Often associated with metal. It indicates that another element has been added to a molten metal (such as carbon added to steel).

Arc: The flow of electrical current through a gas, such as air, from one pole to another.

Boxing: Intermixing the coating material if more than one container of a coating will be used.

Cathodic protection: The use of an electric current or a sacrificial anode to protect a metallic object from corrosion.

Coalescence: Curing property of acrylic, epoxy, and latex based coatings. The particles of resin pull together as the coating dries to form a hard coating.

Confined space: A work area large enough for a person to work, but arranged in such a way that an employee must physically enter the space to do work. A confined space has a limited or restricted means of entry and exit. It is not designed for continuous work. Tanks, vessels, silos, pits, vaults, and hoppers are examples of confined spaces.

Convertible coatings: Coatings that cure by a chemical reaction.

Corrosion: The deterioration of a material, usually a metal, that results from a reaction with its environment.

Corrosion inhibitors: Substances that can be added to the environment to retard or prevent corrosion.

Cutting in: Painting corners and the perimeter of windows, doors, and corners with a brush or trim roller before roller application of coating to walls or ceiling.

Diluent: Something added to a coating for the express purpose of diluting it. Solvent and water can be diluents when used to thin coatings.

Efflorescence: A fluffy white deposit on the surface of concrete that is caused by moisture that carries salt in it to rise to the surface.

Electrostatic spray: Coatings that use electrostatic charges of particles to be sprayed to coat an object.

Emulsify: To dissolve a fat or oil, usually with a detergent or other chemical, so that it can be washed away with water.

Feathering: Tapering off the edges of the coating by gradually pulling the brush bristles off the painted surface.

Fireproofing coatings: Coatings that, due to their composition, are fireproof or fire-retardant.

Flammable: Capable of igniting easily and burning rapidly; used to describe a fuel with a flash point below 100°F.

Flash rust: A type of rust that occurs after a steel surface has been exposed to moisture, either humidity in the air or water.

Galvanic corrosion: The electrochemical deterioration of a metal when two dissimilar metals are placed in contact with the presence of a liquid that can carry a current.

Ground: The conducting connection between electrical equipment or an electrical circuit and the earth.

Intumescent materials: Fireproof materials that turn into foam when exposed to heat.

Laitance: The accumulation of fine particles (usually salt) on the surface of freshly poured concrete. It is caused by water rising out of the concrete.

Lower explosive limit (LEL): Lowest concentration of a gas or vapor in air capable of producing a flash of fire in presence of an ignition source (arc, flame, heat). Concentrations lower than LEL are too low to burn.

Material safety data sheet (MSDS): A document that must accompany any hazardous material. The MSDS identifies the substance and details the exposure limits, the physical and chemical characteristics, the kind of hazard it presents, precautions for safe handling, storage, and use, and specific control and disposal issues.

Mill scale: The black or dark blue coating on steel that is sometimes scaly in appearance. It is a result of hot steel cooling in the presence of oxygen.

Glossary of Trade Terms

Mitting: A coating application technique in which the coating is applied with a hand mitt.

NACE International: An organization whose mission is to protect people, assets, and the environment from the effect of corrosion.

NACE-SSPC standards: Written industry codes that guide blasters, applicators, supervisors, and owners in the use of minimum best practices for successful protective coating systems. The written codes are voted on and approved by hundreds of industry experts.

NCCER: An organization whose primary goal is to standardize construction craft training throughout the country so both employers and employees will benefit from the training, no matter their location.

Non-convertible coatings: Air-drying coatings that dry by solvent evaporation with no change in the chemical nature of the binder. The coating remains soluble in the original solvent after drying.

Orifice: A small, specially shaped opening or hole in a spray or blast nozzle.

Overspray: Atomized coating particles that deflect from or miss the surface being sprayed. Also, spray particles that are not wet enough to fuse when they reach the surface.

Permissible exposure limit (PEL): Regulatory limits on the amount or concentration of a substance in the air. They may also contain a skin designation.

Permit-required confined space: A confined space that has real or potential hazards including atmospheric, physical, electrical, or mechanical, defined by OSHA.

Personal protective equipment (PPE): Equipment or clothing designed to prevent or reduce injuries, and/or contact with or inhalation of hazardous substances.

pH: The alkalinity or acidity of a substance.

Plural component spray: Coatings using several components mixed at the nozzle.

Polymerization: A chemical reaction that joins molecules to create long chains.

Porous: Indicates a material has many small holes that allows liquid (usually water) to pass.

Pot life: The limited amount of time available for thinning (when spraying) and/or application of a coating.

Powder coatings: Coatings applied as a powder and then treated to produce a film of coating.

Process control: A means of employing established work procedures, methods, and tools for improving quality and reducing variation.

Protective coatings: A coating defined by its protective, rather than aesthetic, properties. Protective coatings can include sacrificial, barrier, or inhibitive coatings.

Qualified person: A person who is trained to perform a critical or dangerous task. For example, a safety technician may be a qualified person to certify the safety of scaffolding.

Saponify: Turning to soap. For example, when added to grease, some caustics turn the grease into a soap that can be washed away with water.

Short term exposure limit (STEL): The highest concentration of an airborne material that workers may be exposed over a prescribed period of time.

Stripe coat: A coat of paint the covers the edges of something, often used in overcoating operations to paint over the edges of the old coating before the final coating is applied.

Subliming materials: Fireproof materials that change when heated from a solid directly to a gas without going through a liquid phase.

Surface profile: Graphic representation of a surface condition.

Sweat-in time: The time allowed for the catalyst, converter, hardener, or activator to start a chemical reaction before thinning and/or application.

Thermal spray: Coatings applied as a heated spray.

Thermoplastic: Characteristic of a coating that is pliable when cool and hardens as it is heated.

Glossary of Trade Terms

Thermosetting materials: Convertible coatings where heat causes the coating to harden by chemical reaction.

Threshold limit value (TLV): The level it is believed that a worker can be exposed to chemical substances day after day for a working lifetime without adverse health effects.

Time weighted average (TWA): Limit of amount of time a worker can be exposed to a material over a certain period, usually given in hours per day or week.

Transfer efficiency: The amount of coating that goes onto the surface compared to the total wasted due to overspray or bounceback.

Traps: Devices used to capture something. A water trap traps water.

Upper explosive limit (UEL): The highest amount of vapor concentration in the air that can explode. Vapor concentrations above the UEL displace so much oxygen that there is not enough oxygen in the air to explode.

Viscosity: The thickness of a fluid.

Vitreous coatings: Coatings that have the appearance of glass.

Volatile organic compounds (VOC): Organic chemical compounds that have high enough vapor pressures under normal conditions to significantly vaporize and enter the atmosphere.

Water-miscible: Indicates that a substance can be mixed with water.

Weathering: To expose the unprotected surface to weather conditions.

Index

Index

Note: Figures are indicated with *f*.

B

C

M

N

O

P